Changing Planet, Changing Health

Changing Planet, Changing Health

How the Climate Crisis Threatens Our
Health and What We Can Do about It

Paul R. Epstein, MD, and Dan Ferber

Foreword by Jeffrey Sachs

UNIVERSITY OF CALIFORNIA PRESS
Berkeley · Los Angeles

University of California Press, one of the most distinguished university presses in the United States, enriches
lives around the world by advancing scholarship in the
humanities, social sciences, and natural sciences. Its
activities are supported by the UC Press Foundation and
by philanthropic contributions from individuals and institutions. For more information, visit www.ucpress.edu.

University of California Press
Berkeley and Los Angeles, California

Library of Congress Cataloging-in-Publication Data
Epstein, Paul R.
    Changing planet, changing health : how the climate
crisis threatens our health and what we can do about
it / Paul R. Epstein and Dan Ferber ; foreword by
Jeffrey Sachs.
    p. cm.
    SUMMARY: "Spotlights the threats of global
warming and offers a systems approach for possible
treatments. Decades spent as a physician and public
health scientist have allowed Dr. Epstein to examine
and now comment on the dynamics of global politics,
climate change, and global health. Together with
journalist Dan Ferber, he expresses a fundamental
need for communities (of all scales) and industries (of
all kinds) to reach together for a low-carbon economy.
They make their argument by combining personal
accounts with accurate histories and industry case
studies. What enfolds is a prescriptive narrative for
repairing an ailing planet"—Provided by publisher.
    Includes bibliographical references and index.
    ISBN 978-0-520-26909-5 (hardback)
    1. Medical climatology. 2. Climatic changes—
Health aspects. I. Ferber, Dan. II. Title.
    RA793.E67 2011
    616.9'88—dc22   2010051311

Manufactured in the United States of America

19 18 17 16 15 14 13 12 11 10
10 9 8 7 6 5 4 3 2 1

This book is printed on Cascades Enviro 100, a 100%
post consumer waste, recycled, de-inked fiber. FSC
recycled certified and processed chlorine free. It is acid
free, Ecologo certified, and manufactured by BioGas
energy.

Contents

Illustrations

Foreword

This is a book about complexity—the complexity of humanity's predicament at the start of the twenty-first century. The predicament is easy to state, hard to understand, and perhaps hardest to solve. With nearly seven billion people on the planet producing and consuming at an unprecedented rate, humanity is deranging the world's ecosystems and threatening the survival of our own species and millions of other species with which we share the planet. How we are doing that is a scientific detective story of the first order, told with brilliance and relish by one of the world's great ecological detectives. What we are going to do about this remains hidden in the fog of complexity. The answers here are incomplete and may be wrong in many specifics. However, this fascinating book is important because it is a cry to take our survival seriously and to address it from the vantage points of science, ethics, and a moral commitment to the most vulnerable among us.

Paul Epstein is a pioneer of the new complexity. His intelligence, commitment, and perseverance have helped to change the way we employ science to serve society, and to alert the world to our harrowing risks. Readers will be entranced, edified, and inspired by Epstein's journey, which began when he was a young medical doctor learning the basics of tropical health in Mozambique and has carried him to the forefront of earth-systems science. Ever since that early start, Epstein has displayed the traits of a great scientific problem solver. He is both practitioner and theorist. He brings his erudition to problems but allows the problems themselves to challenge his ideas. He is ready to take great risks, intellectual and personal, to find truth and solutions. And he simultaneously leads his colleagues and perseveres on his life mission. In the way that energy flows through a complex ecosystem, Epstein's lifetime scientific journey flows across the disciplines of public health, ecology, climatology, and social change. And he leaves a crucial mark in each area, inspiring new insights and scientific efforts.

Earth-system change is a new and still emerging field. As Epstein and coauthor Dan Ferber make clear, the Earth is changing much faster than our ability to understand the changes or to control them. While the basic idea of climate change has been known scientifically for well over a century, the needed intensive and large-scale scientific exploration of climate change in all its crucial dimensions—including climatology, ecology, public health, engineering, and economics—is a much newer enterprise of just the past quarter century. And while Epstein and others have skillfully put the pieces of the jigsaw puzzle into place, nature has not been kind enough to wait for humanity to catch up in our understanding. The evidence is clear that the rate of climate change is accelerating. We are reaching thresholds of monumental danger, and according to many leading scientists, we've already surpassed some of those thresholds.

This wonderful book brilliantly illustrates many of the complexities of ecosystem functions, such as the ways that slight changes in temperature can fundamentally alter the evolutionary battleground between predators and prey, or between crops and pests or pathogens. The authors' breadth of knowledge and experience is vast. The case studies are fascinating. The detective work needed to explain the emergence of new diseases, or the spread of new pests, or the implications of climate change for forest fires, extreme storms, and other threats, is exceptionally and clearly described. Scholars and students will benefit enormously from these cases, which include the inspiring background stories of the scientists who unlocked the mysteries.

Readers will also share the authors' mix of exhilaration and dread: exhilaration that careful study can unlock complex mysteries and dread that the biggest mystery of all—whether humans will cooperate to secure their own survival—is still mostly unsolved. The story starts roughly a quarter century ago, when issues surrounding climate change were first reaching the public's awareness and when new epidemic diseases were beginning to appear with a frightening frequency. During the quarter century to follow, we would achieve much progress in scientific knowledge, increase our modeling capacity, gain insights toward solutions, and engage political, business, and social leaders. But what has not yet happened is any real societal problem solving, certainly not at a scale and magnitude that begins to address the scale and magnitude of the threats.

The past quarter century has been filled with treaties and protocols, global meetings, declarations, campaign speeches, and of course, tens of thousands of scientific studies; yet the sum of this enormous effort has not resulted in a true global commitment to needed solutions. It is fair to say that since ratifying the UN Framework Convention on Climate Change in 1992, the U.S. Senate has taken almost no single constructive step to honor the treaty or provide U.S. leadership toward global solutions. Epstein and Ferber explain the paradox with insight and fervor. Corporate propagandists have filled the airwaves and media with lies and confusion. As one tobacco company infamously wrote in a secret memo during the battles over tobacco and lung cancer, "Doubt is our product."

Nor are the economic and engineering solutions easy. Fossil fuels are at the very heart of the world economy. They have powered the escape from poverty achieved by most of the planet. As the authors make clear, many low-carbon energy alternatives—such as ethanol for driving, nuclear power, or carbon capture and sequestration—raise serious unsolved concerns. These are not reasons for inaction, to be sure, but they do exemplify the complexity of the economic challenges on top of the ecological ones. There is as yet no convincing pathway to a global solution.

I agree fully with the authors that such a pathway can be found and that our efforts need to be invested urgently and fully in that direction. While I have some doubts about some of the specific economic proposals put forward in this book, they point in the right direction: finding ways to further economic prosperity, to end extreme poverty, and to make the rapid transition to ecological sustainability, all within the boundaries of our market economy. If we are to find real solutions, it will be by deploying the scientific talents and humanistic efforts so richly on display in this book: serious research in full rigor; the capacity to work across disciplinary lines; a perseverance in the face of enormous obstacles and uncertainties; the highest of intellectual integrity; and a breathtaking love of humanity and nature, knowing that it is our responsibility to be stewards of this glorious planet and the generations to come.

Jeffrey D. Sachs, Quetelet Professor of Sustainable Development and Director of the Earth Institute at Columbia University

Introduction

On January 31, 2002, ominous crevasses widened in the Larsen B ice shelf, a mass of sea ice the size of Rhode Island located along the east coast of the Antarctic Peninsula. Then, over several weeks, it shattered, sending thousands of icebergs into the sea.

The ice shelf, thicker than the length of two football fields, had formed 11,500 years ago during the last ice age, and for most of the time since then, it had been melting ever so slowly from below. By the twentieth century just a few dozen meters of ice had melted, and the ice shelf remained solid. Over the past century, however, the air over the Antarctic Peninsula has warmed some 6°C (11°F). Beginning in 1992, the top layer of the Larsen B ice shelf began melting at unprecedented rates; then in the particularly warm Antarctic summer of 2002, extensive pools of meltwater formed on its surface. The water poured into the crevasses in the ice sheet and cascaded over its side, splintering the shelf and setting thousands of smaller pieces adrift.

The breakup of this ice shelf was by no means an isolated incident. In 2008, a 3,500-year-old ice shelf in northeastern Greenland collapsed. In the summer of 2007, the expanse of Arctic sea ice thinned and shrank to a quarter of the size it had been for thousands of years. Many scientists believe that, in just a few summers, the North Pole will be in open ocean.

Such anomalous climate-related events are hardly limited to melting ice. Since 2002, the worst megadrought in recorded history has crippled agriculture in southern Australia. In December 1999, three freak windstorms swept through central Europe, felling large swaths of forests in France. In 2004, the first hurricane ever recorded in the southern Atlantic Ocean made landfall in Brazil. In 2008, Cyclone Nargis raised a tidal wall four meters (twelve feet) high and a remarkable forty kilometers (twenty-five miles) long that slammed into Myanmar (Burma). The Earth's climate is changing—fast.

For more than ten thousand years, the planet's climate has been relatively stable. Not long after the Larsen B ice shelf formed in Antarctica, Earth entered the Holocene epoch, a relatively mild period that nurtured the rise of human civilization. For thousands of generations before and during the Holocene, humans developed ingenious ways to survive the elements—to stay cool in the heat, to keep dry when it's wet, to find warmth in the cold. When the wood, peat, and whale oil used to fuel the fires of our burgeoning population ran low a few centuries ago, we discovered abundant new sources of fuel—the coal, oil, and gas that began forming more than three hundred million years ago from the buried remains of ancient plants and animals.

For the past few centuries, we've mined and burned these fossil fuels to cook our food, heat our homes, and transport ourselves with increasing speed around the world. Because fossil fuels are rich in carbon, they combine with the oxygen in air to form carbon dioxide when burned, and, in powering the human enterprise, we've pumped more than a half trillion tons of this gas into the atmosphere. This has raised the atmospheric carbon dioxide concentration by almost 40 percent over that of the preindustrial era, to a level the planet has not seen for thirty million years.

Like the glass roof of a greenhouse, carbon dioxide in the atmosphere traps heat. Higher levels of gas act like thicker glass, trapping more heat. Over the past century, atmospheric temperatures have risen 0.8°C (1.4°F) on average. But the oceans have stored twenty-two times as much heat as the atmosphere. The warmer atmosphere and oceans have altered humidity and wind patterns in some regions and generated more extreme weather in others. These changes have brought a myriad of outcomes, including record floods in Iowa, a dwindling snowpack in the Sierras, and drowning polar bears in the Arctic.

As the Earth's climate changes, it will do far more than maroon polar bears amid melting ice. Climate change has already contributed to the remarkable 2003 heat wave in Europe that melted 10 percent of the ice in the Alps and killed more than fifty-two thousand people. It has contributed to an extraordinary single day of drenching rains in Mumbai that killed one thousand people, contaminated water supplies, and sickened hundreds. It has contributed to the spread of malaria-carrying mosquitoes, to the disappearance of mountain glaciers, which threatens drinking water supplies on five continents, and to at least 150,000 additional deaths worldwide each year and five million years of healthy life lost to disability. Clearly, climate change is hazardous to our health.

. . .

This book is about how climate change harms health now, how it could devastate public health by midcentury, and how we must transform the way we power society and organize our economy to preserve a livable planet. But it is also about the incredible opportunities that will arise once we do.

By the time climate change first entered public consciousness in the scorching summer of 1988, I'd had the privilege of working as a physician for almost two decades, mostly in low-income communities. My patients' symptoms stemmed from microbes or chronic ailments, to be sure, but also from larger social and environmental factors—many of which could have been prevented. During a two-year stint in Mozambique, for example, I treated patients with tropical parasitic diseases that could have been eliminated by simple public health measures. In and around Boston, I helped treat children suffering from lead poisoning caused by high levels of lead in house paint and gasoline, decades after the hazards of each had been identified.

The suffering I saw in my patients fueled my passion to prevent such health problems rather than treat them after they'd developed, and that desire led me to complement my clinical work with formal training in public health. This dual perspective gave me a useful vantage point. From where I stood, it seemed clear that the best way to prevent health problems was to identify their root causes, then work to change them.

By the late 1980s, it was evident to many scientists that humans, by destroying forests and burning fossil fuels, were warming and altering the Earth's climate. The United Nations had just created the Intergovernmental Panel on Climate Change (IPCC), which brought together scientific experts from around the world to assess the condition of the planet. It was clear even then that climate change was happening on top of a host of other human-induced changes in the global environment, including thinning of the ozone layer, forest loss, destruction of plant and wildlife habitat, and pollution of freshwater, oceans, air, and land. Leading scientists were just then beginning to ask how these global environmental changes might threaten human health. It has been the great challenge of my career to help find the answers.

In pursuing these complex issues, I've been fortunate to collaborate with talented scientific colleagues in fields as far-flung as oceanography, epidemiology, ecology, and atmospheric science. Over the years, our studies and those of other key researchers have painted an alarming picture. Climate change threatens health directly, causing lethal heat waves, spreading insect-borne disease, worsening air quality, and more. But it also threatens health indirectly by the long-term and pervasive decline of the world's ecosystems.

As the scale of these threats became clear in the early 1990s, I broadened my focus as a physician and public health scientist. By then, I'd spent more than a decade conducting research in developing nations and thinking broadly about how to prevent disease. As I began to focus on global issues, I began engaging in the public discourse to help educate my colleagues and the broader public about how we could protect ourselves from the ravages of climate change. I reached out to colleagues in other nations and to those in business and international environmental organizations. My goal was to better understand the real-world workings of those realms and help to change them to promote healthy development.

In the last twenty years, the public debate on climate change has moved, slowly but inexorably, from denial and disinformation to acceptance, consensus, and the first small steps of the necessary societal transformation. This book will trace that journey and, I hope, further it.

. . .

In planning for a healthier world, we can draw from history. As the industrial revolution gathered steam, urban populations swelled, and between 1790 and 1850, the population of London grew sevenfold. By the 1830s, the city was overcrowded, with congested factories filled with child laborers, garbage-strewn streets, and contaminated water. These conditions—captured in so many novels by Charles Dickens—bred a seemingly never-ending series of epidemics: tuberculosis, smallpox, cholera, more smallpox, more cholera. The epidemics spawned epic riots, as protesters streamed onto the streets to demand better living conditions; one such riot claimed more than four thousand lives.

During another cholera epidemic two decades later, a British scientist named John Snow mapped cholera cases in London's Soho district, and by overlaying the data on a map of the water system, he traced the source of the scourge to drinking water drawn from a single pump. It would be three decades until Louis Pasteur discovered that bacteria could cause disease, so Snow could not have known what exactly it was about the neighborhood's water that was causing cholera. But he knew enough to turn off the Broad Street pump, saving countless lives.

In the latter half of the nineteenth century, society changed course dramatically. Civic leaders in London and throughout the Western world invested heavily in infrastructure. They built new purification plants and laid new piping to supply clean drinking water, they installed sanitation systems to remove human waste, and they paved the streets and enacted laws to protect workers. These efforts systematically addressed the root causes of disease. Indeed, public health became the raison d'être for development, dramatically boosting the health of the population.

Today, climate change poses health risks even more profound than those posed by poor sanitation and living conditions in nineteenth-century London. Will we grasp the interconnected nature of these threats and seek solutions that get to the root of the problem?

Dan Ferber and I have written this book to call attention to the perils climate change poses for human health and welfare and to help chart a path toward a cleaner and healthier future. It is no longer a luxury to make our economy low-carbon and sustainable. It's a matter of preventing harm to the species who dwell on the Earth, including our own. Just as an ailing patient can recover, so can an ailing planet. But we must act now.

Paul R. Epstein, MD, MPH
Boston, MA
October 2010

1

Mozambique

My first epidemic began quietly, as most epidemics do. It was May 1978, and I was working as a physician at the Central Hospital of Beira, Mozambique, which was the only hospital for hundreds of miles. One morning I was summoned to attend to a dangerously debilitated man in his thirties. The man's family had brought him a great distance from the *mato,* or countryside. The ailing man was so severely dehydrated that when I gently pinched his skin, it tented, meaning it retained the profile of a small tent where I'd pulled it away from the underlying tissues. His eyes were sunken, his gaze terrified. He was clearly near death.

Three months earlier, I had arrived in Beira with my wife Andy (short for Adrienne), who's a nurse, and we had begun caring for patients in the city's Central Hospital. Beira is a major port city in southeastern Africa and Mozambique's second largest urban center. It lies some seven hundred miles north of Maputo, the capital, in a flood-prone rice-growing region on the Mozambican coast. Although it is not a spectacular city, it does have its share of natural beauty, with long, curving white sand beaches rimmed by the warm waters of the Indian Ocean. The beauty of Beira's beaches, where well-off white Rhodesians once played, could have blinded the casual observer to the existence of ancient diseases rife in the population, diseases scarcely known in Western societies.

Beira was then an impoverished city of about three hundred thousand in a nation that had wrested its independence from Portugal just three years earlier. Andy and I were part of a small troop of profes-

FIGURE 1. Drs. Paul Epstein (right) and Steve Gloyd with a Mozambican colleague in Caia, Mozambique, in 1978. In the background is the Zambezi River, one of Africa's four largest rivers. (Photo courtesy of Paul Epstein)

sionals, of all stripes and from numerous countries around the world, known as *cooperantes,* or international aid workers, who converged on Mozambique to help the fledgling nation rebuild its health care system, its economy, and its society (figure 1).

To prepare for my work in Africa, I had audited a course on tropical diseases taught by top experts at the Harvard School of Public Health. Upon our arrival in Mozambique, I had also undertaken six weeks of on-the-job training in the sprawling 1,600-bed complex of the Central Hospital of Maputo, the country's most modern city. In the vast, open wards of the Maputo hospital, a collegial group of local and foreign doctors had given me a hands-on crash course in recognizing and treating the many afflictions common in southern Africa, including well-known diseases like malaria and tuberculosis, along with a multitude of life-sapping diseases caused by worms of all sorts and sizes. These included hookworm, which lives on blood and causes anemia, weight loss, and stunted growth, and schistosomiasis, or snail fever, a debilitating ailment endemic in Mozambique that causes urinary tract and kidney disease.

My suspicions about what was ailing the desiccated, frightened man in Beira did not derive from the training I'd received in Maputo. Instead, the deathly ill man recalled images in tropical disease textbooks I'd studied in Boston. His sunken eyes and dehydration presented the classic picture of cholera, a waterborne disease capable of blossoming into a raging epidemic.

Completing the diagnosis required microscopic examination of the patient's stool for the cholera bacillus. Other strains of bacteria, as well as some viruses and parasites, can cause diarrhea that results in extreme dehydration, and we needed to rule out these infections. I placed a drop of the man's watery stool on a glass slide and peered at it under the microscope. I saw hordes of wriggling, comma-shaped microbes dancing on the slide—telltale signs of *Vibrio cholerae*. The diagnosis was established when the organisms later grew in petri dishes containing agar made with small amounts of sheep's blood.

Andy and I had become familiar with the host of serious but preventable diseases that afflicted Mozambicans and those in neighboring nations, resulting from poor nutrition, inadequate sanitation, and poverty. But cholera—with the exception of a brief appearance in 1973—had not been among those ills.

. . .

When we applied to work in Mozambique in 1976, the former Portuguese colony had been independent for a year. Mozambique's revolutionaries in the Front for the Liberation of Mozambique (FRELIMO) had fought a successful thirteen-year war for freedom from colonial rule, a war that ended when colonial soldiers returned to Portugal and overthrew their dictator, in turn freeing the nation's African colonies. The revolution held the promise of a better life for Mozambicans, but its immediate aftermath had major repercussions. Upon independence, the Portuguese fled en masse—more than a quarter million Portuguese left the city of Maputo alone.

The fleeing Portuguese packed up their riches as they exited, and, in some instances, they sabotaged development projects on the way out. But the losses to the country weren't solely material. The exodus included virtually the entire professional class of Portuguese settlers, including teachers, foresters, mining specialists, engineers, and doctors. Under Portuguese rule, education for most Mozambicans had ended after fourth grade, with the exception of students sent off to seminaries. When the Portuguese left, most Mozambicans were illiterate.

Because Mozambique needed so many kinds of experts to build its new infrastructure, Mozambique's first president, Samora Moisés

Machel, and his government reached out around the world for assistance. My wife and I were part of the wave of international *cooperantes* who responded to that call. Indeed, during our time in Beira, Andy and I worked alongside *cooperantes* from, among other places, England, Holland, Sweden, Russia, Bulgaria, Cuba, Zambia, Brazil, and Chile.

The new government had many concerns to address. Perhaps the most immediate had to do with medical care for their newly liberated citizens, who were overwhelmingly rural farmers. Many nurses had departed, and almost all of Mozambique's Portuguese doctors had abandoned the country, leaving just a handful of physicians to care for a population of twelve million.

Fortuitously, President Samora Machel, whom Mozambicans knew simply as Samora, was a trained and experienced nurse, and he assigned health care a high priority in those early heady days. Samora's aim was to develop a well-distributed health care system that integrated public health services. Achieving these worthy goals was difficult with so few resources and health practitioners at hand. At Samora's insistence, neighborhood health clinics were opened throughout the country and stocked with a carefully selected list of imported generic medications. These efforts led the World Health Organization to recognize the Mozambique health care system as exemplary. But that was only a beginning. One of the important tasks for medical *cooperantes* was the training of Mozambican nurses, nurse practitioners, and new doctors in the nation's one medical school in Maputo.

From 1976 to 1978, we waited while our applications to work in Mozambique wound their way through the fledgling, byzantine Mozambican bureaucracy. At last we were informed that we would be welcome for a two-year posting. Although we were officially employees of the Mozambican Ministry of Health, our sojourn had been arranged by the American Friends Service Committee, the Philadelphia-based Quaker organization that supports humanitarian aid and peacemaking efforts around the world.

Upon our arrival in the country in February 1978, we were relieved to discover that we could speak our rudimentary Portuguese haltingly with Mozambicans. (It would be six months before we could hold our own in a dinner conversation.) Assured that we were capable of taking a medical history, we got to work immediately. Our children adapted in a different fashion. They played silently for the first six weeks with the children of our Mozambican, Chilean, Swedish, and Portuguese neighbors and coworkers. Then, suddenly, they began to speak the lingua franca fluently, interjecting the word *coisa* (thing) for whatever object they could not yet name. It was an exciting time.

With FRELIMO's explicitly race-blind policy, we felt welcomed and accepted in this beautiful port city of a million people, with its vistas of the radiant Indian Ocean. With our children we walked downtown and through shantytowns, exploring the spicy and delicious victuals. Fish and shrimp and chicken were abundant—and cold, cold beer *(bem gelado)* as well. In our hotel, there was only the occasional mosquito, but crickets, birds, and roosters could be heard everywhere.

During our six weeks of training at Maputo's Central Hospital, a delightful, well-trained Cuban hematologist befriended me, easing my transition from Western to tropical medicine. When Andy arrived in the surgical ward, her first patient was a man who'd been bitten almost in half across his abdomen by an alligator while crossing a nearby river. With surgery and good medical care, the man survived intact.

Most of the diseases afflicting Mozambicans presented with signs almost as obvious as an alligator bite; thus diagnosis was generally easy. Prevention, on the other hand, was difficult to implement, given the country's low level of development, which made even simple preventive measures for many common maladies hard to come by. Few women or children wore shoes (figure 2), which would have spared them the anemia caused by hookworms that enter the body through the skin of the foot and then line the sides of the intestines. (Men, who benefited more from the country's limited prosperity, were more apt to be shod.) From our vantage point on the ground, the shape the nation's development would take was not obvious. Even today, the questions of how to develop and to power that development remain central issues for Mozambique, as they are for many underdeveloped nations.

Shortly after arriving, we were posted to Beira, the second largest city in Mozambique. After completing our training at the seven-hundred-bed Central Hospital in Maputo, Andy and I launched our family's four-day trip on the northerly road to Beira in a spirit of adventure seeking and with a sense of purpose. We traveled a hardscrabble highway, which was intermittently paved, and ascended onto the vast African savannah dotted with wide-crowned acacia, mango, and cashew trees. If Maputo had seemed exotic, the landscapes and civilization we encountered on our journey to Beira were even more so. We stayed overnight at a hotel in the beach town of Vilancoulos, where we were the only guests. We were served by a Mozambican staff that otherwise stood quietly behind nearby palm trees, shoeless but clad in white colonial-era uniforms, as

FIGURE 2. Mozambican boys near *canisos* (houses made of sugarcane stalks) in Buzi, Mozambique, a sugarcane-growing area west of Beira, in 1978. Many women and children then lacked shoes, which put them at risk for parasites, such as hookworm, that enter the body through the feet. (Photo by Paul Epstein)

if the revolution had never happened. We swam for the first time in the surprisingly warm sea. That evening, the four of us stood surprised and transfixed, watching a lunar eclipse from the hotel balcony. The moon seemed just a few feet away as it dropped into the sea.

Two days later, following a stop in Xai-Xai to visit a Mozambican friend who had studied in Boston and taught us Portuguese and Mozambican history, we reached Beira. The city had been a major trading center for goods from Salisbury, Rhodesia (now known as Harare, Zimbabwe), in the heyday of colonialism. By the time we arrived in 1978, however, it was a diminished outpost populated by a few Indian shop owners and many unemployed Mozambican men. Women were barely in evidence in the modern part of Beira, which the locals called the "cement city." The town's center seemed almost abandoned. The real life of Beira, we discovered, could be found on its outskirts. Dirt roads and narrow paths weaved among houses made from sticks and stones, cement, or sugarcane stalks. Outside, children played and women sewed and sold goods from stalls, keeping one another company. Women tended the cooking fires, and the air was infused with the sweet scent of

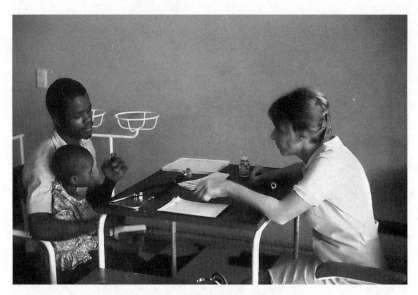

FIGURE 3. Adrienne (Andy) Epstein interviews a father and son in the emergency room at the Central Hospital of Beira, 1978. She was performing triage—treating some patients then and there, admitting others to a hospital bed, and sending others to an outpatient clinic. (Photo by Paul Epstein)

burning acacia. We lived nearby, and from our cement house we could hear *batuki* (drumming) well into the wee hours on the weekends.

I was soon appointed chief of medicine at the Central Hospital of Beira, and I set about working with the thirty other international doctors and one dentist. Each medical *cooperante* put in one twelve-hour work shift a week in the hospital's emergency ward. Andy worked there each day as well, performing a form of triage, determining when patients required hospital admission, simple packets of pills, or treatment in a neighborhood health center (figure 3).

In the evenings, our home was a magnet for this multicultural crowd of international coworkers, Portuguese friends and coworkers, and Mozambican locals and their families. The children, lacking television to mesmerize them, kept us entertained with their improvised theatrical and musical productions.

Conditions varied on the hospital's wards. A common problem on the women's ward was severe anemia from hookworm, acute malaria, or both. The problem was exacerbated when women were pregnant, as the body's blood volume rises from five to seven quarts and already scarce iron is stretched even thinner. Other patients had one or more of a range of debilitating diseases, including mosquito-borne elephantiasis, amoebic dysentery, and schistosomiasis. Mosquitoes infected and reinfected our patients with malaria, and they infected Andy and the children with malaria as well, despite the chloroquine pills they took to prevent it. Then there were the diseases imposed on Mozambicans solely as a result of inadequate sanitation, poor nutrition, or both, among them rickets and gastroenteritis from many sources. "It's a wonder there are so many people walking about," I wrote in one of my earliest letters to our sponsors in Philadelphia. On the other hand, the pervasive diseases of industrialized societies—diabetes, hypertension, and heart disease—seemed nonexistent.

Our work was demanding but rewarding. After working all morning in the hospital, we'd head out to the shantytowns—the *bairros*. Andy worked in the health clinic of a *bairro* called Inhamudima near our home, while I worked in another clinic out in Munhava, the largest of the *bairros* ringing the cement city. The locals introduced us to locally available foods, and we taught them about nutrition and other healthy practices. We saw patients—usually about twenty-five each day—throughout the afternoon. The medical and social needs were enormous.

. . .

An epidemic is defined as an unusual occurrence of disease—an unexpected number of cases occurring in a particular time and place. Understandably, epidemics are often not immediately recognized. But in the case of the cholera epidemic that erupted in Beira in 1978, the breadth of the outbreak was evident in a matter of days.

That the disease was cholera made the episode all the more remarkable. Cholera had spanned the globe in seven pandemic waves since the 1800s, including the London cholera epidemic in the 1850s that crusading epidemiologist John Snow had helped stop. Epidemiologists knew that cholera was circulating in Asia in the late 1970s and was, in fact, considered a permanent blight in the many countries bordering the Bay of Bengal, reaching from India all the way to Thailand. But the disease had been absent from Africa for most of the twentieth century, so its appearance in Mozambique was surprising. The deathly ill man from the *mato* was the first, or "sentinel," case of the epidemic, and his illness indicated that the seventh pandemic wave had now spread to East Africa. In the weeks that followed his arrival, hundreds of people in

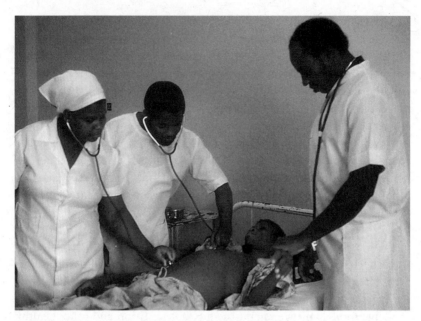

FIGURE 4. Nurse-practitioners-in-training examine a patient at Beira Central Hospital, Beira, Mozambique, in 1979. The hospital's facilities were barely adequate to provide modern medical care to Mozambicans. (Photo by Paul Epstein)

all stages of the disease found their way to us, to Beira's neighborhood clinics, and to clinics in surrounding towns and villages. In addition to those presenting as ill, we knew there were thousands more in the Mozambican countryside who were carrying the bacteria but showing no symptoms. It was truly a major epidemic.

When the cholera epidemic first hit, we were fortunate to have the staunch support of the province's health officer, the capable Dr. Pascoal Mocumbi, a practicing obstetrician and gynecologist who would later become the nation's minister of health and then its foreign minister. Working together, in short order we organized a thirty-bed ward staffed by four doctors and eight nurses.

Treatment of cholera demands immediate administration of fluids, a therapy that is the only real cure for the disease. (Antibiotics are generally useless as treatment, though they can be preventive.) Without treatment, a patient might lose up to seven quarts of fluid a day from blood and tissues as a result of severe diarrhea, and such extreme dehydration results in a fatality rate of up to 50 percent. Treated properly, cholera can have a case fatality rate of less than 1 percent. We developed protocols, which included giving all but the most severe cases of dehydration an oral salt-and-sugar solution. We knew from work in Bangladesh that such a regimen could treat the majority of patients. Patients who were too weak to drink or who were vomiting required intravenous administration of this simple fluid mix.

Doing medical work in a poor developing country like Mozambique presented many unexpected challenges. The water in the salt-and-sugar solutions had to be free of pathogens, and that meant that every liter of water had to be boiled, using heat derived mostly from burning wood and charcoal. Plastic catheters and other disposable supplies readily available in developed nations for administering intravenous fluids were nonexistent in Beira. We needed steel needles to deliver fluids directly into veins, but they were dulled from repeated use. A one-time shipment of plastic disposable IV equipment from our sponsors arrived, but when it ran out, Mozambique could not afford to order more (figure 4).

The cholera epidemic tried our medical and nursing skills, it taxed our capability to rapidly train other care providers, and it stretched the resources of this young nation. At the time, though, we were just trying to save the ailing Mozambicans who were flooding into our clinic. The epidemic had become a trial by fire for us all.

. . .

The man from the *mato* stayed at the Central Hospital in Beira for more than a week, and a combination of intravenous solutions followed by oral rehydration gradually restored his strength. Then one day he walked out of the hospital with his family. We saved many others as well. But we couldn't save everyone. Many of the ailing people lived far from the hospital, and slow access to transport and care meant that an estimated 5 percent of those stricken with cholera died, many of them deep in the countryside.

Epidemics had previously been events I'd studied in historical and medical textbooks, but now they were a dreadful reality. However, after the cholera epidemic abated, I was struck by just how much could be done to stem the tide of an outbreak with early warnings, rapid diagnoses, and relatively basic training of health care personnel. During the epidemic, the services we set up to rapidly diagnose and treat cholera patients had saved hundreds of lives.

We had far less success stopping a devastating man-made scourge. Just a few weeks after we began working at the Central Hospital of Beira, the first of what would be thirty-five young Zimbabwean refu-

FIGURE 5. Dr. Paul Epstein conducting a public health survey in a refugee camp at the border area of Mozambique and Zimbabwe, 1985. (Photo courtesy of Paul Epstein)

gees, most of them under twenty-five years of age, wound up at our hospital, bleeding from multiple orifices.

At the time, Ian Smith's all-white Rhodesian army was fighting to retain control of that country, and they occasionally pursued fleeing Zimbabweans, some of whom were fighting for their country's independence, across the border into Mozambique. The young men and women we encountered had been staying in a refugee camp several hundred miles from Beira. Given how much they were bleeding, we initially suspected they were suffering from one of the viral hemorrhagic fevers—Ebola, Lassa, or Marburg—that were newly emerging in Africa. But that was not the case, and we—and other infectious disease experts we consulted—were at first bewildered about the cause of their ailments.

We organized a ward for these desperately ill young people and infused them with blood provided by other Zimbabweans and Mozambicans, but over the next two weeks, fifteen of the thirty-five died. After the last victim, a handsome young man, succumbed in front of me, I excised a small sample of fatty tissue from his abdomen. A toxicological investigation revealed the cause of this macabre disaster. The man's fatty tissue had been contaminated by warfarin, a blood anticoagulant commonly used as a rat poison. It turned out that members of the Rhodesian intelligence forces, aided by foreign mercenaries, had deliberately impregnated the Zimbabweans' clothing with rat poison.

That event was our initiation into the darkest reaches of biological and chemical warfare. Through both the epidemic and the poisonings, however, Andy and I gained a new appreciation of the interwoven environmental and social determinants of health (figure 5). The lessons learned have stayed with me and helped shape my views on how to cope with and prevent the mounting health problems associated with a changing climate.

THE ROOTS OF DISEASE

I returned from Mozambique in 1980 with a strong desire to better understand the field of public health. I was acutely aware that, while individual health care was essential and needed to be distributed to all corners of the globe, public health measures and social interventions were needed to prevent the ills afflicting large populations. After returning to Massachusetts, I worked as a primary care doctor at a neighborhood health center in Boston for a while, then at Cambridge Hospital, a teaching hospital affiliated with Harvard Medical School. Most days I worked at the associated East Cambridge Neighborhood Health Center, located in a neighborhood populated by Portuguese-speaking people from the Azores, Portugal, the Cape Verde Islands, and, later, Brazil.

In 1981, I enrolled in a master's program at Harvard's School of Public Health. Over the next two years, I was exposed to ideas and people who influenced me deeply and helped shape my approach to medicine, science, and international public health policy. One significant encounter occurred in a classroom in an eleven-story brick building at the Harvard School of Public Health. There, Richard Levins stood at a blackboard, speaking quickly and sketching a diagram as a small group of graduate students, most from or with experience in developing countries, leaned forward in our seats and focused intently. Levins was and still is a professor of population sciences at the school. A self-described ex–tropical farmer turned ecologist, biomathematician, philosopher of science, and political theorist, Levins was one of the founders of modern ecology. Of solid build and with a bushy beard, Levins resembled a modern-day Charles Darwin. And like Darwin, Levins was a scientific leader far ahead of his time.

The diagram Levins had sketched on the blackboard represented a scenario involving a factory worker and his foreman. The diagram consisted of a pair of circles with arrows pointing from one circle to another; one circle was the factory worker and the other was the foreman. The foreman was pressuring the employee to work harder and faster and threatening to fire the employee if he didn't. The worker's adrenaline surged, causing his blood pressure to rise. Chronically elevated blood pressure could have placed this worker at risk for heart and kidney disease.

Then Levins drew another circle—the union organizer—who stepped forward to protect the worker's rights. The foreman backed off, and the man's blood pressure settled back to normal. Then Levins modified the diagram again. The union had been weakened by social and political factors; the worker was under chronic stress and experienced high blood pressure repeatedly. By using examples like this, Levins taught how both social and environmental conditions can profoundly influence human physiology and health. In laying out the specific pathways and mechanisms by which this occurs, Levins provided a powerful intellectual tool that sheds light on the root causes of individual and global distress.

I had signed up for Levins's seminar soon after beginning my training at Harvard. I'd heard high praise for the man and soon joined a fellow physician classmate who'd also worked in Mozambique to seek out Levins's counsel and teachings. Levins attracted many students like us who had lived and worked in developing countries, an experience that creates a breed unto itself. In the seminar, we discussed emerging issues such as integrated pest management, an approach to farming that minimizes the use of toxic chemicals. We discussed sustainable development—half a decade before an influential United Nations report, *Our Common Future,* popularized that term. For graduate students like me, Levins's methods offered useful tools to help tackle complex real-world systems, and he set an example as a thoughtful, rigorous, ethical, and independent thinker. Dick became a friend and a scientific mentor—a central figure who helped me understand the world and what it would take to transform it.

In the 1950s, Levins and his wife, Rosario Morales, a poet, had farmed for five years in rural Puerto Rico, growing vegetables that included lettuce, bush beans, snow peas, and Chinese cabbage. There he learned firsthand through hard work how the richness of the soil and a diversity of plants provided natural resistance to pests and microorganisms that plague farmers worldwide. After returning to the mainland, Levins earned a doctorate in evolutionary biology at Columbia University. His early work established the importance of maintaining corridors that link wildlife habitats in order to sustain large and genetically diverse populations of animals—an idea that is applied widely today in designing wildlife refuges and conserving rare and endangered species.

While Levins was making his mark as an ecologist and biomathematician in the 1960s and 1970s, infectious disease declined in North America and in Europe, having been successfully suppressed with vaccines, insecticides, and antibiotics. Scientists concluded that we had passed through what was termed "the epidemiological transition," in which the main scourges of society went from acute infectious disease to chronic illness. "Public health assumed that we'd licked it," Levins recalled recently. Harvard's health experts were so convinced of this that in the 1980s its public health school abolished its department of microbiology and decided to focus almost exclusively on the chronic diseases of developed modern societies, including heart disease, stroke, and cancer.

But Levins never believed the battle was won. His evolutionary biology training had taught him that pathogens such as bacteria and parasites are highly resourceful creatures. For example, some one-celled parasites switch their overcoats—their outer membranes—several times as they course through our bodies, thereby evading our immune responses. Many bacteria have evolved resistance to antibiotics. By the late 1970s, such knowledge led Levins to suspect that fast-growing, easily adapting pathogens that cause infectious diseases would not give up without a fight.

By the 1980s, and in spite of the epidemiological transition, it was clear to most epidemiologists that dozens of infectious diseases, including HIV/AIDS, tuberculosis, and Ebola, were emerging or reemerging. Levins searched for clues to why so many U.S. and European researchers had gotten it wrong. He concluded that they had based their assessment on just several decades of disease trends in North America and Europe. They had neglected to consider how infectious diseases have waxed and waned through history. They had also neglected to consider how microbes regularly evolve antibiotic resistance, or how mosquitoes and other insect carriers of disease evolve pesticide resistance. They had assumed wrongly that the vaccines and antibiotics that vanquished infectious disease in the developed world would be available widely in the developing world. And they hadn't appreciated that sunny projections for economic development in many countries masked widening inequities and growing impoverishment, which often lead to malnutrition, which in turn lowers immune defenses. The source of the error was a "narrowness in the field," Levins concluded.

A SCIENCE OF THE WHOLE

The narrowness of thinking that Levins encountered pervades much of modern science and leads to inaccurate assessments and prescriptions in many fields. The narrowness itself stems from a perennial challenge with which every scientist must grapple: many phenomena we'd like to understand are highly complex and have multiple, interacting causes. For the last three centuries, many scientists have tried to overcome that challenge by simplifying, using an approach called reductionism. The codification of this approach is often credited to the seventeenth-century French philosopher René Descartes, whose work helped launch the rise of modern science. In his *Discourse on the Method* (1637), Descartes provided reductionism's central metaphor. The natural world works like a machine, he wrote. More broadly, the central tenet of reductionism and the Enlightenment that followed was that by studying pieces of the world and reducing them to basic laws, we could unravel the marvels of the whole.

Over the centuries, this reductionist approach to science has revealed a great deal about complex phenomena, and that understanding has led to technologies that have transformed our lives. In medicine, William Harvey, a seventeenth-century doctor and physiologist, dissected animals and discovered that veins had valves and that the heart pumped blood through our arteries—insights that revealed, with remarkable accuracy, how blood circulated through the human body. In biology, Rosalind Franklin isolated DNA from cells and bombarded it with X-rays, creating images that helped James Watson and Francis Crick determine that DNA is shaped like a double helix—a finding that paved the way for modern biotechnology. And today's nanoscientists conduct experiments on individual molecules and use what they learn to build microscopic machines that promise to revolutionize batteries, solar energy collectors, and medicine. Reductionism has taken hold so widely that, for many people, its analytical, mechanistic approach has become synonymous with science.

There is another powerful but less well-known scientific approach, however, that plays a vital role in addressing some of today's global issues, including climate change. It has its roots in biology rather than physics, and it uses concepts derived from the study of life rather than the study of machines. The approach is known as systems theory, and it was pioneered in the first half of the twentieth century by a renowned biologist and philosopher named Ludwig von Bertalanffy.

Bertalanffy was born in 1901 in Atzgerdorf, Austria, a small village near Vienna, to a distinguished and scholarly family that traced its roots to sixteenth-century Hungarian nobility. He trained in both philosophy and biology at the University of Vienna in the early 1920s, then launched his career as a biologist by investigating how animals develop from a few simple embryonic cells into complex creatures with many types of interacting tissues and organs.

When Bertalanffy began his work, a debate was raging in biology. In the late nineteenth and early twentieth centuries, one school of biologists proposed that every process of living organisms, from metabolism to development, could ultimately be explained completely using only the laws of physics and chemistry. The way to gain that understanding was via a reductionist approach: take organisms apart into their components, literally or conceptually, and then understand the parts and use that knowledge to comprehend the whole. For example, reductionists believed that complete knowledge of the human body could ultimately be obtained by completely understanding the workings of human cells.

Other reputable biologists believed this approach would never work in living beings because they were fundamentally different from machines. To support that conclusion, these biologists cited several properties of life that no machine possessed. For example, organisms, unlike machines, become progressively more complex as they develop from embryos to adults and as they evolve over millennia into complex life-forms. And organisms, unlike machines, can grow, adapt to circumstances, reproduce, and even regenerate missing parts. Organisms were so different from machines that some of these biologists invoked an ancient idea first attributed to Aristotle: that each living creature possesses an intangible life force that directs its mechanistic forces.

Ludwig von Bertalanffy didn't buy either explanation. Classical physics and chemistry did not seem able to explain all the properties of living organisms. But the idea of an intangible life force was vague and useless, he believed, and would do nothing to advance a scientific understanding of living organisms.

Instead, Bertalanffy believed every living organism could be thought of as an open system. In science, the term *system* has a specific meaning: it's a collection of interrelated parts that function together via driving processes. An open system takes in matter and energy from outside itself, uses that flux of matter and energy to maintain itself, and releases waste matter and energy to the environment. Such systems are omnipresent. The flame of a lit candle is a classic example: it takes oxygen from the air and matter from the wick and the wax, then releases heat to the environment and uses this flux of matter and energy to maintain itself. Similarly, all life-forms take in matter (nutrients from food, water, oxygen from air) and energy (the calories stored in food), and they release matter (bodily wastes, exhaled carbon dioxide) and energy (heat) to their surroundings.

In a crucial insight, Bertalanffy realized that every open system, living or not, has properties characteristic of life. He sought to define and describe those common properties with a unified theory that explained open systems. To do that, he drew ideas from philosophy, several fields of biology, and physics.

From the then-new science of ecology, Bertalanffy borrowed the essential concept of networks. In the 1920s, pioneering ecologists had realized that most real-world forests and tide pools, indeed all biological communities, actually consist of food webs: interconnected networks of species whose relationships are determined by whom one eats, whom one helps, and with whom one peacefully coexists. This led to a view of systems as "an integrated whole whose essential properties arise from the relationship between its parts," as Fritjof Capra put it in his book *The Web of Life,* an illuminating exposition of nonreductionist science. This emphasis on relationships and connections differentiates systems theory from a mechanical world view. Both are useful.

Bertalanffy also understood that the whole of a biological community or any complex system is more than the sum of its parts. A system at any level of organization—whether a cell, a tissue, an organ, or an entire biological community—exhibits properties and behaviors that its component parts do not.

From physiology, Bertalanffy realized that open systems could maintain a steady state by regulating the functions of their parts. Physiologists call this concept homeostasis. For example, the body keeps blood sugar levels relatively constant by adjusting the levels of two hormones, insulin and glucagon, that control the release of sugar from the liver and its uptake into muscle and other tissues. Similar stabilizing mechanisms keep blood pressure, body temperature, and the saltiness of our body fluids relatively constant, within healthy limits. Since all open systems maintain themselves in steady states, Bertalanffy posited that they all had analogous stabilizing mechanisms.

From cybernetics, a science developed in the 1940s to understand and control how machines process information, Bertalanffy and others added the idea of self-regulation by feedback—the act of sending information about the outcome of a process or activity back to its source. Feedback mechanisms are described as *positive* if an increase in one part of the system stimulates an increase in another. A suckling baby, for example, stimulates its mother to produce more milk. *Negative* feedbacks, in contrast, are self-correcting, dampening, and balancing; thus they help to maintain homeostasis. A thermostat provides a negative feedback: when it gets too cool, the heater kicks on; when it gets too warm, it shuts off.

From cybernetics, Bertalanffy and others drew the pivotal idea of feedback loops—circular sequences of feedback mechanisms in which component A influences component B, which influences component C, and so on, until the final component in the chain influences component A. Negative feedback loops in our endocrine and nervous systems steady our warm body temperature, our blood sugar level, and more, thereby maintaining our health. Positive feedback loops, in contrast, amplify change, quickly turning small changes into big ones. Examples include a nuclear chain reaction and a mob that's escalating out of control. The terms we use for positive feedback loops reflect their volatility: *chain reactions* or *vicious cycles*.

The Earth's climate system contains both negative feedback loops that help stabilize the climate and positive feedbacks that destabilize it. Our global economic system also contains both types of feedback loops: the financial sector integrates feedback information about the state of our world and responds by deciding where to best invest its capital, which maintains stability. But investors also feed on each other's irrational exuberance, creating bubbles, or feed on each other's panic, accelerating a crash.

As the example of a financial crisis illustrates, the central concepts of systems theory—input and output, networks, homeostasis, and feedbacks—apply equally well to social systems, including those of our families, congregations, corporations, communities, and the community of nations. Each of these systems, taken alone, is composed of a web of relationships among its members, each has its own ways of maintaining stability, and each receives input from the world outside itself.

Bertalanffy realized this early on, which in the 1940s led him to propose systems theory as a universal scientific framework, a way to unify the natural and social sciences. Other scientific heavyweights, including Massachusetts Institute of Technology mathematician Norbert Wiener and anthropologist Margaret Mead, contributed to its development in the mid-twentieth century. Today systems thinking is influential in the fields of biology, education, psychotherapy, technology, and economics—and climate science. Thanks in large part to decades of work by Levins, it is also on the rise in epidemiology, ecology, and public health.

Using systems thinking, scientists now have tools to consider entire systems too complex to analyze using reductionism alone. "Someone must dare to look at the whole," writes physicist Murray Gell-Mann, founder of the Santa Fe Institute, in the introduction to his fascinating book *The Quark and the Jaguar.* Systems science allows us to do just that, and it has framed my thinking along my scientific and experiential journey.

AN EPIDEMIC OF EPIDEMICS

After 1976, forty diseases new to medicine surfaced, including HIV/ AIDS, Ebola, and a new hantavirus. Other long-quiescent diseases like cholera resurged and spread, as I had witnessed in Mozambique. The resurging infectious diseases began invading every medical practice. Why, in the middle of the epidemiological transition to diseases of affluence, were microbes reasserting themselves?

As Levins suggests, we can look to medical history for clues. Major epidemics have often followed periods of decaying infrastructure and dissolving social fabric. In A.D. 541, for example, as Emperor Justinian was presiding over the dissolution of the Holy Roman Empire, the bubonic plague swept through Constantinople (today, Istanbul), killing an estimated ten thousand people per day at its peak—so many that survivors ran out of grave space and stacked bodies in the open. The plague then spread throughout Europe to China, killing forty million people in numerous waves over the following two centuries. The plague died out, then reemerged in the early 1300s as peasants from the European countryside pushed into cities that were built for far fewer of them, leading to overcrowding, shortages of drinking water, and poor sanitation. The Black Death, as the plague was then known, killed 30 percent of Europe's population in half a decade.

Such pandemics (epidemics that spread through multiple regions) have transformed history. The Plague of Justinian led people to abandon cities en masse and settle into isolated fortresslike rural communities, thus helping to usher in the Feudal era. The Black Death led ultimately to protests against the old church order, to Protestantism, and to the Reformation. Smallpox, plague, typhus, and other diseases, sometimes introduced intentionally, contributed to the decimation of native peoples throughout the Americas between the sixteenth and nineteenth centuries, helping European colonists take control over new territories.

Astute observers of health and history have perceived the deep connections between disease and social order. Dr. Rudolf Virchow was a nineteenth-century pathologist and activist who pioneered the study of social medicine, which examines how social, cultural, and economic conditions affect health and disease. "Epidemics," he wrote, "are like sign-posts from which the statesman of stature can read that a disturbance has occurred in the development of his nation."

Virchow, a contemporary of John Snow, was active in the protests spurred by the rash of urban epidemics. These protests had a positive result, as civic leaders responded by installing sanitation systems and drinking water plants and pipes. These changes to the infrastructure helped quell infectious diseases, which made cities safer and became the critical driver of economic development. Infectious disease declined steadily, with a bit of help from antibiotics, which were first employed in the late 1930s. The decline continued until the mid-1970s, when the first sign of an uptick in more than a century was seen.

By the 1980s, it was clear that infectious disease had returned with a vengeance. We had not just two or three simultaneous epidemics but dozens of new diseases, occurring in humans, other animals, and plants: an epidemic of epidemics. As a physician and a public health practitioner, I had begun to study the environmental and social origins of individual diseases. But was this new raft of infectious ills—occurring worldwide—a symptom of a more fundamental global change?

In 1991, a devastating cholera epidemic gave me an epiphany. On January 29 of that year, the first cases of severe diarrheal disease appeared in the coastal district of Chancay, Peru, a port city near Lima. Health officials raced to the city to investigate, soon isolating the bacterium *Vibrio cholerae* from the stool of those who'd fallen ill. These were the first cases of cholera in the Americas in more than a century, and the sickness simultaneously appeared in two other port cities along Peru's 1,200-mile coastline. From there, it spread to other port cities and inland up streams and rivers, including the Amazon, throughout South and Central America. A massive epidemic was underway.

CLIMATE AND CHOLERA

My experience confronting cholera in Mozambique made me particularly curious about the causes of the Latin American epidemic. I soon came across the groundbreaking research of Rita Colwell, a tenacious and innovative microbiologist then at the University of Maryland, Baltimore County, who later became the head of the National Science Foundation. By then Colwell had been investigating the underlying causes of cholera epidemics for more than two decades.

When she began her work in the 1960s, several harsh realities were apparent to modern research scientists. Cholera was caused by the bacterium *Vibrio cholerae,* and people were infected when they drank water contaminated with the bacterium. Cholera bacteria passed into the drinking water when fecal discharge from cholera-infected people was spread by poor sanitation and mixed with water supplies. Many microbiologists considered the case closed.

But a great puzzle remained. Cholera epidemics waxed and waned, raging some years, then disappearing for decades or centuries at a time. Where did the cholera bacteria hide between outbreaks? Or, in the scientific vernacular, where was the "reservoir"?

From the beginning, clues pointed to the sea. The seven major cholera epidemics that have occurred in history all began on or near continental coastlines. Likewise, the first victims have tended to be fishermen and other seafaring people. Even more telling, historical accounts of the outbreaks often linked them to the arrival of ships from areas where cholera was endemic. In the early 1960s, when Colwell began her work, most scientists had put aside this "ancient" history. Colwell, however, had not.

For her doctoral research at the University of Washington in the early 1960s, Colwell had studied the salt requirements of several harmless bacterial cousins of *V. cholerae* that thrive in open oceans as well as in the less salty waters of estuaries, bays, and salt marshes. Colwell's early career took a turn for the better when a colleague suggested that she look for *V. cholerae* in the same bodies of water in which she was searching for the non–disease-causing bacteria. In 1969, the young microbiologist searched for and found *V. cholerae* in, of all places, the Chesapeake Bay.

"This was not something people wanted to know," Colwell recalled wryly. Infectious cholera bacteria were not supposed to be floating in the Chesapeake Bay, or any other large body of water—certainly not one frequented by commercial fishermen. The National Marine Fisheries Service of the National Oceanic and Atmospheric Administration (NOAA), which was providing her grant money at the time, promptly cut off her funding.

Colwell rebounded with financial support eight times larger from

National Sea Grant, another NOAA program, which supports coastal research. Colwell went on to pursue her cholera research for years, ultimately opting to work in the most cholera-prone region of the world: the low-lying Ganges River Delta in Bangladesh. There, large cholera epidemics erupt as if on cue from September through December, following the seasonal monsoons, and smaller epidemics occur with equal regularity between March and May.

Colwell chose to study Ganges delta water near Matlab, a small city just forty-five kilometers (twenty-eight miles) from the sprawling capital city of Dhaka. In one now-famous experiment, she began by pulling fifty-two water samples from the river delta. Into those samples, she added antibodies that had a fluorescent chemical attached and that bound specifically to *V. cholerae*, thereby helping to visualize the bacteria under the microscope. The results were stunning. All but one of the fifty-two water samples had tiny glowing particles in them, indicating the presence of *V. cholerae*. Perhaps most astonishing was the fact that the particles were less than one-fifteenth the size of the typical *V. cholerae* bacteria. And yet, Colwell was able to culture *V. cholerae* from seven of the positive samples.

Colwell concluded from this and other confirming experiments that *V. cholerae* could exist in the ocean and in estuaries in a tiny and dormant but hardy state. Colwell's colleagues were not swayed—a situation common for those who make paradigm-shifting discoveries. Most refused to accept that these miniature microbes were actually alive and capable of causing disease. One called her work "rubbish"; others mocked her, dubbing the dormant microbes "Colwell's ghosts."

Despite such derision, Colwell pressed on. She knew that benign, or noninfectious, marine *Vibrio* bacteria hitched a ride on minute marine animals called zooplankton that have shells made of a hard, carbohydrate-based substance called chitin. By sampling water from ponds and rivers in and around Matlab for three years, Colwell proved that infectious *V. cholerae* also hitched a ride on zooplankton and sustained themselves by feeding on the chitin in the zooplankton's shells. Had Rita Colwell discovered cholera's reservoir?

To see if cholera bacteria from seawater could cause disease, Colwell undertook another remarkable study—done with extreme care because of the toxic potential and conducted with medical treatment available promptly if needed. She enlisted human volunteers to swallow water containing the tiny dormant form of *V. cholerae*. The results were, once again, stunning. Colwell was able to isolate infectious cholera bacteria from her volunteers' stool samples. Seemingly, in one fell swoop, Colwell had solved one of the great mysteries of this ancient disease: the location of the *V. cholerae* reservoir. She bolstered that study with later work that showed that *V. cholerae* could hide out in a dormant state for years in the plankton in estuaries, swamps, and coastal oceans yet remain fully capable of infecting people when they drank water or ate shellfish that harbored the plankton.

Colwell's team obtained other evidence that bolstered the case for the ocean–cholera connection. When seawater warmed and its nitrogen and phosphorus levels were heightened, dormant cholera bacteria emerged from hibernation and became infectious. Under the same conditions, algae bloomed, followed after a lag by zooplankton, which fed on the algae. Meanwhile, monsoon-driven floods could flush cholera-carrying zooplankton into Bangladesh's inland waterways. The result was a kind of perfect storm, if in microcosm: as zooplankton were carried inland, the cholera bacteria they carried became infectious and would then infect people who drank unpurified water, which led to epidemics. In fact, annual algal blooms in coastal waters near Bangladesh were followed, after a lag, by cholera epidemics.

Almost as if to vindicate Rita Colwell's years of scientific dedication, the Peruvian cholera epidemic of 1991 began almost simultaneously in three distant port cities. If the disease had appeared in a single city, its appearance might have been explained by a traveler carrying the bacteria. But since the outbreaks began in widely separated harbors nearly simultaneously, cholera was most likely coming from the sea. This fact, and the fact that the cholera epidemic occurred during an El Niño, a large and periodic climatic event that warms nearshore waters, suggested a link between climate and cholera.

Over fifteen months, half a million people in nineteen South American countries contracted cholera and almost five thousand died. The death rate was much lower than in the Mozambique epidemic I experienced, because of accessible public health services and the coordination of the Pan American Health Organization, but cholera has persisted in Peru ever since.

My appreciation for Rita Colwell's revolutionary cholera research, and the drama of the Latin American cholera epidemic of 1991, against the backdrop of my time in Mozambique, set me on a new path. Global warming, I suspected, could threaten human health in ways that scientists had yet to fully anticipate.

2

The Mosquito's Bite

In 1991, several Harvard faculty members, including Dick Levins, Uwe Brinkman, Mary Wilson, Andy Spielman, Richard Cash, and I, invited other colleagues to join what we called the New Disease Group, which met weekly to discuss the causes of emerging and reemerging diseases that had begun to afflict the world. We sensed that changes in the global environment might be contributing to the rise in disease; thus we cast our net wide in deciding who we should include. Participants and guest speakers were experts in fields as diverse as ecology, entomology, epidemiology, infectious diseases, population biology, mathematical modeling, international health, evolutionary biology, climate, environmental analysis, and marine ecology, among others. For more than two years, I was energized by the intellectual brilliance and creativity of these colleagues.

By the time we launched the New Disease Group, I'd been encountering new and reemerging diseases in my medical practice for more than a decade. I'd begun work as a primary care doctor at Cambridge Hospital, a Harvard-affiliated hospital, in 1981 and had joined the Harvard faculty as an instructor that year. It was while working at Cambridge Hospital that I encountered the epidemic of HIV/AIDS. In the United States, HIV/AIDS spread farthest in low-income communities, and I focused on treating patients from East Cambridge, many of whom were poor and working-class immigrants from the Azores and other Portuguese-speaking countries.

At first, no drugs were available to battle the human immunodeficiency virus, which meant that an HIV-positive diagnosis was most often a death sentence. Together with a dedicated team of doctors and nurses, we established the hospital's first AIDS clinic. We treated disadvantaged patients with the best therapies then available. We also provided acupuncture, social services, psychological care, housing referrals, and nutrition counseling. Our clinic was part of an early vaccine trial—one that proved unsuccessful in stemming the march of this devastating illness. Most patients succumbed. It was difficult, sad work, but our community of caring caregivers enabled us to be supportive.

Drawing on my public health training, we started a community outreach and support program for those afflicted. (The nonprofit organization Cambridge Cares About AIDS grew from this initial vision and continues to work to prevent AIDS and assist AIDS patients today.) We also pioneered an educational program geared to preventing transmission, which took us into schools in Cambridge, South Boston, and other Boston neighborhoods to educate students about safe sex, intravenous drugs, and HIV. Our work treating AIDS, caring for AIDS patients, and increasing AIDS awareness among an urban population deepened my appreciation of the social forces and inequities that contribute to health, health care, and disease.

In the early 1960s, I had visited Haiti and seen abject poverty up close, and I'd taken two months off from medical school in 1967 to study health and nutrition in a small town in southern Mexico. Throughout the 1980s, I continued to work in the international arena, primarily with Physicians for Human Rights. On a 1986 mission to Nicaragua, where U.S.-backed Contras were fighting the left-leaning government, I accompanied public health students to study the health of civilians living in war zones. Despite the physical ills and injuries they suffered, we discovered that, while anxiety was universal, those active in community organizations were not depressed. The lesson—that being part of a community and engaging in solutions enhance psychological well-being—continues to influence my outlook and my work.

On a 1991 mission to refugee camps in a mountainous region of Turkey that borders Iraq, our team carried out a rapid assessment to gauge the illness and mortality of three-quarters of a million largely middle-class, professional Iraqi Kurdish refugees who had fled northern Iraqi cities during the first Gulf War. They had no access to potable water, and many children were dying daily of diarrheal diseases. Our assessment helped evacuate the camps, relocating the refugees to better and protected conditions in the western Iraqi plains. A 1985 mission—this time sponsored by the fledgling American Jewish World Service—took Andy and me back to Mozambique. By 1985, South African–backed Contras were waging a war to undermine Mozambique's progressive health care system and destabilize its government. We accompanied an airlift of medicines and flew up-country to visit war victims in a rural hospital.

The international work in these war zones gave me a keen sense of the physical and emotional damages that result from armed conflict and the crucial role of peace in ensuring the delivery of essential public health measures. These experiences continue to drive my work for clean development as a force for building peace and healthy, equitable societies.

. . .

By the mid-1980s, the international scientific community was beginning to recognize the magnitude of the world's environmental problems. Biologists were reporting startling declines in frog populations around the world, even in what seemed like pristine surroundings. The ozone hole had just been recognized, and, in 1987, the United States and twenty-three other nations signed the Montreal Protocol to phase out the use of chlorofluorocarbons and other chemicals that caused it. (Almost two hundred nations had signed it as of 2010.) During the same period, the scientific community began to seriously discuss global warming.

In the summer of 1988, more than five thousand people, an extraordinary figure by any measure, died in the United States as a result of record-setting heat waves. Blistering temperatures withered crops in the Midwest and Great Plains and sparked wildfires that consumed a third of a very parched Yellowstone National Park. That summer, a scientist named James Hansen sounded the first major public alarm on global warming. On a 101-degree day in the nation's capitol, Hansen, a physicist and atmospheric researcher at NASA's Goddard Institute for Space Studies in New York City, testified to a U.S. Senate committee that "the earth is warmer in 1988 than at any time in the history of instrumental measurements."

Hansen's overhead transparencies showed ominous data to back up his assertion. It showed that the Earth's temperature had warmed 0.4°C (0.7°F) in the seven years between 1980 and 1987, compared with the thirty-year mean of annual temperatures between 1950 and 1980. Moreover, the 0.4°C increase was almost exactly what NASA's state-of-the-art computer models had predicted would occur as a result of the greenhouse effect. "There is only a one percent chance of accidental warming of this magnitude," Hansen told the intently listening senators and journalists in the Dirksen Senate Office Building.

When Hansen's testimony came to an end, Senator Timothy Wirth, a Colorado Democrat who had organized the hearing before the Senate Committee on Energy and Natural Resources, concluded, "Now, the Congress must begin to consider how we are going to slow or halt that warming trend and how we are going to cope with the changes that may already be inevitable."

The assembled press treated the NASA scientist's comments with gravity. Hansen's credentials were first-rate and his data impeccable. He was widely respected in the scientific community. He was also the first climatologist who refused to equivocate on the definitive nature of his findings. The following day, June 24, the *New York Times* reported on Hansen's testimony in a story that carried the headline "Global Warming Has Begun, Expert Tells Senate."

. . .

For most Americans, Hansen's testimony was the first bona fide confirmation of the climate change phenomenon. Indeed, he had managed that summer to focus attention on global warming in a way scientists before him had been unable or unwilling to do. Other scientists had been investigating the subject: some tabulating temperatures at various latitudes, longitudes, and altitudes, which indicated that the planet was warming, and others building computer simulations to project how the greenhouse effect might alter the world's climate in the decades to come.

But scientists had veered toward excessive caution, emphasizing the uncertainty in their data, even though their cumulative results suggested overwhelmingly that CO_2 levels in the atmosphere were rising ominously and that global warming was under way.

A few prescient scientists had become concerned about rising atmospheric CO_2 levels decades earlier. In 1958, climate scientist Charles David Keeling established a CO_2 monitoring station on the isolated, blustery peak of Mauna Loa in Hawaii. When Keeling began measuring CO_2 in that rarified atmosphere, the gas was present in the atmosphere at levels of 315 parts per million (ppm). Measurements over three decades had demonstrated a steady rise of CO_2 year after year. By 1988, CO_2 levels had reached about 350 ppm. (Today, they are approaching 400 ppm.)

Even before 1958, however, CO_2 levels had been rising. By examining air bubbles trapped in ice cores, scientists have sampled atmospheres of the past. Those studies show that around 1800, before the rise of polluting industries, the CO_2 level was 278 ppm. In short, by 1988 confirming data had been accumulating for decades. Not only did Hansen's courageous and groundbreaking testimony establish a new benchmark; his presentation persuaded formerly reticent scientists and policy makers to step forward and speak out.

Soon after Hansen's testimony, the United Nations General Assembly produced a resolution calling for "protection of the global climate for present and future generations of mankind." They went on to establish the Intergovernmental Panel on Climate Change (IPCC), a UN body representing more than one hundred nations. The UN charged the IPCC with producing a definitive report to evaluate the science and to offer policy options to stabilize the climate and ease the impacts of ongoing climate change. The panel met for the first time in November 1988, five months after Hansen's watershed testimony to the Senate.

GLOBAL WARMING IS REAL

At first nations responded with an urgency commensurate with the severity of the climate crisis. In 1990, the IPCC produced the first of its influential and sobering assessments. The report was the product of an unprecedented level of international cooperation. One of the authors' striking predictions was that global mean temperatures would climb between 0.2°C and 0.5°C (0.36°F–0.90°F) per *decade* in the twenty-first century, a faster climb than the Earth had experienced in more than ten thousand years. They also projected that sea levels would rise 0.6 meters (2 feet) by 2100. (Today, with glacial melt accelerating, a rise of up to 1.8 meters, or 6 feet, by century's end is possible.)

The IPCC authors concluded that "human activities are substantially increasing the atmospheric concentrations of greenhouse gases [that] will enhance the greenhouse effect [and result in] an additional warming of the earth's surface." The report called for a 60 to 80 percent reduction in greenhouse gas emissions.

In 1990, nearly six billion tons of carbon was being pumped into the atmosphere each year as a result of burning fossil fuels and felling forests. Today it is ten billion: approximately eight from fossil fuels and two from deforestation. Meanwhile, the Earth's natural "sinks"—forests and oceans—together remove half of what is generated, leaving five billion tons to accumulate in the atmosphere each year (figure 6). These sinks, too, are destined to become saturated, causing yet more CO_2 to accumulate in the atmosphere.

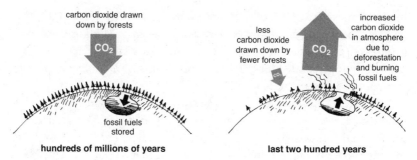

FIGURE 6. Fateful fuels. Vast forests helped draw down CO_2 from the Earth's early atmosphere, creating underground stores—today's fossil fuels. By burning fossil fuels and destroying today's forests, we're sending more and more carbon to the atmosphere, reversing the evolution of the planet's atmosphere.

As the IPCC was preparing its report, the medical profession received its first warning that a warming climate might pose health risks. In 1989, Dr. Alexander Leaf, a professor of medicine at Harvard Medical School, presented the spectrum of health issues in a groundbreaking article in the *New England Journal of Medicine*. But the UN-sponsored scientific group made little mention of the potential health impacts of climate change. At the time, few in the health community, let alone the general public, were aware of these threats. But some of us began reaching out across the globe to one another to share these concerns.

It was against this backdrop that the New Disease Group convened. Our weekly meetings were filled with informed, energetic, free-form discussion. As Dr. Mary E. Wilson, a professor of medicine at Harvard Medical School, wrote several years later, "We threw ideas into a crucible for testing and remolding. We marveled at concepts and ideas that were at the fringe of our understanding. We strained to look at data and events simultaneously through the prism of other disciplines." It was an exciting period of discovery and exchange.

It was clear that most emerging diseases could—like most outcomes of a disease—be traced to multiple causes. One resurging disease, tuberculosis, thrives when the host is malnourished or is weakened by infection with HIV, and it is most easily transmitted in close quarters, such as a New York City crack house in the early 1990s or the mines of South Africa. And while AIDS is caused by the HIV virus, genetic, economic, and behavioral factors influence its development as well. Its very emergence as a virulent pathogen may have been connected with the heavy burden of disease and malnutrition that weakened immune defenses among Africans oppressed by colonialism.

Ecological changes under way were clearly also contributing to the spread of some diseases. Fertilizer runoff into streams and estuaries promoted the growth of red-tide algae that produced dangerous neurotoxins. Removal of top predators from food webs allowed the proliferation of pesky animals—like mice and rats—that could carry disease.

After exploring how warming seas might spread cholera, I began to consider how a warming climate might spread other infectious ills. It was known that certain infectious diseases occurred only in geographic areas that were neither too cold nor too hot and only when weather conditions were right. We began to examine diseases transmitted by vectors—animals that transmit but do not suffer from a disease. For many diseases, the vector is a flying insect. Mosquito-borne disease poses the greatest danger—a danger illustrated by the case of a seven-year-old girl named Elena from Karatina, Kenya.

THE MOSQUITO'S BITE

Elena Githeko was normally energetic and chatty. But on a Tuesday morning in 2003, Elena's mother, Anne Mwangi, found her daughter quiet and listless, her forehead warm with fever. Anne thought it was just the flu, so she did what any concerned mother would do: she stayed home from work to care for her daughter.

At age seven, Elena had her mom's mischievous almond eyes, her dad's chubby cheeks, neatly braided cornrows, and a broad smile. Until that Tuesday, she'd been perfectly healthy.

But she was still sick on Wednesday morning. Anne stayed home yet again. On Thursday, Elena looked a little better, Anne thought, so she left her younger daughter at home with the family's maid, an older woman who had raised several children of her own. Anne then headed off to work at Mathaithi Secondary School, a girls' high school where she teaches history and Christian religious education. Her husband, Peter Mwangi Githeko, who goes by "Mwangi," went off to his job as a chemistry and biology teacher at a local boys' high school.

The phone rang late that afternoon. As the maid spoke, Anne's fine-boned face knotted with worry.

Elena could not keep her food down. She had a horrible headache, and she was burning with fever. The worried maid, who had no car, had carried the sick girl about a mile on her back to the nearest community clinic. The doctor there gave Elena medicine to cool her fever and stop her vomiting. But it didn't work, the maid told Anne. Elena was getting worse fast.

Anne called her husband. Within minutes, Mwangi Githeko arrived to pick up Anne in the family vehicle, a blue 1970s-era Toyota pickup. At 5 P.M., they sped home over winding, hilly roads. By the time they arrived at their house, Elena was crying. Her feet were cold, she was dehydrated, and her forehead was on fire.

They raced to another clinic for a second opinion from another doctor. Take Elena straight to the hospital, he told them. It was 6 P.M. At Jamii Hospital in Karatina, a doctor quickly took Elena's blood pressure and temperature, listened to her symptoms, and did some tests. Malaria, he said. She needed to be admitted—immediately. That horrible headache might be a sign of cerebral malaria, a condition in which the malaria parasites burrow into the cerebrospinal fluid that bathes the brain, and sometimes into the brain itself. Anne Mwangi had lived in Kenya her entire life, long enough to know what cerebral malaria could mean. "I thought my daughter was condemned to death," she says.

. . .

Each year, malaria sickens one of every twenty people on the planet—some three hundred million people, a total roughly comparable to the population of the United States. In many ways, Elena Githeko's case was typical: intense fever, sweats, shaking chills, and extreme weakness. Many of those who recover suffer long-lasting anemia, periodic fevers, and chronic disability. The World Health Organization estimates that malaria kills more than a million people a year, most of them children. In Africa, where 75 percent of all cases occur, a child like Elena Githeko dies of malaria every thirty seconds.

Malaria harms so many in part because the malaria parasite, a complex single-celled microbe called a protozoan, dodges the immune system more effectively than most bacteria or viruses. That's one reason why no effective vaccine has been developed, despite decades of effort. The antimalarial drugs that work reliably have nasty side effects, or they are too expensive for most African governments. What's more, the Kenyan government, like most African governments, often lacks the manpower and money to get the drugs to enough people in time to stop the disease. In many places, the parasite has evolved resistance to once-effective drugs such as chloroquine. And bed nets—lightweight plastic nets that cover the

FIGURE 7. Anne Mwangi, a schoolteacher in Karatina, Kenya, holds a photo that includes her daughter Elena Githeko (front, center), who suffered a life-threatening bout of malaria in 2002. In Karatina and other villages in the foothills of Mount Kenya, malaria, once absent, has become commonplace as warmer temperatures have allowed malaria-carrying mosquitoes to live at high altitudes. (Photo by Dan Ferber)

sleeping person like a tent—can help protect people from malaria-carrying mosquitoes while they're sleeping but not when they're out and about.

In much of sub-Saharan Africa, Elena's diagnosis would have been sadly routine. Here, in the foothills of Mount Kenya, it was remarkable. Mount Kenya is a massive, long-extinct volcano, more than fifty kilometers wide at its base, with snow-capped peaks that graze the sky at an altitude of more than five kilometers. Despite being just fifty kilometers (thirty-one miles) from the equator, Karatina sits in Mount Kenya's foothills at an altitude of 1,600 meters (almost a mile), high enough to have a distinctly cooler climate than the low-lying tropical areas of the country. When the first British colonists settled in Kenya in the late

1800s, the central Kenya highlands, like highlands in Tanzania, Uganda, and Ethiopia, were considered a place to escape the *mal aria*—the bad air—that was thought to cause the disease. When Anne Mwangi and Mwangi Githeko were growing up in Karatina in the 1960s, malaria was unheard of. A 1970 national atlas had deemed the region "malaria-free." "We never had this problem," Mwangi Githeko says.

Elena, however, definitely had a problem with malaria, and possibly cerebral malaria. As her mother watched, the doctor approached Elena with a syringe full of antimalarial drugs. "Doctor, do not touch me! Don't inject me!" she cried. But the doctor did what he had to do, putting an intravenous line into Elena's hand. A powerful antimalarial drug coursed directly into her blood, where it could do battle with the parasites then wreaking havoc in her frail body.

All Thursday night, Anne lay with her daughter on a narrow bed at Jamii Hospital as other patients and their mothers lay nearby. From inside the white mesh mosquito net, Anne listened to the slow whirring of the overhead fans and held her daughter. She slept with one eye open, ready to call a nurse in case of trouble. Because the antimalarials made her ears ring loudly, Elena couldn't hear. She didn't eat, she didn't talk, and she hardly moved.

On Friday, the epic battle of drug versus parasite continued in Elena's body. On Friday night, Anne again slept in the hospital bed, holding her daughter close. On Saturday morning, Anne left her daughter briefly, then returned with family friends. Another little girl, a friend of Elena's, called out to Elena. Elena answered! Soon she was sitting up and smiling. She remained in the hospital for another day before she was released (figure 7).

Elena is now an energetic girl of fourteen, but for years the experience haunted her. Months later, she spoke of a dream she'd had while her body was in the grip of malaria. Three clouds were coming for her—one for mom, one for dad, and one for herself. "I saw a white cloud coming down, and I wanted to go in it," she said. And again and again in the months and years that followed, Elena would ask her mom, "Do you remember the day when you walked into the hospital and carried me because I was dead?"

CLUE FROM A CATTLE TROUGH

Elena Githeko's close call shocked more than just her parents. It stunned Andrew Githeko, Mwangi Githeko's brother and Elena's uncle. Andrew also happens to be a friend and colleague and a world-renowned malaria

FIGURE 8. Andrew Githeko, shown here in front of his family's ancestral tea farm in Ihwagi, Kenya, directs the Climate and Human Health Research Unit at the Kenya Medical Research Institute in Kisumu. His research has documented a link between climate change and malaria in the country's central and western highlands. (Photo by Dan Ferber)

expert who directs the Climate and Human Health Research Unit at the Kenya Medical Research Institute (KEMRI) in Kisumu. When I first met Andrew in the late 1990s, he was a rising star in the field of medical entomology, the study of insects that carry disease. Andrew Githeko's is the tale of a scientific underdog who, through patience and persistence, triumphed against formidable opposition. And it's a tale of discovery that has helped settle a contentious question: Can global warming contribute to the spread of disease?

Now a compact man of fifty-two with an oval face, close-cropped hair, and a steady gaze, Githeko listens patiently when asked a question, then speaks without hesitation or hurry, with the air of someone who's used to being heard. That sense of authority may come from growing up as the eldest of five brothers in a prominent and relatively well-off tea-growing family in Ihwagi, a rural village near Karatina, approximately fifty miles north of Nairobi. In a country where many scratch out an existence from tiny farms, they lived in a sturdy, four-bedroom stone house on their five-acre hillside farm (figure 8). Across a verdant valley from their home was the edge of a vast forest—home to elephants, bongos, and buffalo—that in those days continued unbroken to the slopes of Mount Kenya.

On the farm Andrew, his four brothers, and their parents tended pedigree cows and worked fields of maize, potatoes, and vegetables irrigated by water from the nearby Ragati River, one of several rivers to come coursing down from the then plentiful snows of Mount Kenya. Despite being just a few miles from the equator, Ihwagi had air that was healthy and cool; in the winter months of June through August, the temperature dropped as low as 4°C (43°F). Each evening, as the sun dipped over the hills, Andrew's job was to bring in firewood and light the family's hearth.

Githeko's father was a pious man, a church elder as well as a successful businessman, and he poured much of his excess profits back into the local community. Such success and good works led others in the local community to revere the Githeko name, and they came to expect the same from Andrew and his brothers. The country was young then, having attained independence from Great Britain only in 1963, and all things were possible for a young Kenyan. At the University of Nairobi, one of Kenya's top universities, Andrew majored in chemistry and zoology, then worked for a few years as a junior scientist at KEMRI in the western Kenyan town of Kisumu, a port town on Lake Victoria. The ambitious young scientist next nabbed a prestigious fellowship to the world-famous Liverpool School of Tropical Medicine in England, where he trained with a highly regarded medical entomologist and investigated the biology of malaria-transmitting mosquitoes in the rice-growing areas of western Kenya. In 1991, before returning to England for his final year of training, he witnessed his first intense malaria epidemic in the hills of western Kenya.

It was pandemonium. "Patients come in. Some need a blood transfusion," Githeko recalls. Others develop cerebral malaria. "They're mad, and you need to strap them to the bed. . . . Others, there's nowhere to put them; you put them outside, and it's raining." Githeko shakes his head. "You've got people under the bed; others are yelling. . . . You can't go home. You can't get tired. The morgue is full." As KEMRI's resident mosquito biologist, Andrew went from village to village, hunting for malaria-carrying mosquitoes. "They were everywhere, breeding in large numbers," he recalls. The malarial parasites responsible for the epidemic had evolved resistance to chloroquine, the most common antimalarial drug. "We couldn't control this thing," he says. "It was very scary."

Runaway epidemics like the 1991 western Kenya outbreak occur only at the edge of malaria's range, typically at higher elevations. At lower elevations in the tropics, steamy weather maintains mosquitoes and malaria year-round, and people have developed an uneasy balance of power with malaria parasites. Most residents are exposed as children, and some succumb to the disease. But the majority of malaria patients survive, and those who do develop partial immunity. That reduces the intensity of later malaria infections to that of the flu.

But the story is different at higher elevations, including the western Kenya highlands, the Usambara Mountains of Tanzania, the highlands of Ethiopia, the Ruwenzori Mountains of Uganda, and the mountains of Indonesia and New Guinea. The good news in such areas is that most residents have not suffered from malaria; the bad news is the same. Because their bodies have never been exposed to the parasite, their immune systems respond too slowly to stop the infection during its early stages. And they, unlike lowland residents, usually lack two genetic traits that make people less susceptible to malaria. This genetic and immunological vulnerability mean that when malaria does hit highland residents, it hits harder, creating waves of disease that lay waste to vulnerable Africans, just as smallpox laid waste to vulnerable Native Americans during early colonial times. People in highland areas account for 12.4 million cases each year—just 2.5 percent of the global total—but the 150,000 annual malaria deaths in highland areas are 12–25 percent of the annual worldwide total. One in five residents of East Africa, or about 125 million people, live in highland areas susceptible to malaria epidemics.

For several years following the 1991 epidemic, Githeko continued to monitor epidemics near his home base in western Kenya, doing the careful, rigorous work that contributes to the steady march of public health science. Then, one day in 1993, Githeko looked into a cattle trough and got a glimpse of the future.

It was around Christmas, and Githeko was visiting family in Ihwagi, high in the Mount Kenya foothills. While out for a stroll, he peered into the water the cattle were drinking on a neighbor's farm and saw a wriggling, wormlike creature the size of a grain of rice. He immediately recognized it as a mosquito larva. He saw the way it lay flat on the water's surface, and he noted its brown body and the white-colored band near its head—all telltale signs of the genus *Anopheles,* which includes the three species that transmit malaria in Kenya. But was it really an *Anopheles?*

To find out for sure, he would have to have the mosquito larva analyzed by entomologists in a laboratory. But it was a holiday, and Githeko's laboratory was halfway across the country. So he made do. He didn't have the correct chemical—a mix of 90 percent alcohol that he'd typically use to preserve mosquito larvae—but he did have some vodka. "I put it in a vial and hoped," Githeko recalls.

He had not yet begun studying climate change in 1993, but he was curious enough about the misplaced mosquito to ship the vodka-pickled larva all the way to Atlanta, Georgia. From his lab in Kisumu, he called Frank Collins, a colleague at the U.S. Centers for Disease Control and Prevention, who had the reagents and expertise needed to confirm the genus and species of the mosquito. He said, "Frank, can you have somebody look at this thing and tell me what it is?" Collins and some colleagues did as Githeko asked them. But vodka hadn't quite worked as a preservative, so by the time the larva arrived in Atlanta, it had seen better days. Collins's team could tell only that the mosquito was in the genus *Anopheles*—possibly, but not definitely, one of the *Anopheles* species that transmits malaria. Githeko, busy with unrelated research, let the matter drop for a time.

For several years afterward, Githeko finished up research he'd started in graduate school on the biology of malarial mosquitoes, publishing paper after paper in entomology and public health journals. Then, in 1996, he found himself at loose ends, unable to do the field studies he had planned. "I didn't have a research grant. I had no money," he says. "What I had was my laptop."

Again Githeko made do. He retooled an established mathematical model that takes weather measurements and biological traits of *Anopheles* mosquitoes and uses them to predict mosquito population growth. He started with predictions from top climate scientists about how climate would change in the twenty-first century; then he plugged those data into the model. "It was a nice way to spend my time," he says. "I needed little grant money. I could go to the literature, find data, plug it in, then simulate all kinds of things." One series of simulations led to an unsettling prediction about the impact of climate change on the spread of malaria in the highlands of East Africa.

ROUSING THE PARASITES

To understand the reasoning behind Andrew Githeko's projections, it's important to take a close look at the small insect with a sharp proboscis that causes a disproportionate amount of human suffering. Only some of the three thousand known species of mosquitoes transmit disease to humans, while others feed exclusively on rodents, birds, or even snakes. And only female mosquitoes suck our blood, which they do to

FIGURE 9. Andrew Githeko samples water from a ditch in a farm field and former swamp near Kisumu, Kenya. The widespread presence of larvae of a malaria-carrying mosquito species in such standing water augurs deadly malaria outbreaks. (Photo by Dan Ferber)

obtain the nutrients required to allow their fertilized eggs to survive and develop. But they still feed on us often enough to transmit more disease to humans than any other animal vector does.

They can transmit disease, however, only when the climate is right. The reasons for this are biological. Depending on the species, female mosquitoes lay their eggs on flood-prone soil or in sun-warmed puddles, open containers, ditches, streams, or swamps—any place rife with water (figure 9). The eggs hatch into larvae, such as the rice-sized wrigglers Andrew Githeko had spotted in the cattle trough. The larvae feed on microbes, algae, and other nutritious muck for one to three weeks, shedding their skin four times as they grow into bigger versions of themselves. Then they morph into a resting stage called a pupa before miraculously emerging whole as an adult mosquito, much as a caterpillar morphs into a pupa before emerging as a butterfly.

Mosquitoes, like all insects, are cold-blooded animals, which means that the temperature of the air outside determines the temperature inside their bodies. When it's too cold, mosquito larvae simply can't develop.

The mosquito in sub-Saharan Africa that transmits malaria most efficiently, *Anopheles gambiae,* cannot breed below 16°C (61°F). The other two species of malaria-transmitting *Anopheles* mosquitoes in the region shut down below 18°C (64°F). The warmer it gets, however, the more quickly *Anopheles* larvae develop, up to about 35°C–40°C (95°F–104°F); above that, they're cooked. What's more, as it gets warmer, adult female mosquitoes feed more often and digest blood faster. As long as the temperature is within the insect's range of viability, warmer weather means more mosquitoes.

Malaria begins when a female mosquito settles on a human, inserts its proboscis, and withdraws blood for its future brood. If the mosquito is infected with one of four species of single-celled malaria-causing parasites of the genus *Plasmodium,* the parasites can move from the mosquito's saliva into our bloodstream, infecting us. From there, it's on to our liver, where they replicate, morph into a different form, and emerge to infect red blood cells. They multiply inside those cells, creating armies of progeny. Those progeny then rupture their host cells and escape to reinfect still more red blood cells, creating a vicious cycle that kills blood cells by the millions, causing the anemia associated with the disease. Some of the protozoans morph again inside red blood cells into a new life stage and become gametocytes—the reproductive form of the parasite—which then circulate in the human bloodstream. When a new mosquito settles onto our arm and takes a blood meal, gametocytes may end up in its stomach, where they grow and morph again, ultimately landing in the mosquito's saliva in a form that's ready to infect a human.

Along with the mosquitoes that carry it, the malaria parasite also develops faster when the weather is warmer. It is this stage of the life cycle—parasite development inside the mosquito—that is most exquisitely sensitive to temperature. The most dangerous malaria parasite, *Plasmodium falciparum,* takes fifty-six days to mature inside the mosquito at 18°C (64°F), nineteen days at 22°C (72°F), and just eight days at 30°C (86°F). But a female *Anopheles gambiae,* the mosquito that transmits *P. falciparum* in Kenya, lives only about two to three weeks. Therefore, at 18°C, the mosquito will likely die and take the immature parasite with it. However, when the average temperature is just two or three degrees warmer, there's a strong possibility the parasite will mature before the mosquito dies, increasing the chances for malaria transmission, which can lead to epidemics. And epidemics were exactly what Andrew Githeko wanted to prevent.

MALARIA CREEPING UP

Public health workers are taught to think systematically about how to prevent disease. If an epidemic has already begun, the first imperative is to treat and rehabilitate those affected, to minimize death and disability. This is called tertiary prevention, and it's what the workers in the health clinics of Kenya's malaria-stricken regions were already doing. Secondary prevention means reducing the number of people being harmed by an epidemic, and it is better than treating patients after they've been stricken. One good way to reduce the number of people harmed is to predict the onset and intensity of an epidemic, then take measures to prevent it and mitigate its effects.

That's what Andrew Githeko was doing. To predict an epidemic, first one has to understand how epidemics tend to unfold. Typically, cases trickle in at first, then skyrocket, then peak and fall off. Plotted over time, the result is a bell-shaped curve—the same curve that's used to plot the distribution of student grades in a class. Health officials typically detect an epidemic only when the first rash of cases has been diagnosed and numbers are already climbing. By the time they can institute protective measures, many of the people who will be sickened are already infected, and it's too late to stop the spread. During malaria epidemics in the East African highlands, the disease can infect up to 60 percent of a village population, overwhelming local health workers. It's even worse when the parasite is drug-resistant: thousands get sick; hundreds can die.

Githeko wanted to use local weather data to predict epidemics early enough to ship medicine and health workers to villages that would be affected, to distribute bed nets, and to spray insecticides that kill malarial mosquitoes. But predicting epidemics of infectious disease is no routine task.

Armed with his laptop computer, Githeko began modeling the effect of climate and weather on malaria transmission. He started with a mathematical equation that's proven to help predict potential outbreaks of malaria and other mosquito-borne diseases. To apply this equation, one first needs to know which species of mosquito and which species of parasite are prevalent in the area. One also needs to know how often a mosquito bites humans in a twenty-four-hour period and the mosquito's odds of surviving during that period. Finally, one needs to know how long it takes for the parasite to develop and become infectious inside that particular species of mosquito. Decades of malaria research have established reliable estimates of these quantities for major malarial parasites and the mosquitoes that transmit them. When one plugs these quantities into the equation, the result is a single number that represents the capacity of a specific species of disease-carrying mosquito—the vector—to transmit a specific type of malaria. The higher that number—called the vectorial capacity—the greater the chance of an epidemic.

Githeko knew that warmer temperatures—within a range—made it more likely that mosquitoes would survive and bite more often, and parasites would then multiply and mature more quickly. He also knew that rainfall created the puddles and standing water that are ideal breeding spots for mosquitoes and could quickly make them more abundant. By predicting temperatures and rainfall over a period of months, he could, in theory, project where and when an epidemic was most likely to arise.

Without any of the high-tech capabilities grant money would have provided, Githeko was able to use a simple spreadsheet program—and historical temperature and rainfall data for the western Kenya highlands, as well as data from published studies of mosquitoes, malaria parasites, and past epidemics—to develop a mathematical model that correctly predicted past outbreaks in the region, thereby verifying the accuracy of his model.

The model also made a startling new prediction. By the late 1990s, when Githeko did his analysis, the IPCC already had reported that the Earth had warmed 0.6°C (1.1°F) through the twentieth century and had predicted that temperatures would continue to rise by at least several degrees more through the twenty-first. Githeko's model indicated that a temperature change of just 2°C could trigger malaria epidemics in many once-safe highland regions of East Africa. This meant that global warming would eventually make millions of people in malaria-free parts of East Africa—including his family in the highlands of central Kenya—vulnerable to the disease. Though alarming, the threat seemed distant; climate scientists thought it would take decades for the region to warm that much.

In 1997, Githeko was invited to give a talk at a conference on the potential health effects of climate change. The conference was held in Nairobi at the International Center of Insect Physiology and Ecology, a respected research institute, and the sponsor was an organization called the International Geosphere–Biosphere Programme, which provides scientific knowledge to advance sustainability and respond to global change.

Githeko struggled to keep up with technical talks on climate and atmospheric science, fields then new to him. Toward the end of the meeting, the discussion veered back to biology. The young entomologist stood up. "I said, 'Gentlemen, I've been doing some modeling. Please give me five minutes,'" Githeko recalled. He put up his transparencies. He told them about his modeling studies and about his predictions that global warming would spread malaria in Kenya. Soon after, an invitation arrived: would he be interested in joining the Intergovernmental Panel on Climate Change, the one-thousand-member body of experts whose mission was to issue definitive scientific reports on the science of global warming?

"I had no clue what the IPCC was, but I was ready to do anything," Githeko recalls. "I didn't have a grant. I was so frustrated. I said, 'Whatever. I'm ready!'"

CLIMATE SENTINELS

By the time Andrew Githeko joined the IPCC in 1997, a debate was coming to a boil in the public health world. A growing number of public health scientists were concerned that global warming would alter the distribution of infectious disease and, in some cases, expand its range. Still, at the time, there was not much hard evidence supporting that belief, and climate skeptics in the scientific community didn't hesitate to say so.

That year I pulled together a team of eight climate scientists, ecologists, and public health researchers to search for early signs that climate change was altering global patterns of disease. To study global change, it's necessary to employ methods that differ from the controlled experiments used by laboratory researchers, who discern cause and effect by subjecting otherwise identical samples to treatments that differ in a single variable and then assessing the outcomes. For example, to determine conclusively whether warmer temperatures cause mosquitoes to reproduce faster, laboratory researchers might monitor reproduction of identical groups of mosquitoes, one group at room temperature and one in a warm room. If the mosquitoes in the warm room produce progeny more quickly, the researchers could then conclude that warming causes mosquitoes to reproduce more quickly. To be confident in that conclusion, a good scientist would also reproduce the experiment several times to see if the result was the same each time, thus making sure it could be trusted. If it couldn't, then the hypothesis would be disproved—and rejected.

In studying the causes and consequences of global change, such tightly controlled experiments are impossible. We can't control the

Earth's atmospheric conditions, and the experiment humanity is conducting with our planet can't be replicated. But we can examine trends and patterns, and test whether predictions from computer models match real-world observations. When they do, we say we've found a fingerprint of global warming.

The IPCC bases its conclusions about the existence and effects of climate change in part on such fingerprint studies. By 1995, three clear and ominous fingerprints had already appeared: warm winters, more extreme weather events, and disproportionate warming at high elevations.

In our fingerprint study, we investigated the effects of high-elevation warming on glaciers, plants, and mosquitoes. Mountains are excellent study sites precisely because of their steepness. In fact, moving up just four meters (thirteen feet) on a mountain is the equivalent in temperature change to moving 2.4 kilometers (1.5 miles) in latitude. Because of this, one can move from desert to tropical to tundra conditions simply by traveling a few miles.

For these reasons, mountain environments provided us the perfect petri dishes to study climate change and glimpse a snapshot of our future. By the late 1990s, we found, mountaintop glaciers were retreating rapidly worldwide—throughout the Andes, in the African highlands, the Alps, New Zealand, Asia, and Indonesia, and on the Tibetan Plateau. For example, the edge of the Qori Kalis glacier, the major outlet glacier of the Quelccaya Ice Cap in the Peruvian Andes, the largest glaciated area in the tropics, had retreated four meters (thirteen feet) a year throughout the 1960s and 1970s and by 1995 was shrinking by thirty meters (ninety-eight feet) annually. As we moved into this century, these trends continued to accelerate, and many mountain glaciers may soon disappear.

We also found that plants were responding to the changing mountain climate. The plants had moved upward on thirty Alpine peaks, and they were moving up at four meters (thirteen feet) a decade, according to the investigation by one member of our team, a German plant ecologist named Georg Grabbher.

Insects were following the warming climes as well. Populations of Edith's Checkerspot butterfly, a gorgeous tapestry of orange, black, and white, had disappeared more than twice as often at low elevations than at elevations above 2,438 meters (1.5 miles) and more than twice as often in Mexico than in Canada. They were surviving at higher, cooler elevations and dying out where it was getting too warm. As we scrutinized the scientific literature, the story repeated itself in species after species.

Mosquitoes and mosquito-borne diseases were moving up, too. For more than a decade, reports had been rolling in of malaria epidemics in the high-elevation areas of the tropics. In East Africa, these included the Usamabara Mountains of Tanzania, the Ruwenzori Mountains of Uganda, the Ethiopian highlands, and the hills of western Kenya. Dengue fever, which used to appear only below about 975 meters (3,200 feet) in Latin America, had broken out at an elevation of 1,676 meters (5,500 feet). The mosquito that carried dengue and yellow fevers, *Aedes aegypti,* had been spotted high in the Colombian mountains, above 2,134 meters (7,000 feet).

Taken together, the temperature, glacier, plant, insect, and disease trends provided a clear and compelling picture: it was getting warmer in the mountains, the mosquitoes were moving up, and they were bringing infectious disease with them.

A BATTLE FOR THE BOOKS

By 1999, Andrew Githeko's work on malaria had earned him a prestigious appointment as one of the two lead coauthors of the health chapter of IPCC's influential report. In 2001 that report, IPCC's third assessment report, was released, reminding readers that a million people die from malaria each year, most of them in Africa and most of them children.

What's more, Githeko and his colleagues reported, models from several scientific teams had predicted that climate change would increase the number of regions in which the disease could spread. The regions at risk included some surprising locations, such as Australia, Europe, and the United States. In the 1950s and 1960s, all of these developed regions had successfully eradicated malaria, but not the mosquito vectors that carry it. Local transmission of malaria had occurred occasionally since then, but well-developed public health measures had kept the disease at bay. Now, Githeko and his colleagues had spotted what could be a massive problem. "Malaria could become established again under the prolonged pressures of climatic and other environmental-demographic changes if a strong public-health infrastructure is not maintained," they wrote in the report. In other words, global warming raised the risk that malaria could return to the developed world, and those nations should prepare to head it off.

Predicting that diseases could spread is one thing. Determining whether they already have is another. Showing that climate change was playing a role is yet another, far more difficult task. In part, that's because many factors can set the stage for an epidemic:

- A deteriorating public health infrastructure allows otherwise preventable diseases to emerge and spread.

- Microbes and parasites acquire resistance to the drugs meant to kill them, and insect vectors can evolve resistance to pesticides.

- Massive migration of people from the countryside to the cities in developing countries can result in overcrowding and poor sanitation, which breed disease.

- When people move into forested areas, they can find themselves in close proximity to wildlife harboring new microbes (Ebola and AIDS are suspected to have arisen this way).

- Dams can create new sources of standing water, creating favorable conditions for breeding mosquitoes and other organisms, such as snails, that spread disease.

- People travel, bringing their microbes with them.

Any one or a combination of these factors can contribute to an epidemic.

Given the many possible causes, can we say with confidence that global climate change is contributing to the spread of disease? It's a challenging problem, but for a growing number of researchers, the answer is an emphatic yes.

In 2001, however, not everyone agreed. As soon as the third IPCC assessment report was released, blowback on the health chapter began. The handful of climate change "skeptics" (more aptly called deniers and naysayers in the face of new evidence, for scientists are naturally skeptics) who worked in public health included biologist Simon Hay, of the University of Oxford, and Paul Reiter, then at the Centers for Disease Control and Prevention. They asserted that climate change played no role in the spread of malaria in the East African highlands. The debate, although phrased in the muted and technical language of the scientific literature, turned into a slugfest. Much was at stake: professional reputations, the scientific consensus, and—most critically—societal consensus about the changes needed to deal with climate change. In the hallway at a scientific conference, one of the climate change deniers took Andrew Githeko aside. "We're going to dent you," he said.

East Africa and Kenya, in particular, had become the focus of the climate and malaria debate, and of the available studies, Githeko's work offered some of the strongest evidence of a climate–malaria connection. He had years of experience in western Kenya tracking malaria caused by *Plasmodium falciparum,* the most dangerous kind, and had published much of his research in leading public health journals. He had set up weather stations to obtain his own climate and weather data in or near the towns where epidemics had occurred, giving him the best data available. And his models were sophisticated enough to take into account how *P. falciparum* and anopheline mosquitoes respond to temperature and rainfall. It was the sort of rigorous science that won over impartial experts. But the deniers of the climate–health connection were poised to tear it apart.

In 2002, Simon Hay joined two Oxford colleagues and several others and published a paper in *Nature* that claimed the western Kenya highlands had not gotten any warmer over the last few decades. Therefore, they asserted, it couldn't have been climate change that had caused malaria to resurge. They called the claims of such a link "overly simplistic"—fighting words for scientists. Later that year, Hay and his Oxford colleagues published a review article aimed specifically at discrediting Githeko's models. The Oxford team blamed malaria resurgence in the East African highlands not on climate but on diminished efforts to kill mosquitoes and the malaria parasite's growing resistance to chloroquine and other drugs. The paper's title captured its message and its tone: "Hot Topic or Hot Air? Climate Change and Malaria Resurgence in East African Highlands."

Githeko's critics were well-credentialed biologists from the Centers for Disease Control and Prevention and from one of England's foremost universities, all publishing in the world's leading scientific journals. Githeko's reputation was beginning to take a beating. Then he and his colleagues pushed back.

First, they found a fatal chink in the Oxford group's armor. In their *Nature* paper, the British scientists had used seventy-five years of data on average monthly temperature and rainfall from four weather stations—one each in western Kenya and the nearby highland areas of Uganda, Rwanda, and Burundi—and concluded that the average temperature and rainfall hadn't changed. This was the basis of their critique against Githeko. However, the Oxford group had made a crucial error, which can be explained best by analogy.

If the average monthly temperature in January in Kansas City is 4°C (39°F) and the average temperature in Los Angeles is 20°C (68°F), you might assume that the temperature high in the Rocky Mountains, halfway in between, would be 12°C (54°F). But of course it's much colder high in the Rockies in January; the model would fail because it doesn't take altitude into account.

The Oxford team had made a slightly more sophisticated version of the same mistake. They had taken data from a grid of weather stations and calculated the temperatures in East African towns between them. Those towns were in hilly regions, so the method, which scientists call interpolation, was inappropriate for the task. They asserted that they had disproved a link between malaria and global warming because the region hadn't warmed. But it was they who had been too simplistic. "It was statistical smoke and mirrors," says Githeko. He and his IPCC colleagues were invited by *Nature* to reply to the paper, and, in restrained scientific language, they said as much.

The naysayers' reaction was swift. "Oh my God, they went bananas," Githeko says.

The battle raged for several years. Githeko published a paper describing his early modeling studies. The Oxford group published a paper saying Githeko's data were no good. The Oxford group attended Githeko's conference presentations and tried to shoot him down with hostile questions, though he consistently fielded them well.

In 2004, Githeko and a U.S. malaria researcher named Guiyun Yan, then at the State University of New York at Buffalo, published a paper in the prestigious *Proceedings of the National Academy of Sciences* that made a key advance based on a crucial but often overlooked facet of climate change. Scientists agree that climate change not only can lead to a warmer climate but also can destabilize the climate, leading to more extreme weather—more heat waves, more cold spells, more drought, and more drenching rains that leave pools of standing water. By looking at malaria and meteorological data from seven East African towns, they found that the more the climate yo-yoed in a town, the more likely there was to be an increase in malaria. The finding meant that to predict an outbreak, one needed to take into account weather fluctuations and not just warming. Indeed, Githeko's study was pivotal in the assessment of climate change's health impacts, reminding scientists that weather fluctuations, and not just warming, could induce illness.

In 2006, a group of U.S. researchers unconnected with either camp repeated and improved upon the Oxford group's climate analysis, reaching a very different conclusion. They used the same data Hay's team had, plus an additional five years' worth, which took the climate record up to 2002. Using a more sophisticated and effective statistical method, they were able to mathematically separate the actual temperature trend from the random noise in the data. Their results, reported in *Proceedings of the National Academy of Sciences,* showed that each of the four East African sites the Oxford team had examined had remained at about the same temperature until about 1980, at which point, over the subsequent twenty-two years, they warmed by 0.5°C (0.9°F). Not coincidentally, the tropical Indian Ocean sea surface temperatures, which are known to powerfully affect weather conditions in East Africa, had been climbing since 1976.

What's more, by plugging temperature trend data into a mathematical model like Githeko's, the American researchers had demonstrated that climate warming of just 0.5°C (0.9°F) over two decades would more than double the simulated mosquito population. And in the highlands, where there had been too few mosquitoes to transmit the disease, the more mosquitoes, the more malaria. "That was fantastic," Githeko says. "It really changed opinions." And Hay "kept quiet at that point," he adds.

MOSQUITO HUNT

As these predictions reached the scientific community, Githeko was already back at work, leading several field studies to document where in the western Kenya highlands the most dangerous strain of malaria parasite and its mosquito vectors were actually found. But Githeko was also thinking about another, higher part of the country—the central Kenya highlands, where Anne Mwangi and her husband and daughters lived and where Andrew Githeko had grown up. This region was considered a safe haven, high and cold enough to be free of the disease. A 1970 national atlas had called the region "malaria-free." There'd been scattered reports of malaria there since the mid-1990s, but Andrew Githeko and other scientists thought travelers had brought it there from a rice-growing lowland area called Mwea a few dozen miles away, where the disease was well established. Based on his modeling studies, using global warming rates at that time, Githeko predicted that warming temperatures would eventually bring malaria to the Mount Kenya highlands, but not until 2020 or 2030. That prediction had been met with skepticism from the same scientists who doubted Githeko's conclusions about warming temperatures and malaria in western Kenya. But all that was before Hong Chen came to Kenya and got lost.

Chen was a postdoctoral scientist who worked for Guiyun Yan, the U.S. malaria researcher. Yan had collaborated with Githeko on several projects, investigating the ecology and epidemiology of highland malaria, which occurs when the disease spreads at high elevations where people haven't previously been exposed. Chen's immediate goal was to map the locations around Kenya where malaria-carrying mosquitoes withstood the insecticides meant to kill them.

"He came here to study insecticide resistance in mosquitoes. I told him to look at mosquito larvae because they are easier to collect and to show resistance than adult mosquitoes," Githeko says.

Chen planned to crisscross the country, from Kisumu in the west to Mombasa by the sea. Everywhere he went, he'd look for standing water and take samples of mosquito larvae. As a mosquito biologist, Chen could routinely spot the sorts of puddles, streams, and swamps that were likely to harbor them. Githeko recommended that Chen sample around Mwea, and in April 2005, during the rainy season, he gave Chen directions there. Mwea is located between Nairobi and the Mount Kenya highlands, and, for as long as anyone could remember, there had always been mosquitoes and malaria there. Githeko outfitted Chen with a car, some equipment, and a driver named Mohammed.

Chen and Mohammed, who'd been on this sort of field trip before, headed north from Nairobi. They drove for about an hour before they saw a swampy area with some streams—a perfect place to find mosquitoes. Chen got out and saw the white-banded brown larvae floating on the surface of the water. *Anopheles!* He collected them and transferred the wriggling larvae to vials. He asked Mohammed to drive farther. The terrain became higher and drier, and Mount Kenya drew closer. They stopped again to look in some puddles and a stream. They found more *Anopheles*. They had what they'd come for, and they were pleased. But then they realized that they hadn't passed any rice fields. They stopped and asked some locals, as Githeko recounts the tale. The locals said, "You guys are so lost."

Eventually, they found their way to Mwea, did their sampling, and returned to Kisumu. They later discovered they'd actually been sampling in Karatina, twenty miles north of the rice fields, and in Naro Moru, high in the rain shadow of Mount Kenya. Mohammed was embarrassed. "You know, even if we got lost, we still found mosquitoes in Karatina," he told Githeko.

Githeko's eyes widened. "It's one thing to get lost. It's another thing to find mosquitoes in Karatina. There are no malarial mosquitoes in Karatina," Githeko said.

"No, no, I'm sure. I saw them," Chen insisted.

"It's impossible. That's where I was born. There was never anything like that."

But that's where they'd been, and they had GPS readings to prove it. Githeko told Chen to send the larvae to another lab and determine their species using a sophisticated molecular biology method called PCR that obtains unique fingerprints of their DNA.

Chen was still concerned that he'd gotten lost. "Don't worry, Hong," Githeko said. "You've just made a major, major discovery."

And he had. Everyone had assumed that the scattered cases of malaria in the central Kenya highlands since the mid-1990s, including cases like Elena Githeko's, were carried there by travelers from nearby malaria-ridden lowlands—from Mwea. But if malarial mosquitoes were breeding in the highlands, that meant the parasite had what it needed to breed locally—mosquitoes, warmth, moisture, and human hosts. Chen, Githeko, Yan, and their colleagues published the results in *Malaria Journal* in 2006, concluding that conditions in the central Kenya highlands were ripe for local malaria transmission. In 2007, Githeko studied the temperature trends of the area and observed that around 1993 a change had occurred that allowed malaria transmission. "In my whole life, that's my best contribution to science," Githeko said.

But in public health, determining the cause of an epidemic often requires a series of studies to rule out other possible causes and build up a sturdy body of evidence. Githeko was one step closer, but he'd need still more research to prove that climate change was spreading malaria in the foothills of Mount Kenya.

GATEI HEALTH CENTRE

The village of Ihwagi, Kenya, sits in the crowded countryside about fifty miles north of Nairobi, a few miles from Karatina. In March 2008, it was a patchwork of green—small plots of vegetables, light green tea fields, broad-leaved banana palms that leaned over small grassy yards. Men in work clothes on the roadside pushed sturdy bicycles up hills, with bundles of vegetables or tan *nipia* grass tied to the back rack. A middle-aged woman in a worn but colorful dress trudged uphill, balancing a white plastic bucket on her head with one hand. Small groups of goats wandered the roadside.

Githeko was returning to his hometown in the central Kenya highlands to do the work of a disease detective. Because he lives and works hundreds of miles away in western Kenya, Githeko had not yet had the chance to visit clinics and hospitals here to talk to health care workers. Such visits, he says, are the best way to uncover clues about emerging diseases.

Here, more than 1,600 meters up in the Mount Kenya foothills, malaria had never been a problem. If malaria was spreading even here, then global warming was contributing, Githeko believed. His plan was to visit dispensaries (community health clinics staffed by nurses), school clinics, and regional hospitals, where he'd speak with nurses and doctors to find out if malaria had taken hold.

But first there were some stops to make. Githeko weaved a diplomatic-plated Land Cruiser down a barely paved road, careening from one brown-dirt shoulder to the other, dodging potholes big enough to swallow a tire. He turned off the paved road and drove down a wide brown-dirt one through Ihwagi's business district: a block of weather-beaten, low-slung buildings with names of businesses—Jekima Hotel, Patrose Animal Feeds and Kerosine, Nyawira's Tailoring House—neatly hand-lettered on brightly painted walls.

Looking northeast from Ihwagi, he could see a trio of peaks mingled with the clouds. For people here, Mount Kenya is the only mountain that matters (figure 10). Massive, tall, and snowcapped, the mountain's amazing biodiversity—882 plant species, rare elephants, black rhinos, giant forest hogs, and bongos—has made it a Kenyan National Park and National Reserve, an International Biosphere Reserve, and a UNESCO World Heritage site. Mount Kenya is the headwaters for five rivers, including the Tana River, which provides water to hydroelectric plants that produce half of Kenya's electricity; the Ewaso Nyiro River, the only river that sends water to the arid north of Kenya, home of the colorful Samburu people; and the Ragati River, which passes through Ihwagi and once provided the Githekos and their neighbors with all the water they needed. The Kikuyus, the people who live in these fertile foothills, call it the mountain of God.

On a bumpy dirt road, Githeko stopped on a bridge over a steep valley. Below was a riverbed dozens of feet wide filled with boulders the size of basketballs. A small stream dribbled along at the bottom—the Ragati River. Githeko shook his head. "This was a huge mountain stream," he said sadly. He swept his arm to indicate the whole valley and the grassy banks above it. "It used to flood here all the time." But since he was a kid, there has been a population explosion in the region, and ten times more people live in the Mount Kenya region now than a half century ago. They're drawing the river down, but that's only part of the problem.

FIGURE 10. Mount Kenya, an International Biosphere Reserve and a UNESCO World Heritage site, nurtures a tropical forest of incredible biodiversity, and it serves as headwaters of five rivers that supply millions of Kenyans with water for drinking, irrigation, and hydroelectric power. A warming climate has melted about 80 percent of the mountain's ice cover. (Photo by Dan Ferber)

Eighty percent of the glaciers on Mount Kenya have also melted. "The mountain is drying up," he said.

Githeko soon turned onto an even narrower dirt road through some woods and stopped at a blue wooden gate. He honked his horn. There was no answer, so he opened the gate himself. He parked on a small lawn in front of a one-story stone house overlooking a verdant valley. A rooster crowed, a cow bellowed, and the warm air filled with the tweets, whistles, and calls of tropical birds. Two trim men in muddy work clothes walked over, smiled, and shook hands with Githeko. They were John and Philip, brothers who tended the Githeko family's property, raising chickens and growing their own corn, potatoes, tomatoes, bananas, and coffee. The three chatted softly in Kikuyu. The workers told Githeko that there were mosquitoes in the area during the rainy season. "There were none when I lived here—not a single one," Githeko said.

At the edge of a small plot rough with corn stubble was a strip of dirt where a shallow one-meter-wide canal once ran. The Githekos' father had dug the canal in the 1970s to draw water from the Ragati River to irrigate the family farm. The canal used to go around to all the farms, but now, with the river drying up, it was gone. The two workers told Githeko that they were into the third week of water rationing in the area, which forced them to dig the property's first-ever well. Githeko took a long look at the house, at the valley, at the two large white wood crosses in the cornfield, his parents' graves. Then he moved on.

For Githeko, the trip up the dirt road prompted memories and stories. Here was a sloping field with bright-green tea bushes, still owned by the family; there was the high school his father built. Here was the former edge of the forest, where the elephants and buffalo would come; there was the spot on the dirt road where the teenage Githeko, trying to impress a girl, gunned the engine of the family car, promptly drove off the road, and crashed. On the side of the road sat a stretch of small huts, constructed entirely of sticks the width of baseball bats, still with their bark on. The huts provided homes to a tribe that until recently had lived in the shrinking Mount Kenya forest.

Not far from that makeshift village, Githeko spotted a stone entrance-way on the left side of the dirt road. A large sign sat next to the entrance-way, hand-painted in large red letters: Gatei Health Centre. He parked on the lawn in front of the clinic. A few feet away was a small one-story building with rust-stained cinderblock walls and a corrugated tin roof.

Githeko got out of the Land Cruiser as two women appeared at the door. One woman was big-boned, in a lavender skirt, matching top, and beaded necklace; the other was shorter, with almond-shaped eyes and smooth skin the color of dark chocolate. A bespectacled man in a white lab coat joined them. Githeko walked over and greeted them. "I'm Dr. Githeko," he said, then paused. "From KEMRI."

The woman in lavender, Margaret Kariuki, identified herself as the district public health officer; the shorter woman, Susan Wangiki, was the clinic's lab technologist; the man in the white coat, Bernard Gikandi, was a nurse. The Gatei Health Centre was the first stop for sick locals. Githeko chatted with the health workers for a few minutes in English, one of two common languages in Kenya, about Githeko's roots in the area, the brand-new university campus up the hill, the famous cool weather of the area. "This place used to be very cold," Githeko said. "Mmm," Margaret murmured in acknowledgment.

Githeko explained his intention to investigate the impacts of global

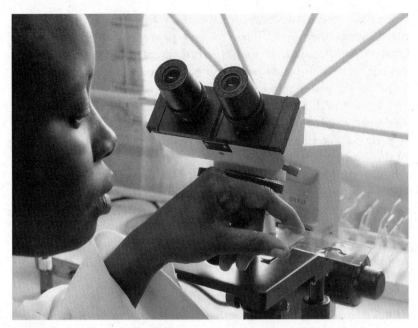

FIGURE 11. Susan Wangiki, laboratory technologist at the Gatei Health Centre in Ihwagi, Kenya, in the Mount Kenya highlands, examines a slide containing a blood sample from a patient that she has stained to detect malaria parasites. (Photo by Dan Ferber)

warming. "You know global warming?" he asked. They all nodded. He told them he had come to see what global warming is doing to malaria. Githeko told them of the malaria mosquitoes his team found in nearby Naro Moru in 2005. "We were told that Karatina had a lot of malaria," he said.

Oh, yes, they saw it all the time. "Actually there is a lot of it," added Margaret, the district officer. "Malaria is here."

"Local people?" Githeko asked. If people who hadn't traveled out of the area had gotten the disease, that meant the disease was almost certainly being transmitted locally, Githeko believed.

"Yeah, local people!" exclaimed Susan, the medical technologist.

"Wow," Githeko said. He shook his head. He asked to come into the clinic. "I have to explain to you why this is going on," he told them.

In the clinic's laboratory, a room the size of a small bedroom, a microscope sat next to a rack with drying glass slides on the clinic's lab bench, a waist-high counter facing a window. Here, Susan Wangiki, the laboratory technologist, used her microscope and biochemical tests to analyze blood samples from patients to determine what was making them sick (figure 11). Taped to the wall were well-thumbed photocopies from a medical textbook with diagrams of malarial parasites. Githeko placed his laptop on the lab bench, booted it up, and launched a PowerPoint presentation.

As the health workers crowded in to watch the screen, Githeko explained the biology of malaria mosquitoes—how at 18°C (64°F), malaria parasites reproduced too slowly to mature during the lifetime of the mosquito, but at 20°C (68°F) they reproduced quickly and easily. "That's why you might not think there's a big climate change, but, in the mosquito world, two degrees makes a big difference." He showed a slide with an analysis they did of historical weather records. It showed that in the central Kenya highlands—right where they were—the mean temperature warmed past 18°C, the critical threshold for malarial mosquitoes, in 1994. The health workers listened intently and nodded. Bernard, the nurse, took notes.

Githeko contrasted endemic malaria, in which people have been exposed and have some degree of resistance, with infections in highland areas, where the disease was not historically endemic and where residents would lack the immunological defenses to hold the parasite at bay. That's been the situation in the western Kenya hills over the last two decades, Githeko told the health workers. "We call it unstable transmission, and it's very, very dangerous."

The busy health workers at Gatei Health Centre saw about five hundred patients a week with all sorts of ailments; about fifteen of them, or 3 percent, had malaria. They often saw malaria in babies and young children who came in with diarrhea, vomiting, headaches, and high fever. Bernard spotted the symptoms and took the blood samples, Susan did tests to confirm malaria parasites in the blood, and then they prescribed medicine. Almost none of the families had cars, and the sickest kids often needed to go seven kilometers on a bus to the regional hospital. The clinic workers saved many, but some children died before they could get there.

In an epidemic, one in three people walking into the clinic would have malaria, Githeko told them, and the numbers could overwhelm the clinic's capacity. That was not yet the case. But a 3 percent infection rate indicated that there was unstable transmission in Ihwagi, Githeko explained.

On his laptop, he pointed out graphs from the Kisii district of western Kenya that showed weather extremes plotted over time. Epidemics regularly follow the extreme weather. With enough data, Githeko said, he could develop a model that would predict epidemics in the central Kenya highlands and that would help them prepare to counter the epidemic. But then Githeko cautioned the health workers. "This," he said, "is what could happen to you."

3

Sobering Predictions

In the first two weeks of June 1992, thousands of people, including 108 heads of state, converged upon Rio de Janeiro for a scientific conference, formally the United Nations Conference on Environment and Development, that ultimately came to be known simply as the Earth Summit. The conference was unprecedented for its size and ambitions. The goals of this international body were to find new ways to halt the destruction of the Earth's natural resources and reduce climate-changing pollution. Part of that agenda, by necessity, included a need to examine and reimagine world economic development.

Eric Chivian, a Harvard colleague who was also immersed in the science of global change and medicine, attended this historic conclave with me. Eric was a cofounder of International Physicians for the Prevention of Nuclear War, an organization with a membership of a quarter million doctors. In 1985, he and his fellow cofounders were awarded a Nobel Peace Prize for their efforts to alert the world to the dangers of nuclear war. By the end of the 1980s, as a result of Mikhail Gorbachev's emergence in the Soviet Union and the end of the Cold War, the threat of nuclear war had seemingly begun to ease, and Chivian turned his attention to the catastrophic threats to humanity posed by climate change. After all, most people had a clear understanding of the threat to all of civilization posed by nuclear war, but their grasp of the slow-moving environmental catastrophe under way was weak.

"It was even more important to translate abstract and technical.

scientific terms into concrete, personal health issues," Eric recalled recently.

My colleague and I were stunned, ultimately, by the absence of any discussion of health issues in Rio. We attended many research presentations on climate change, trade, and endangered species but could find no one talking about human health consequences. We moved from speech to speech and spoke to one official delegation after another. When they found out we were physicians, delegates stared at us blankly and asked, "What are you doing at an environmental meeting? You're doctors." The disconnect was glaring.

Chivian and I agreed that something needed to be done to educate not just the general public but policy makers as well. Although we were just attendees and not delegates to the conference, we decided to organize an ad hoc press conference. (Impressive display of chutzpah, for sure.) About forty people showed up, including other conference attendees and a handful of curious reporters. Chivian addressed the issue of climate change and biodiversity and the ways in which the loss of species threatened human health. I chose to talk about the debate over the link between climate change and the Latin American cholera epidemic, which was still raging.

. . .

The year of the Earth Summit—1992—was a time of high hopes for those of us concerned about the health of the planet. In June, the same month we were in Rio, Senator Al Gore published his international bestseller, *Earth in the Balance,* in which he explained the world's ecological predicament and proposed a series of solutions. Gore had by then evinced a record of dedicated and persistent environmental leadership. As a junior congressman from Tennessee in the late 1970s, he was the most vocal member of Congress on the dangers of toxic waste, and in 1981 he convened the first-ever congressional hearing on climate change.

During the 1992 presidential campaign, George H. W. Bush, who made his fortune in the oil industry, famously slammed vice presidential candidate Gore as "ozone man" and claimed Gore was "so far out in the environmental extreme we'll be up to our necks in owls and outta work for every American." When Bill Clinton defeated the elder Bush, ushering the environmental champion Gore into the White House policy apparatus in 1993, many of my fellow scientists and I hoped and believed that the new administration would offer serious policy responses to the planetary catastrophe that was unfolding in slow motion.

We had a lot of company in the scientific community. Just two weeks after Clinton's victory, 1,600 scientists from seventy countries, including 104 Nobel laureates, signed the joint statement "World Scientists' Warning to Humanity." It minced no words:

> Human beings and the natural world are on a collision course. Human activities inflict harsh and often irreversible damage on the environment and on natural resources. If not checked, many of our current practices put at serious risk the future that we wish for human society and the plant and animal kingdoms, and may so alter the living world that it will be unable to sustain life in the manner that we know. Fundamental changes are urgent if we are to avoid the collision our present course will bring about.

SOUNDING THE ALARM

After the Earth Summit, my colleagues and I decided to bring our concerns directly to our medical peers. In Western medical education, doctors are trained to focus on the disease or injury, sometimes to the neglect of the patient as a person in a family and as a member of a community. That focus can lead practitioners to miss the connections between a patient's health and his or her physical and social environment. Medical students, even today, receive minimal training about these connections.

In an effort to raise awareness of these vital links, I proposed an idea to an editor named David Sharp at *The Lancet,* the venerable British medical journal. The idea was to publish a series of articles, each focusing on specific aspects of the climate change–human health nexus. These articles would address the science of climate change, ozone depletion, degradation of marine ecosystems, and loss of biodiversity. The *Lancet* editors responded enthusiastically, and the articles were assigned to prominent experts. *The Lancet* published the eight-part series over the fall of 1993.

Cholera researcher Rita Colwell, Harvard water specialist Tim Ford, and I teamed up to report on the infectious disease risks linked to the deterioration in coastal marine ecosystems, including toxic red tides, cholera, and marine viruses. Andrew Dobson of Princeton and Robin Carper of Johns Hopkins described the ways in which the depletion of tropical rain forests—itself a driver of climate change—was driving the extinction of hundreds of species of tropical plants. (This is especially tragic because plant-based medicines constitute nature's pharmacopoeia, and 80 percent of the world's population relies on such medicines, which are drawn from twenty-five thousand plant species.) Martin Parry of

Oxford and Cynthia Rosenzweig of Columbia and NASA projected that warming from doubling carbon dioxide could decrease food production worldwide.

In our introduction to these reports, we presented the view from twenty-five thousand feet:

> Climate change—alterations in temperature, rainfall, and the pattern of extreme weather events—has an important role in the distribution and impact of disease. The world's ecosystems—terrestrial, marine, aquatic, even atmospheric—have proved remarkably robust over millennia. Can that last as one species, a consumer, energy waster, overpopulator, and polluter, does its best to upset those delicate balances? For many years, persons of "green" or Gaia-esque persuasion have answered "No." Now climate scientists are joining them. And physicians also, as the health implications of global climate change become more apparent. Reluctantly and nervously, politicians too.

In hindsight, we were overly optimistic. The dawning recognition among doctors, scientists, and politicians we described would take longer than anticipated. But the 1993 *Lancet* series helped to bring the climate–health connection to the attention of the international environmental and public health communities.

. . .

In 1995, I received a call from Rudi Slooff, a Dutch public health scientist with the World Health Organization. Slooff had read our *Lancet* series. Would I be part of a team that would comprehensively evaluate the science on climate change and human health? I joined climate–health researchers from the London School of Hygiene and Tropical Medicine and others from Africa, Latin America, Australia, and Germany who were investigating various aspects of the problem.

Over the course of a year we met, combed the scientific literature, shared the results of our own research, talked with dozens of top infectious-disease specialists worldwide, and offered our informed assessment of whether climate change would encourage the spread of ten infectious diseases that plague the tropics and subtropics. It was clear from the science that regional climate—the average humidity, temperature, and precipitation—influenced the range of each of these diseases. We also considered food-borne ills and the emerging threat of harsher weather and more heat waves. We pulled all this research together and published our findings as a scholarly book: *Climate Change and Human Health*.

In 1995, we presented our findings to the IPCC, which was then close to completing its eagerly awaited second assessment report. We managed to persuade them that our findings should be included. They assigned to our group of a dozen the task of condensing our book down to a single chapter. In our chapter, we concluded, "Climate change is likely to have wide-ranging and mostly adverse impacts on human health, with significant loss of life."

More specifically, we stated that global warming was "highly likely" to spread malaria. In the precise terminology of the IPCC reports, *highly likely* meant that the odds of spreading were greater than 95 percent. Dengue fever, a mosquito-borne viral illness that infects more than 10 million people a year, was "very likely" to spread, which meant the odds were over 90 percent. So was schistosomiasis, the tropical parasitic disease transmitted by freshwater snails that infects more than 200 million people per year, and river blindness, a parasitic disease transmitted by black flies that infects 17.5 million Africans and Latin Americans. The odds were 67 percent for the spread of six other diseases: African sleeping sickness, transmitted by tsetse flies; leishmaniasis, transmitted by sand fleas; Chagas' disease, transmitted by the kissing bug; and three mosquito-borne diseases—lymphatic filariasis (elephantiasis), Rift Valley fever, and yellow fever.

Not all the predicted changes in infectious disease were for the worse. Some areas, we found, could become too dry and too hot for disease-carrying mosquitoes to survive. In southern Honduras, for example, the combination of climate change and deforestation have already made it too hot for malarial mosquitoes to exist (although many southern residents have also moved north into milder, malaria-ridden areas, so warming has failed to improve their health).

As our study of changes in the mountains would indicate, a warming climate can open up new territory for insect vectors, creating conditions there that favor insect-borne disease. The IPCC was projecting that, as the tropics expanded toward the poles and to higher elevations, the odds of tropical diseases would rise in those areas. Over the last decade, as models have improved and data on various diseases have accumulated, this idea has held up.

INSIDE THE GREENHOUSE

The urgency we felt about climate change stemmed from more than a century and a half of scientific work that relied on fundamental principles of physics and chemistry, as well as on observations and mathematical models. In 1827, the French mathematician and physicist Jean-Baptiste Joseph Fourier recognized the likely warming effect of some gases on the atmosphere and coined the term *greenhouse effect*. In 1861, John Tyndall, a British physicist and atmospheric scientist measured absorption of infrared radiation by CO_2 and water vapor. In 1896, a Swedish physicist and chemist named Svante Arrhenius, who later won the Nobel Prize in chemistry, calculated—laboriously and by hand—that doubling CO_2 would heat the planet by about 5°C (9°F), a number that squares roughly with today's best estimates. And in 1940, an English steam engineer and inventor, Guy Callendar, calculated that the warming due to CO_2 was coming from burning fossil fuels. We have understood these basic connections for a very long time.

In the early twentieth century, some scientists thought that heating the planet might be a good thing but that it would take thousands of years of industry to raise levels of the main greenhouse gas—carbon dioxide—appreciably. In foretelling the future, however, they greatly underestimated our impact. (Underestimation may be more common than exaggeration in judging how things change.) As we made and used more and more automobiles and jet engines, as we burned more and more coal in hundreds of huge coal-fired power plants, as our industrial economy picked up speed through the twentieth century, and as we cut and burned more and more CO_2-absorbing forest, we added vast quantities of CO_2 to the atmosphere each year. Today, fossil fuel burning and deforestation are responsible for about thirty-six billion metric tons of CO_2 entering the atmosphere every year, enough to push its concentration in the atmosphere to about 390 parts per million (ppm) by 2009, a figure that's rising by between 2 and 3 ppm per year. At the rate we're going, we'll double preindustrial levels by midcentury.

Like the panes of glass on a greenhouse, CO_2 allows visible light to shine through but prevents infrared light, or heat waves, from escaping. Unlike a solid glass ceiling that holds in warm air, though, CO_2 accomplishes this in a more subtle way, by interacting differently with various wavelengths of light.

Incoming solar rays contain visible light of various colors, plus two types of invisible light: ultraviolet light (which can cause skin cancer) and infrared light. Carbon dioxide and other atmospheric gases are transparent to most visible light; thus light shines through the atmosphere, striking land and penetrating the top layers of the ocean. Some of that light is absorbed, adding energy to the planet, and some of it is reflected or reemitted from the land, sea, and air in the form of infrared light. (This is the form of energy we see rippling up from asphalt roads on a hot summer day.) Before the industrial revolution, when the planet's climate was fairly stable, the incoming and outgoing light canceled each other out. But by adding CO_2 and other greenhouse gases, such as methane, we've tipped the balance: more heat is being absorbed by the earth, oceans, and atmosphere than is departing out to space, causing the globe to warm.

The extra CO_2 we've added to the atmosphere heats the Earth by about 1.5 watts for every square meter—roughly equivalent to having a planetwide grid of small Christmas lights, each spaced one meter apart, lighting and heating the surface. Other greenhouse gases contribute to global warming as well. Methane—a chemical from natural gas, cow flatulence, and other sources—is less abundant but powerful, contributing half as much warming as CO_2. We now know that black soot—the gritty black smoke emitted from diesel truck and bus exhausts (but not from vehicles run on cleaner fuels like natural gas)—contributes almost two-thirds as much warming, particle for particle, as CO_2. And chlorofluorocarbons—the ozone-damaging chemicals used in antiperspirants and refrigerants—contribute about one-fifth as much. In contrast, aerosols such as sulfur have a cooling effect by directly reflecting the sun's rays and creating reflective clouds. The sum of all this: gases emitted are, in aggregate, trapping heat in the atmosphere and warming the globe.

This heat has warmed the atmosphere by about 0.7°C since 1900, enough to start melting ice sheets in Greenland and Antarctica, and alpine glaciers like those on Mount Kenya. But most of the heat of last century's global warming was absorbed by the oceans. Since the 1950s, the oceans have accumulated twenty-two times as much heat as has the air, and the oceans' surface waters have warmed significantly. Even the deep sea is getting warmer, down to three kilometers (almost two miles) below the surface. The ocean takes longer to heat up and longer to cool off than does the air and is thus the repository of last century's global warming (figure 12).

More water vapor steams off a warming ocean, just as it would from a warming bath. Moreover, warmer air can hold more water vapor (7 percent more for every 1°C [1.8°F]), making the air more humid. These effects combine to rev up the global water cycle, which circulates water vapor from oceans to clouds to precipitation and back through rivers to the sea. These changes to the global water cycle, combined with changes to the ocean heat engine, are changing our climate. A revved-up water cycle makes wet places wetter, and rain tends to come in more intense downpours. In the United States, for example, the amount of rain falling

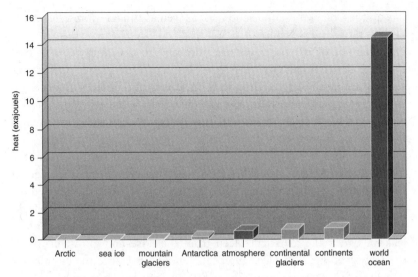

FIGURE 12. Soaking up the heat. The ocean holds far more of the heat of global warming than anywhere else, including about twenty-two times as much as the atmosphere. (From S. Levitus, et al., "Warming of the World Ocean, 1955–2003," *Geophysical Research Letters* 32 [2005]: L02604)

has not changed much since 1970, but the heaviest downpours occur 27 percent more frequently. Meanwhile, higher temperatures over land are also driving more evaporation, causing arid lands to experience longer droughts. "It is well established that droughts are increasing and heavy rains are increasing" worldwide as a result of climate change, says Kevin Trenberth, a leading climatologist at the National Center for Atmospheric Research in Boulder, Colorado.

The rise of such extreme weather is itself a symptom of an unstable climate. Moreover, the variance around the long-term warming trend has begun to influence biological systems. Indeed, two main effects of climate change—warming and greater weather variability—mean that millions of people worldwide face a higher risk of infectious disease.

THE RISE OF BREAKBONE FEVER

Just as rising temperatures drive the spread of infectious diseases like malaria, so too can changes in precipitation patterns. For example, ticks, mosquitoes, and other insect vectors need enough humidity to keep from drying out, so droughts can sometimes eliminate diseases, while rains can trigger them. Strong winds can carry insect vectors, including mosquitoes, blackflies, and sand flies, as well as fungal spores, hundreds or even thousands of miles.

Epidemics of dengue fever are now on the rise, and there's good evidence that climate change is contributing to its spread. (Other factors, including unplanned urbanization, travel, and failures of mosquito control, also play important roles.) Dengue fever is caused by one of four related viruses that are transmitted by mosquitoes in the genus *Aedes*. Infection leads to a rash, a sudden fever, severe headaches (often with stabbing pain behind the eyes), and muscle and joint pain so harsh the disease is nicknamed "breakbone fever." A minority of people infected with a dengue virus develop a more serious illness called dengue hemorrhagic fever, which causes abdominal pain, vomiting, and excessive leakage from the capillaries, the body's smallest blood vessels. The capillary leakage can make a patient bleed from the nose and gums, bruise at the merest bump, and bleed internally enough to cause the circulatory system to fail. Dengue fever infects fifty million people worldwide, and about five hundred thousand are hospitalized with dengue hemorrhagic fever each year, the vast majority of whom are children. When they're treated with nonaspirin pain relievers, bed rest, and plenty of fluids, most recover, but more than twelve thousand die. There is no vaccine available to prevent the disease.

The most important mosquito vector of dengue fever, *Aedes aegypti*, breeds in standing water that collects in rainwater catchment containers, discarded plastic and metal containers, and used tires. In the tropics and subtropics, *A. aegypti* thrives in urban shantytowns, where trash accumulates, sanitation is uncontrolled, and there are plenty of people whose blood the mosquitoes can feed upon (figure 13). Over the past several years, Brazil has suffered multiple epidemics in shantytowns called *favelas* at the edges of cities, often after heavy rains. The IPCC projects that continued warming could spread *A. aegypti* to higher latitudes and altitudes, just as rising temperatures have spread the *Anopheles* mosquitoes that carry malaria. Longer warm and rainy seasons could give mosquitoes more opportunity to breed, and increased drought could do the same.

Dengue is on the rise everywhere it's endemic, including Africa, the eastern Mediterranean Sea, Southeast Asia, and the western Pacific. But it has spread most rapidly in the Americas, where it was under control as recently as 1970. Now explosive outbreaks have been occurring in tropical and subtropical regions throughout the western hemisphere. In 2007, eighty thousand cases of dengue fever were reported in Venezuela, including six thousand cases of dengue hemorrhagic fever. More than

FIGURE 13. The mosquitoes that carry dengue fever are abundant in slums surrounding cities in the developing world, where there are plenty of people to feed on. Warming climate helps spread the mosquitoes that carry dengue fever; heavy rains and droughts can trigger epidemics. Pictured here is Jose Caballero Giron, then 14, in his home in Colonia El Porvenir, a slum on the outskirts of Tegucigalpa, Honduras. (Photo by Jason Lindsey/Perceptive Visions)

thirty-eight thousand dengue cases occurred in Mexico that year, six times as many as in 2001. There, the disease once appeared only in the hottest, wettest months but now strikes almost all year round and is more common in northern states, where it used to be rare.

Nor is the United States immune. Although there have been no major epidemics in decades, the potential is there. A dengue epidemic in Texas in 1922 infected a half million people, and occasional dengue cases, including cases of dengue hemorrhagic fever, have occurred in southern Texas since 1980. What's more, another mosquito that can carry the dengue viruses—the Asian tiger mosquito *(Aedes albopictus)*—has already spread throughout much of south Texas and the southeastern United States and even north to Chicago. Fortunately, though this mosquito is capable of transmitting dengue fever, it seems to do so very inefficiently.

SOBERING PREDICTIONS

To most effectively prevent climate-linked disease, we also need an early warning system that forecasts epidemics months before they occur. Some

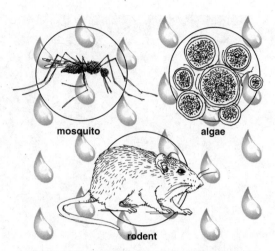

FIGURE 14. Biological indicators. Populations of disease-carrying mosquitoes, rodents, or algae spike in environments disturbed by downpours, drought, or deforestation. By keeping watch on their populations, scientists can help predict outbreaks of diseases like malaria, cholera, or plague.

of the best evidence linking climate and epidemics comes from studies akin to Andrew Githeko's that create such forecasts and cross-check their predictions with observations of where, when, and how much of the disease in question actually occurs. If a model reliably predicts that extreme weather, such as a drought or a period of heavy rains, will bring an epidemic, that's solid evidence that the model and its assumptions are accurate. Predictive climate models have been developed for several diseases. They rely not just on climate or weather data but on biological indicators such as the populations of rodents, algae, or mosquitoes that either transmit disease or indicate an epidemic is brewing (figure 14).

Entomologist Ken Linthicum of the U.S. Department of Agriculture's Agricultural Research Service, Center for Medical, Agricultural and Veterinary Entomology in Gainesville, Florida, has developed one of the best models. He collaborated with colleagues including Assaf Anyaba and Compton Tucker, of NASA's Goddard Space Flight Center, and John-Paul Chretian, of the Department of Defense Global Emerging Infections Surveillance and Response System in Maryland. The team's model successfully combines climate data with satellite data on the greenness of the landscape and uses them to help predict epidemics of Rift Valley fever, a frightening mosquito-borne viral disease that causes a minority of its victims to vomit blood or bleed to death internally.

Rift Valley fever occurs mainly in sub-Saharan Africa, where outbreaks can kill up to 20 percent of infected people and more than 80 percent of an infected herd of sheep, goats, or cattle. There is no effective treatment and no approved vaccine for humans. But predicting epidemics facilitates timely, environmentally friendly protection and prevention in at-risk areas.

In the early 1980s, Linthicum and his colleagues began conducting a series of studies that demonstrated that heavy rains precede Rift Valley fever outbreaks in East Africa. The rains create depressions and puddles where mosquitoes that transmit the disease can develop. They also cause plants in semiarid regions to bloom, and the resulting greenness of a region's landscape can be detected from space using information from orbiting scientific satellites that measure the amount of visible and infrared light reflected or emitted from the surface of the Earth. Each pixel in the resulting false-color maps represents 1.1 square kilometers (270 acres) of land, and the color represents the region's vegetation density, which is quantified using a measure called the greening index.

In the 1980s, Linthicum noted that the greening index spiked upward just before the first signs of a Rift Valley fever epidemic in livestock. In the late 1990s, Linthicum's team noted that every outbreak of the disease in East Africa since 1951 had occurred during an El Niño Southern Oscillation (El Niño for short).

During most years, when warm surface water gathers in the Pacific Ocean off the coast of Peru and Ecuador, relatively constant trade winds near the equator blow this water westward toward Indonesia. As the winds push surface water thousands of kilometers west across the Pacific, the sun heats that water. The newly warmed surface water piles up in the western Pacific as if the ocean were a big bathtub, raising sea levels near Indonesia to more than a half meter higher than off the coast of South America. This warm water in the western Pacific releases more heat to the atmosphere than any other body of water on Earth.

During El Niño events, however, the trade winds from the east weaken, and the piled-up warm water in the Pacific bathtub sloshes thousands of kilometers back east, usually reaching South America around Christmastime (thus the name *El Niño,* meaning the boy or the Christ child). At the same time, a huge seesaw of atmospheric pressures called the Southern Oscillation occurs, with its fulcrum in the equatorial Pacific. El Niños typically change weather patterns on five continents for up to several years at a stretch. Although no two El Niños are identical, they typically bring drought to Indonesia, Australia, and some parts of North and Central America and rainy conditions to Ecuador, Peru, and the U.S. Gulf Coast. Importantly for Linthicum's team, El Niños often warm the Indian Ocean as well as the Pacific, and when they do, they breed storms that move over land to drench East Africa.

In 1999, Linthicum's team reported in *Science* that, by using their model and sensors to assess the greening index and detect Pacific and Indian Ocean sea surface temperatures, they could have given two to five months' advance notice for every Rift Valley fever outbreak since record keeping began. Equally important for a forecasting tool, their model would never have issued a false alarm by predicting an outbreak that never arrived.

In September 2006, Linthicum's team put their model to use. They noticed that sea surface temperatures had risen in both the equatorial Pacific and the Indian Ocean and that it had begun raining heavily in Somalia. "We thought we'd better put out warnings to the area," he said. By October, they'd written their first alert to the World Health Organization, the United Nations Food and Agriculture Organization, and other health authorities to watch for Rift Valley fever.

By November, rains had spread to northeastern Kenya and Ethiopia, and Linthicum's team issued warnings for those countries as well. Major newspapers in the United States and Africa covered the announcement. By early January 2007, there had been 165 human cases of Rift Valley fever reported in Kenya, many in a vast, difficult-to-reach area populated by nomadic livestock herders. The actual number of human cases was thought to be much higher.

In March, Linthicum visited the Sukari Ranch in Ruiru, Kenya, not far from Nairobi. Sheep and cattle had begun dying en masse from the disease. The sheep had been grazing on open ground near wet areas where mosquitoes bred. "I got there and said, 'Don't let the sheep go where mosquitoes are infected by the virus,'" Linthicum recalls. The ranchers also administered a veterinary vaccine to the cattle, which kept them from falling ill. "In a dramatic way we could change the course of what happened on this farm," Linthicum said. That prevented the ranchers from being wiped out economically, and it may have prevented human cases as well, since Rift Valley fever can also be transmitted by handling the blood, meat, or milk of infected animals.

If today's weather fluctuations can trigger epidemics, will changing weather patterns due to climate change lead to more disease? Some of this depends on what happens to the El Niño Southern Oscillation. Rift Valley fever is just one of the diseases that flourishes during El Niño conditions. El Niño delivers epidemics of disease to many regions of the world: malaria to the Punjab region of India and to Venezuela, dengue fever epidemics to Thailand, cholera to Bangladesh, other diarrheal diseases to Peru, and hantavirus pulmonary syndrome to the southwestern United States. The particulars vary, but, in each case, it is extreme weather conditions—prolonged drought, unusually heavy rains, or unusual warmth—that foster diseases.

PAIN, THEN PARALYSIS

As droughts and heavy rains increase because of climate change, millions of people will become more susceptible to infectious disease. Most of them live in tropical and subtropical regions. But people in temperate climate zones of North America and Europe are far from immune to climate-linked disease. Case in point: Chris Ballas's summer from hell. Ballas, who is forty-three, works as an electrician and building mechanic for a state vocational school system not far from his home in Spring Lake Heights, New Jersey. On the job, he climbs and crawls into whatever space he needs to be in, fixing whatever's broken. In his off time, he takes on more fix-it projects to help pay the bills. When he's not working, he lifts weights, rides his bike, and runs around with his twelve- and fourteen-year-old sons and four-year-old daughter. During the spring of 2008, he spent a lot of time fishing in the woods near the Manasquan River Wildlife Management Area in Monmouth County, New Jersey. "I'm the opposite of a couch potato," he says.

Because Ballas's work and play are so physical, he's no stranger to minor aches and pains, which may be why he wasn't alarmed by the shoulder pain he acquired in June 2008, which he attributed to an earlier bout with elbow tendonitis, or by the hip and leg pain he got a couple of weeks later, which he blamed on a strenuous bike ride. He began to get concerned during a stay-at-home vacation the third week in July, when pain between his shoulders made standing, sitting, walking, or lying in any position utter agony. His family doctor diagnosed a muscle spasm and gave him Percocet, an opiate painkiller, and Flexeril, a muscle relaxer. But the pain continued.

On July 31, Ballas and his wife, Laura McKeown, a medical writer and editor, arrived at New Jersey's Giants Stadium for a Bruce Springsteen concert with tickets they'd been thrilled to receive as a Christmas gift.

Ballas was in so much pain they had to leave before the show began—almost sacrilege in New Jersey and a huge disappointment to the two Springsteen fans.

Ballas went back to his doctor, who added a prescription of Valium to further relax his muscles, but the pain persisted. On Saturday, August 2, at 6 A.M., Ballas, beside himself with pain and exhausted from four sleepless nights, asked his wife to drive him to the emergency room.

Over the next several weeks, Ballas's agonizing pain continued. He was diagnosed with a muscle spasm, then a severely herniated cervical disk in his neck. Surgery was scheduled.

Meanwhile, Ballas had missed more than two weeks straight of work, and his worries mounted. He began ticking down the sick days he had left before he thought he'd lose his job. No one knew what was wrong. "I was sweating, nauseous, really depressed; my anxiety was going off the wall," Ballas recalls.

Then Ballas realized he was gradually becoming paralyzed. "From my collarbone to my navel, it felt like I was touching myself through thick clothing," Ballas says.

On August 18, he brushed his teeth, tried to gargle—and drooled. In the bathroom mirror, the right corner of his mouth was drooping. Then his right eye wouldn't blink. By the next day, the entire right side of his face slumped lifelessly. One of his doctors diagnosed Bell's palsy, which made surgery unsafe, and the doctors called it off. He was admitted to the hospital again. The other side of his face began to droop as well—bilateral Bell's palsy, the doctors called it. "It felt like my face was wrapped in a stocking," Ballas recalls. "I couldn't move it, no matter how I tried to blink or squint or smile." By then Ballas could consume only liquids, and he was rapidly losing weight. His twelve-year-old son could hardly bear to look at him. "I was talking like a person without a tongue. My eyes were rolling around in my head. It gave him the creeps. He kept asking, 'When's it gonna get better?'"

McKeown, meanwhile, was terrified and pressed her husband's doctors for answers. "I thought my husband was dying," she recalled. Ballas underwent blood work, a brain scan, and more tests. He became a teaching case for the baffled doctors at Jersey Shore Medical Center, in Neptune, New Jersey, and his hospital room swirled with crowds of residents. A neurologist ordered an MRI to rule out multiple sclerosis, stroke, a brain tumor, or an aneurism, then told Ballas that if the paralysis continued, he might have to go on a respirator.

"That's when I was most scared. I was starting to believe I had a weird, rare disease," Ballas says.

The MRI came back negative, which gave Ballas and McKeown a measure of relief. Finally, an infectious disease doctor ordered a spinal tap. When the results came back, she diagnosed advanced Lyme disease with meningitis.

Lyme disease is caused by a spiral-shaped bacterium called *Borrelia burgdorferi,* a cousin to the microbe that causes syphilis. It's easily treated if it's caught in its early stages, when it can cause fever, aches, and a characteristic large, circular "bulls-eye" rash.

If it's not diagnosed and treated early, however, the initial symptoms may last for some time. Then more menacing symptoms can appear, such as meningitis and Bell's palsy, as Ballas experienced, and painful arthritis of the knees and other joints. Lyme disease is sometimes marked by chronic neurological symptoms, including shooting pains, numbness, or tingling in the hands or feet, and problems with concentration and short-term memory. All these complications make Lyme disease very expensive to treat. A 1993 study showed that each case of the disease cost the health care system $89,000 in 2009 dollars, and a 1998 study showed that it cost the equivalent of $3.4 billion in 2009 dollars to treat all cases of Lyme disease and prevent Lyme-induced arthritis, neurological disease, and cardiac disease.

Unfortunately, Lyme cases like Ballas's have become more common. Each year from 2000 to 2006, between seventeen thousand and twenty-four thousand cases were reported to the Centers for Disease Control and Prevention, making Lyme the most common vector-borne disease in the continental United States, and the numbers climbed to twenty-nine thousand by 2008. Recent research indicates that climate change will increase the incidence of Lyme disease in parts of the United States and especially in Canada.

Lyme bacteria are transmitted in most of the United States via the bite of the black-legged tick, *Ixodes scapularis.* Black-legged ticks live for two years. The first spring and summer after they hatch, tiny larval ticks take blood meals from the white-footed mouse or other small mammals and birds. These animals transmit Lyme bacteria to the tick, which is then infected for life. The larval tick falls off the mouse or bird after a few days and spends the rest of its first year maturing into a life stage called a nymph, which looks like a miniature version of the adult tick, but with shorter legs. The next spring, the nymphs—hungry again for a blood meal—bite large animals such as deer or humans, infecting their hosts with *Borrelia bergdorferi*. The deer don't get sick, but humans can contract a potentially chronic, debilitating disease.

In the twentieth century, former farms in New England regrew into forests. This reforestation allowed deer to spread and become more plentiful, helping set the stage for Lyme disease's spread through the region. Other ecological changes contributed to the disease's spread. As forest habitats were fragmented into smaller tracts, voles and chipmunks declined and mice thrived. Voles and chipmunks do not carry as much Lyme disease as mice, but they were no longer around to dilute the pathogen and reduce transmission. Biologically diverse regions have been shown to dampen infectious disease transmission in several other instances.

Climate also plays a pivotal role in determining where Lyme disease occurs. During its two years of life, the black-legged tick feeds just three times—once each as a larva, a nymph, and an adult—spending between three and seven days on each host animal. The rest of the time it sits in the soil. Instead of drinking, the black-legged tick absorbs water from the air, and when the air becomes too dry, the tick dies. Ticks, like mosquitoes, die when it gets too hot or too cold, which means temperatures and, to a lesser degree, humidity control where ticks can live and where Lyme disease can occur.

In 2003, John S. Brownstein, an epidemiologist who is now at Harvard Medical School, and Durland Fish, a Lyme disease researcher at Yale School of Medicine, devised a model that used temperature and humidity data from several dozen locations to predict where in the United States the black-legged ticks would be found. Then they enlisted teams of undergraduates; sent the students to twenty state parks and forests in Virginia, Maryland, Delaware, New Jersey, and Pennsylvania; and had them repeatedly drag a one-meter-square cloth fixed to a wooden handle through the underbrush in the heat, stopping every twenty meters to reach down with tweezers and pick ticks off the cloth. The model predicted with 89 percent accuracy where the black-legged tick is now found across North America.

A few years later, Brownstein and Fish expanded their model to include Canada and used the best predictions of global warming to predict how Lyme disease would spread in North America. Currently, it's limited in Canada to a few locations in southern Ontario. But by 2080, the model predicted, climate change will more than double the area where the tick, and thus Lyme disease, could reside. Lyme's northward march is already evident in the northeast United States, where from 2001 to 2009, case reports of Lyme disease rose eightfold in New Hampshire and tenfold in Maine. A cousin of the black-legged tick, *Ixodes ricinus,* has already expanded its range northward in Sweden as winters there have warmed, marching northward and spreading tick-borne encephalitis in the Stockholm archipelago.

As climate warms and the black-legged tick spreads, more people may have summers like the one Chris Ballas experienced in 2008. The good news is that Lyme disease, when diagnosed early, can be completely cured.

On a Friday at the end of August, a nurse threaded a catheter into a vein in Ballas's arm, guided it to a larger vein leading to his heart, and began treating him intravenously with ceftriaxone, a powerful antibiotic. For the next month, Ballas or McKeown regularly infused the drug through the catheter at home. Within a few days of starting treatment, all Ballas's shoulder and back pain were gone, and over the next three weeks, he gradually regained movement in his face. By mid-September, Ballas could drive out for his morning coffee. On September 21, 2008, he returned to work, and in the weeks that followed, his paralysis gradually eased.

4

Every Breath You Take

By the time the IPCC issued its second report in 1995, many scientists, including me, believed the evidence was clear that the climate had begun to change, and that the scientific panel should say so.

The first IPCC report, published in 1990, had laid the foundation for the study of climate change but stopped short of stating that climate change had begun. Instead, the report's authors had concluded that certain proof of human-induced climate change "would not be likely for a decade or more."

By the time of its second report, many of my colleagues and I believed the IPCC would be prepared to clearly state the reality of climate change. Few of us, however, foresaw the virulent pressure that would be applied on IPCC members by an unholy alliance of energy interests, with some help from a handful of manufacturers. It was a campaign unlike any since the tobacco industry's cynical public relations campaign to persuade the public that the dangers of smoking were unproven.

By then the oil and coal industries had begun to spend millions of dollars to persuade the public that global warming was still in doubt. The money bankrolled advertising, lobbying, and public relations. It also funded seemingly independent Washington think tanks and a small handful of naysaying scientists who worked tirelessly to undermine the growing scientific consensus. The strategy was right out of the cigarette makers' playbook, as exemplified by this infamous line from a Brown and Williamson internal memo: "Doubt is our product since it is the best means of competing with the 'body of fact' that exists in the mind of the general public." Similarly, fossil fuel interests ramped up to sell doubt about climate change.

The largest companies invested enormous sums with the international public relations giant Burson-Marsteller to sow that doubt. Burson-Marsteller, in turn, created an entity it named the Global Climate Coalition in 1989, shortly before the first IPCC report was released in 1990. Burson-Marsteller's coalition also joined forces with other industry lobbies that were running their own underhanded campaigns. These lobbies included the American Petroleum Institute, which paid Burson-Marsteller $1.8 million in the early days of the Clinton–Gore administration in a successful effort to torpedo a fossil fuels tax, preventing its passage in the U.S. Congress. Between 1994 and 2001 alone, the PR-invented coalition spent more than $63 million on glossy reports, aggressive lobbying of Congress, and widespread advertising that claimed adamantly that global warming was unproven. Burson-Marsteller, through its Global Climate Coalition, also exerted intense pressure on the scientists who produced the IPCC reports. The PR giant leaned on scientists to try to keep them from validating the reality of global warming.

On November 27, 1995, a huge group—scientists, delegates from ninety-six countries, representatives from environmental groups, and industry representatives—convened in Madrid to hammer out the final version of the second IPCC report. British atmospheric scientist Sir John Houghton chaired the meeting. The gathering was contentious from the start, Houghton would later recall in an account in *Nature,* particularly during the drafting of the summary for policy makers. This was the most influential portion of the report—the section the public, policy makers, and reporters were most likely to read.

On the second day, according to Houghton, the meeting slowed to a crawl. Delegates began fighting over the wording of critical sentences. For instance, there was nearly unanimous agreement among those gathered in Madrid to state that "more convincing evidence" for the impact of human activity on climate change was "emerging." However, dissent came, perhaps not surprisingly, from delegates of the oil-producing nations of Saudi Arabia and Kuwait. These delegates consulted with representatives from Burson-Marsteller's Global Climate Coalition throughout the day. They declared that they preferred the phrase *some preliminary evidence* over *more convincing evidence*. Ninety minutes of debate ensued. As the clock ticked, there were more squabbles with Saudi Arabian delegates seeking to weaken the language in other passages. Ultimately, Houghton was able to report, "To my relief, no unsatisfactory compromises had been made." The contentious gathering ended just after midnight.

The intensity of the negotiations and the influence of the Global Climate Coalition failed to mar the final 1996 IPCC report, which was a historic breakthrough. The world's preeminent body of climate scientists had reached consensus that human-induced climate change was underway. Inclusion of a chapter on human health in this important report was a triumph, as well. My colleagues and I believed we had played a part in reaching a critical milestone in the discourse on climate change.

. . .

The landmark IPCC report of 1996 accelerated the worldwide push for policies that would cut greenhouse gas emissions. Plans were under way for representatives of nations to meet in Kyoto in 1997 to finalize a UN-sponsored treaty mandating those cuts. The international community was coming together to address a common threat.

All was not well in the United States, however. U.S. environmentalists found themselves in pitched battle with a conservative Republican Congress led by Newt Gingrich, the powerful speaker of the house. Gingrich had successfully pushed through an aggressive package of legislation he called the Contract with America. This so-called contract amounted to a rearguard action to weaken the nation's most effective environmental laws. If enacted into law, the legislation would have eased regulations on polluters, further damaging the environment and contributing to the human toll.

Given all these events, Eric Chivian and I and several other Harvard colleagues began working to create a center at Harvard Medical School that would provide a credible voice from the world of academic medicine concerning the health impacts of global environmental change. When our Center for Health and the Global Environment opened in June 1996, it was the first of its kind at a U.S. medical school.

The launch of our center was a pivotal change in direction in my life: I stopped treating patients and turned my efforts toward addressing the health consequences of climate change. After nearly three decades of practicing clinical medicine, I looked forward to this new challenge.

. . .

In November 1997, Chivian and I flew to Kyoto where negotiations for the pivotal treaty on global warming were about to begin. Each of us hoped our participation there would help draw attention to the health threats posed by climate change and the urgent need for action. Eric had obtained a grant to place a full-page ad in the *New York Times* that ran in a special supplement on global warming published just as the Kyoto meeting began. It was signed by more than three hundred physicians and health professionals from around the world. In block letters, the ad read, MEDICAL WARNING: GLOBAL WARMING.

The *Times* global warming supplement also provided us with clear indications of what we health and environmental advocates were up against. International Paper and Ford Motor Company bought ads in the supplement, too. There was also a half-page ad that listed thirty-plus business and labor organizations. These groups claimed that mandatory emissions cuts would ruin the U.S. economy by causing huge increases in the cost of gasoline and electricity.

Ads like these were part of the massive industry misinformation campaign on global warming, a campaign that mobilized front groups, created new institutions, used advertisers, and cost millions of dollars. In the run-up to the treaty negotiations in Kyoto, the Global Climate Information Project bought more than $13 million worth of newspaper and television ads, which were produced by Goddard Claussen/First Tuesday, a PR firm whose clients included the DuPont Merck Pharmaceutical Company and the Vinyl Siding Institute. These ads maintained that a treaty to curb fossil fuel emissions would drive up gasoline prices, with a concomitant rise in the cost of heat, food, and clothing. Another group, the Coalition for Vehicle Choice, a front group for oil and gas companies, carmakers, and car dealers, ran ads blasting the climate talks in Kyoto as an assault on the U.S. economy. A third effort, undertaken by the American Policy Center, focused on mobilizing truckers to protest the Kyoto negotiations by parking their trucks on the side of interstate highways; the group also urged farmers to drive their tractors into cities. In damning the possibility of a climate treaty, their prose turned maudlin: "With a single stroke of the pen, our nation as we built it, as we have known it and as we have loved it, will begin to disappear."

In sharp contrast, in Kyoto constructive voices took center stage. Europeans argued for a 20 percent cut in emissions below 1990 levels. Low-lying island nations, increasingly engaged in a race for their very existence given the rising seas that surround them, argued for 30 percent reductions. The Brazil delegation proposed a large global fund called the Clean Development Fund that would have gone a long way toward facilitating the transfer and manufacturing capability of clean technologies.

But the U.S. delegation, heavily influenced by the fossil fuel and automotive industries, nixed the Brazilian proposal, pushing effectively for a provision called the Clean Development Mechanism, which permits companies to buy the right to pollute from those in developing nations that cut their greenhouse gases. The U.S. delegation also lobbied hard and successfully to make the treaty less ambitious than it could have been, setting a goal of reducing global greenhouse emissions just 7 percent by the year 2012.

Ultimately, a less-than-perfect treaty was drafted, but it was one that, if adopted worldwide, would have begun to slow the increase of CO_2 in the atmosphere. However, it mattered little, as opponents in the U.S. Senate pronounced the watered-down treaty dead on arrival. President Clinton decided not to submit it to the Senate for ratification, knowing it wouldn't pass.

GLOBAL WARMING MAKES US SNEEZE

Each spring for over a decade, I cotaught an undergraduate course at Harvard University titled Global Change and Human Health. In the spring of 2000, a student in the class approached me. Susannah Foster was an environmental science and public policy major who was looking for a senior thesis project, and she had become intrigued by the potential health effects of our changing climate.

By then, my colleagues and I were wondering whether the increasing levels of CO_2 in the atmosphere might affect seasonal allergies and asthma. Today, approximately one in ten Americans suffer from ragweed allergies, and allergic diseases are on the rise, now the sixth leading cause of chronic illness in the United States. There was good reason to suspect that higher CO_2 levels might worsen the situation. I suggested Foster might want to look into this, and I teamed her up with Harvard botanist Fakhri Bazzaz as one of her thesis advisers. Bazzaz had done pioneering work on the effects of rising carbon dioxide levels on different plants and operated several research greenhouses on the Harvard campus.

Bazzaz's earlier research had demonstrated for the first time that ragweed grows disproportionately faster than other plants under the high-CO_2 conditions expected in the future. We wondered what the high levels of CO_2 would do to production of ragweed pollen, which is to blame for the bulk of the hay fever North Americans suffer each autumn.

To find out, Foster planted ragweed seeds in pots and placed them in transparent plastic growth chambers in Bazzaz's greenhouses. Half of the plants were placed in chambers with CO_2 levels of 350 parts per million (ppm), 70 ppm more than preindustrial levels. The rest were placed in chambers with twice those CO_2 levels, or 700 ppm.

Over the summer, Foster watched her plants grow, fertilizing them weekly and watering them daily. In September, when they were at the peak of their flowering season, she harvested and weighed the shoots. She shook the pollen-bearing stems into a large funnel to collect the pollen and used a lab instrument to count the pollen grains. It turned out that doubling atmospheric CO_2 concentration spurred ragweed to grow a bit more (10 percent more) but to produce a lot more pollen (61 percent more). Bazzaz and I had expected the plants to grow faster, and we suspected that there might be more pollen, but the magnitude of the change surprised us all.

The results, which we reported in 2002 in the *Annals of Allergy, Asthma and Immunology,* suggested that rising carbon dioxide alone—aside from its impact on the global climate—would significantly raise the risk of hay fever and respiratory afflictions, like asthma, that are exacerbated by allergenic pollen.

As Foster was doing her experiments, another researcher had similar studies under way, we later learned. Lewis Ziska is a plant physiologist with the U.S. Department of Agriculture who also focused his research on the effects of rising CO_2 on plants. As plant biologists go, Ziska is no shrinking violet. Marshaling an impressive array of facts, he speaks with passion and conviction about the importance of plants to our well-being.

Plants evolved in an ancient era when atmospheric CO_2 levels were much higher than today, and, as a result, many of them grow faster at those CO_2 levels. Atmospheric scientists predict that we will reach those levels in the coming decades and centuries. The important question, Ziska says, is this: Under high CO_2 conditions, exactly *which* plants grow faster? Ziska has been investigating this question for more than a decade. The answer, as our study, Fahkri Bazzaz's early studies, and Ziska's studies had all suggested, is the fast-growing, adaptable plants that thrive in disturbed environments—the plants we know as weeds.

In 1999, Ziska conducted an experiment on ragweed that was very similar to ours, and he found strikingly similar increases in ragweed pollen production. Ziska's team planted ragweed in specially rigged growth chambers in his Beltsville, Maryland, laboratory. Those growth chambers allowed them to control the moisture, light, temperature, and CO_2 that the plant encountered. They grew plants at CO_2 levels of 280 ppm, the atmospheric level in 1850; at 370 ppm, the atmospheric level in 1999; and at 600 ppm, the level projected for 2050. The results were clear. "Going from 280 to 370 doubled pollen production," Ziska says. "Going from 370 to 600 doubled it again."

What's more, the high-CO_2 pollen produced was more potent, according to a second study from Ziska's group. A medical student named Ben Singer took the ragweed pollen and measured levels of the protein that's singularly responsible for the plant's allergenicity. Levels of that protein doubled between the simulated 1850 environment and the environment of 2050—a finding that no one had anticipated.

Next, Ziska and his team set up a study to see if ragweed behaved the same way outside, where it mattered. To recreate the atmosphere of the future, they took advantage of an experiment we're unwittingly conducting. Thanks to fossil-fuel burning by vehicles and industry, atmospheric CO_2 levels in major cities are significantly higher than those in the surrounding countryside—so high, in fact, that they mimic the high-CO_2 atmosphere expected by midcentury. For that reason, Ziska's team planted ragweed on an organic farm in rural Maryland, in suburban Baltimore, and on a plot in an urban setting not far from Baltimore's Inner Harbor. With all other factors held equal, the high-CO_2 ragweed plants from the inner city grew three to five times larger than rural ragweed and pumped out ten times more pollen per plant.

Although many factors affect allergies, several of them come from burning fossil fuels. Black soot, formed from the incomplete combustion of diesel and other fossil fuels, consists of miniscule particles, each less than one-twentieth the width of a human hair, that are tiny enough to penetrate into ultrathin airways deep in the lung, causing respiratory disease. But they also glom onto pollen grains, and the combined particles are then carried deeper into the lungs than the pollen would penetrate alone, worsening allergies. Burning fossil fuels also produces nitrogen oxides (NO_x), gases that react with oxygen in the air to form ground-level ozone, or smog, which corrodes the lining of the lungs and further enhances the allergic response. And fossil fuel combustion is causing global warming, which has extended the allergy seasons two to three weeks over the past few decades, depending on locale. Putting the story together, burning fossil fuels produces allergens, soot, and smog, and together these worsen our air. This exacerbates not just allergies but also asthma and other lung diseases such as chronic bronchitis or chronic obstructive pulmonary disease.

Other aspects of our urban lifestyle conspire to make air quality even worse. Cities create microclimates potentially much hotter than the areas would be otherwise. This is called the urban heat island effect, in which the sun heats dark, impermeable asphalt and other urban surfaces to much higher temperatures than shaded, moist, or white surfaces. Industry and vehicle traffic add to that heat. In a city of one million, the urban heat island effect can make the city up to 4°C (7°F) warmer on average and up to 12°C (22°F) warmer in the evenings. Rising temperatures speed up the smog production from vehicle emissions, and this positive feedback amplifies the urban heat buildup. All these factors combined increase the vulnerability of urban dwellers to heat exhaustion and heat stroke. They're trapped in crowded, asphalt-laden inner cities, where escaping to fresh open air and shade can be almost impossible.

For many urban dwellers and visitors, the ongoing assaults to the lungs from the combined effects of fossil fuel addiction can be merely an irritating norm. For some, however, they can pose a mortal danger.

GASPING FOR AIR

Jaquan Doctor loved macaroni and cheese and playing hide-and-seek. He did a mean Soulja Boy dance that made his grandma laugh. On July 29, 2008, when his mother, Latisha Doctor, sped him to the emergency room at New York's Harlem Hospital, Jaquan Doctor was four.

A few days earlier, Latisha had noticed her son sniffling and coughing. She was not particularly alarmed, even when Jaquan's symptoms lingered for four days, and she did not expect anything out of the ordinary the following Monday evening when she put her son to bed. She dozed off herself not long after in her two-bedroom Bronx, New York, apartment. But at 4 A.M., when Jaquan shook his sleeping mother awake, he was gasping for air. "It hurts," he said, pointing to his chest. And he was screaming, "Ma, I can't breathe!"

In a flash, Latisha had her clothes on and her son in the car. The fifteen-minute drive that followed was one of the longest of her life. Terrified for her son, she sped Jaquan to Harlem Hospital's emergency department. There, the doctors on duty that night noted that Jaquan was breathing very fast and that the muscles between his ribs were retracting as he breathed—a sure sign that his body was working extra hard to get the air it needed. They diagnosed Jaquan with acute asthma, gave him oxygen and albuterol through an inhaler, and admitted him as an inpatient.

. . .

Latisha Doctor had reason to be frightened for her son. Some three hundred million people of all ages worldwide suffer from asthma, according to the World Health Organization. Asthma is an equal-opportunity scourge, affecting people in developed and developing countries alike, including the young, old, and in-between. In the United States, emergency departments like the one at Harlem Hospital recorded 1.8 million visits for acute asthma, including nearly 700,000 visits by children, per year earlier this decade. Indeed, asthma is the top cause of serious chronic illness among children worldwide.

Asthma can be controlled with a short-term drug to stop asthma attacks and a long-term drug to reduce the airway inflammation that heightens the risk of an attack. But it's still debilitating for patients, who suffer from wheezing, coughing, and shortness of breath up to several times a week.

Asthma is the third leading cause of hospitalization among children younger than fifteen. It is one of the largest causes of school absenteeism: in 2003, U.S. children with asthma missed 12.8 million school days due to the disease. It requires regular care, forever. And without proper care—and sometimes even with it—asthma attacks can kill. In the early years of this decade, 4,200 Americans died each year in the United States from acute asthma attacks—200 of them children. Latisha Doctor herself had lost an aunt to an asthma attack the year before.

Asthma has also been on the rise. Since 1980 asthma rates have more than doubled in the United States and throughout the developed and developing world. In the United States, about 4 percent of children had asthma in 1980; by 1995 the figure had doubled, and it has remained roughly steady ever since. No fewer than nine million children and sixteen million adults had asthma in the United States in 2004—a remarkable 7.5 percent of the population.

These facts are all too familiar to Dr. Ben Ortiz, a pediatrician and asthma specialist and the medical director of the Harlem Children's Zone asthma initiative. Ortiz helped oversee Jaquan Doctor's treatment in the Harlem Hospital inpatient unit where, he says, "the number one admitting diagnosis is asthma."

ASTHMA ON THE RISE

Now a thirty-eight-year-old with jet-black hair, graying sideburns, a direct gaze, Ortiz has a warm, reassuring manner that he uses with great effect to calm nervous young patients. He is a New Yorker through and through, having grown up in Washington Heights, a neighborhood of Manhattan just a few miles uptown from where he works today. By his senior year at Brooklyn Tech High School, a public high school that specializes in science and engineering, Ortiz had heard enough about medicine from an older cousin who was an emergency room physician to make plans to follow in his footsteps. Big and broad-shouldered, Ortiz also played on the school's baseball team and dreamed of a baseball scholarship to help him pay for college. His talents twice drew professional baseball scouts to watch him play. "I bombed both times," Ortiz recalled with a smile. "That's a sign from above that said this ain't going to be your chosen field."

As a medical student at Mount Sinai School of Medicine in the early 1990s, Ortiz specialized in asthma because it was a treatable disease and because the growing asthma epidemic made it a challenge. After medical school and a three-year pediatrics residency at Mount Sinai, Ortiz took a job in 2001 at the pediatrics department of Harlem Hospital.

By 2001, Ortiz's new colleagues had already realized that a remarkable one in five people admitted to Harlem Hospital were there because of asthma symptoms. To find out why this was happening, in 2000 a group of doctors from the hospital, including Dr. Steven Nicholas and Dr. Vincent Hutchinson, the former and current directors of pediatrics at Harlem Hospital Center, had teamed up with the Harlem Children's Zone, a large and ambitious organization that ran education, social-service, and community-building programs. The staff at Harlem Children's Zone had already made a disturbing observation: all over Harlem, school absences were soaring.

To see if asthma was responsible, the two organizations joined forces to study the problem. The hospital contributed its admitting diagnoses and other clinical data, and Harlem Children's Zone contributed information on school absences. Together they screened nearly two thousand children to see how common childhood asthma really was in Harlem.

To recruit and keep their subjects in the study, they offered free asthma screenings for children, complete with gifts and pizza parties for those who signed up, and followed up politely but persistently with parents. Their goodwill and persistence paid off when they enlisted the large numbers of local families they needed to get a solid estimate of Harlem's childhood asthma rate.

Ortiz joined the study as soon as he arrived at Harlem Hospital, helping screen children for asthma. Because asthma rates were known to be higher in poor urban communities, the researchers expected to find maybe double the national average, or about 250 of the roughly 2,000 children they screened. Instead, almost 1 in 3 kids had asthma—no less than four times the national average. "That surprised even us," Ortiz recalled. Now they'd be getting twice as many new patients as they'd expected. They didn't have the staff to handle it. Their reaction, according to Ortiz—"Holy cow, we need more people fast."

CODE RED FOR ASTHMA

Asthma attacks begin when the airways are provoked. The triggers vary. Many of them occur mostly indoors: tobacco smoke, dust mites in bedding or carpets, cockroach droppings, pet dander, and chemical irritants trigger many asthma attacks. Others are induced when the body is in a particular state: respiratory infections—colds or flu—can kick off asthma, as can exercise or extreme stress. And other triggers occur outside, including air pollution, cold weather, and airborne allergens.

When allergies kick off an asthma attack, as they do in about half of all asthma patients, they start a cellular chain reaction. As an allergen such as pollen is drawn into the airways, they narrow. This happens to all people, but in those with asthma, the airways overreact. Immune cells called T helper type 2 cells become overexcited and coordinate an immune attack. They start by producing immune factors that help crank up production of a special type of antibody involved with allergies. These antibodies signal other immune cells to release a chemical that activates the airway muscle cells. The airways in turn produce thick, sticky mucus. The airway muscles swell and begin to tighten, starving the body of oxygen. By this point, the patient knows what's happening, and doctors can see the signs.

Standing by the nurse's station on the seventeenth floor of Harlem Hospital, Ortiz and Hutchinson explained what happens next. First, as the body starves for oxygen, it compensates by breathing faster and harder. At this early point in an attack, Hutchinson explained, the patient will be breathing fast and often coughing, but not wheezing. "If you give a bronchodilator, you can open up the lungs, get oxygen in, and the respiratory rate goes down," he said.

If doctors aren't there or they don't get there in time, that's when it can get hairy. In relaxed, healthy people, the diaphragm does most of the physical work in breathing, with some help from the muscles between the ribs, in the neck, and in the abdomen. But when the body is starving for oxygen, it compensates by using everything it's got. Hutchinson pointed to a small dent on his own body where the chest bones give way to the neck. When asthma attacks get bad, doctors see that dent retracting, nostrils flaring, and the muscles between the ribs starting to suck in. They can hear wheezing without a stethoscope. "If you put a pillow over your face and you can't breathe, that's how it feels," Hutchinson said.

Ortiz shared what he tells the parents of his asthma patients: "If you're a weight lifter and you're carrying weights and you get tired, you can drop the weight to give your muscles a chance to relax. But if your breathing muscles get tired, then your only option is to stop breathing." This is why asthma patients can go into respiratory failure after hours or days of struggling to breathe. "At that point," Ortiz added, "it's a battle of time."

A few minutes later, a half dozen young doctors, all wearing white coats and stethoscopes, circled around Hutchinson in the hallway near the nurse's station. It was time for rounds, when senior doctors meet with doctors in training to discuss the most interesting or informative cases on the ward that day. As Hutchinson and the young doctors listened, an intern named Dr. Gloria Adjekum, a petite woman in a white lab coat with a lilting west African accent, presented the first case of the day: Jaquan Doctor. Adjekum described Jaquan's family history of asthma and the arc of his illness—how his cough and wheezing had worsened, how his mother had brought him to the emergency room, how he'd been doing in the two days since he'd been admitted. "His oxygen saturation has been normal. . . . He was stable when he came up to the floor. . . . He had a regular diet and slept well, and he seemed to improve with the management that was given him. In the evening, he was running around."

Hutchinson walked into Jaquan's room, followed by the younger doctors. Jaquan was sitting in an easy chair, legs splayed, thumbs flying on a video game. Hutchinson leaned over and listened to his chest and back through a stethoscope as the young doctors watched. Jaquan alternately glowered at Hutchinson and tried to ignore him, his game beeping nonstop. "I wanted to see if he had any of the flaring nostrils or sucking on the sternum, but he doesn't have any distress," Hutchinson told the residents. "His lungs are completely clear." Not long after, Latisha Doctor arrived to take her son home.

NO_x, OZONE, AND SOOT

Several decades before anyone began studying the health effects of climate change per se, scientists were documenting the various ways that air pollution, most of which comes from fossil fuel combustion, harms human health. This research, spurred by horrible pollution events like the Great Smog of London, which killed more than four thousand people in December 1952, helped lead to modern air pollution laws to reduce some of the more commonplace pollutants.

These include NO_x, which is produced by cars, trucks, and coal-fired power plants and reacts in the air to form nitric acid vapor and particles that can damage lung tissue, cause cardiac arrhythmia, and lead to premature death. When NO_x reacts on sunny days with airborne hydrocarbons (also from fossil fuels), they form ozone, the chief component of smog. Ozone can aggravate asthma, raise the risk of pneumonia, and cause permanent lung damage. Carbon monoxide, produced mostly by motor vehicles, can aggravate heart disease, impair vision, and starve the brain of oxygen, making it hard to work or learn. Particle pollution, a complex mix of particles of various sizes, raises risks for a litany of cardiovascular problems including heart attacks, stroke, irregular heartbeats, and an increased rate of death from heart disease. Although the effects are relatively small compared with something like smoking cigarettes, billions of people are exposed. Epidemiologists have calculated that particle pollution causes early deaths in about eight hundred thousand people annually.

One of the most dangerous types of particle pollution is black soot. In 1989, the International Agency for Research on Cancer, the premier cancer research agency in the world, identified diesel exhaust as a likely human carcinogen, and several studies have linked on-the-job exposure to diesel fumes with an elevated risk of lung cancer.

Inhaling soot particles can exacerbate several serious lung ailments: emphysema, chronic obstructive pulmonary disease, and asthma. In a 1999 study, Columbia University researchers analyzed urine samples from Harlem teenagers for a chemical that indicates soot exposure and found that 75 percent of them had been exposed to enough diesel exhaust that its dangerous chemicals had infiltrated their bodies. In a 2006 study, New York University researchers had forty asthmatic children wear backpack-sized air monitors as they attended their schools, which were located near heavily traveled trucking routes in the South Bronx, just a few miles from Harlem Hospital. On days with the most black soot, the children had double the usual number of asthma symptoms.

ALL IN THE FAMILY

On a warm July morning in 2008, the reception room at Harlem Hospital's pediatric outpatient clinic is packed with families from Harlem and surrounding neighborhoods, in clothes with hues of different cultures, speaking languages from three continents. They are all waiting for their turn to see a doctor.

Inside, in the hallway of the clinic area, Ortiz welcomes Ebony Clark and her three children—George, Jerome, and Jazzmyn Miller, ages nine years, three years, and fifteen months, respectively. He leads the family into an examining room. Ortiz has been treating the Clark-Miller children since they were babies and even calls their mother sometimes to check up on them. Ebony Clark, twenty-four, and her three children have lived in Harlem all their lives, and all four of them have asthma.

Clark places Jazzmyn on the examining table, and Ortiz puts a stethoscope to her chest and back, moves it around, and listens. The baby squalls the whole time. When Jazzmyn is finished, it's Jerome's turn. He climbs up onto the examining table, crinkling the white paper (figure 15).

In May, less than three months earlier, Jerome Miller had a runny nose and a very congested chest, and he was wheezing. Ebony Clark took him in for an exam. He was given a few albuterol treatments. His wheezing continued, so they kept him on albuterol for about five days. He missed two weeks of preschool, and finally when his symptoms would not go away, he was diagnosed with asthma.

Now Jerome was in for a follow-up. As Ortiz examined the boy, listening with his stethoscope, Ortiz ran his pediatrician rap.

"What do you like to eat?" Ortiz said.

"Pizza," Jerome answered.

"What kinds of things do you like to drink?"

"Green soda."

"Green soda?!" Ortiz said with mock outrage.

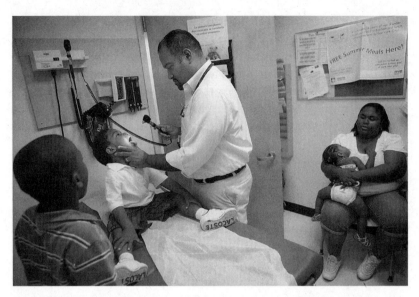

FIGURE 15. Dr. Ben Ortiz, assistant medical director of the Harlem Children's Zone Asthma Initiative, examines three-year-old Jerome Miller as his mother, Ebony Clark, and his brother, George, look on. Clark and her two sons and her baby daughter, Jazzmyn, all suffer from asthma, and the high levels of air pollution in their neighborhood worsen their suffering. (Photo by Jason Lindsey/Perceptive Visions)

Next, Ortiz examined George, the nine-year-old, who'd been diagnosed years before as a young boy, and told Clark her older son's asthma was under control. But George had had an albuterol treatment that day, Clark told Ortiz (figure 16). And just a few days before, he had woken up wheezing uncontrollably. And now the temperature outside was forecast to head into the nineties.

"What do you tend to do on those ozone alert days?" Ortiz asked her.

"Stay in the house," Clark replied.

The Clark-Millers live just a few blocks from the hospital in the Abraham Lincoln Homes, an aging public housing project hard by the Harlem River. The highways on the edges of Manhattan Island are some of the city's busiest, with traffic that rarely stops. The bridges to and from the island of Manhattan are equally busy, with a large volume of traffic and its associated pollution. The Abraham Lincoln Homes sit a block or so from the Madison Avenue Bridge to the Bronx. It's also two blocks from the Amtrak rail line, through which older diesel-powered trains travel regularly. With all this motor vehicle and train traffic nearby, the Abraham Lincoln Homes are getting hammered by air pollution.

As Ortiz finishes examining the Clark-Miller children, he lets the two

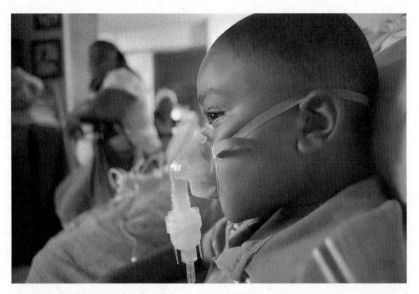

FIGURE 16. George Miller, with a nebulizer. With four asthmatics, the Clark-Miller family keeps a nebulizer handy in their apartment, in case of emergencies. (Photo by Jason Lindsey/Perceptive Visions)

brothers play with his stethoscope as he talks to their mother. George gives Jerome a mock exam, listening to his brother's chest and palpating his belly, just as Ortiz had done. Jerome giggles. And as Ortiz dispenses medical advice to Clark, baby Jazzmyn, the newest asthma patient in the family, closes her eyes and lays her head against her weary mother's chest.

POLLUTION ON THE MOVE

As winds sweep over Africa's vast Sahara and Sahel deserts, they scoop up tons of dust in plumes that are lifted into the upper levels of the atmosphere and carried out over the Atlantic Ocean, following the trade winds thousands of miles to the Caribbean and other endpoints. In 2005, a dust cloud the size of the continental United States blew from Africa out across the Atlantic. Climate change–linked shifts in ocean temperature and salinity are propelling these massive dust clouds even faster.

Dusty desert soil may look dry and lifeless, but it contains a plethora of life that can be very harmful to human health: bacteria, fungi, viruses, plant pollens, heavy metals, and often unknown industrial or chemical toxins from pollution and dumping. Dust clouds blowing from Africa to the Caribbean carry fungal and bacterial diseases that attack trees and crops, and they carry the bacteria that cause highly lethal meningitis. And the particulate matter of dust itself has been linked to increased cardiac and respiratory problems, including a spike in pediatric asthma found in Trinidad and Barbados after dust storms. Dust storms also carry a soil-dwelling fungi called *Aspergillus* that can be lethal to people and livestock and is infecting fan coral in the Caribbean. These infections are becoming so bad, it could affect the islands' tourism industry.

In the deserts of Saudi Arabia, people wrap themselves tightly in blankets to survive dust storms that leave them "covered in a layer of white, like you have suddenly grown old," in the words of an Iraqi who spent two years in Saudi refugee camps plagued by clouds of powdery dirt. In China, it is not uncommon to see lovely ancient temples nearly obscured by sallow yellow curtains of dust. The American Southwest has apocalyptic dust storms of its own—huge brownish yellow clouds spreading above picturesque cities like Tucson. The southern United States has gotten drier in recent years, causing more dust storms there as well.

Such changes occurring in many parts of the world are due to climate change and unwise land use—overgrazing, monocropping, and deforestation. In Asia 2,100 square kilometers (800 square miles) each year turn into desert. This means that where previously wind would have blown cleanly across forest or prairie, it now stirs up gritty pathogen-laden dust from new deserts.

As with other effects of climate change, climate and weather forecasting programs that help anticipate dust storms and warn people to take precautionary measures can minimize adverse health effects. "If we can predict when and where this will occur, we can move people out of the way," said William Sprigg, an atmospheric physicist at the University of Arizona who studies dust storms and who has developed a three-day forecast system for southwestern dust storms using NASA satellite data. He has tailored his system to specific zip codes. A worldwide dust storm prediction system is in the works.

Meanwhile, vast hazes of air pollutants from coal-fired plants and automotive emissions are accumulating over large parts of Asia, affecting respiratory health, visibility, and the local climate. These hazes, along with dust, can severely limit visibility, causing hazards for vehicle and air traffic.

TWENTY-FIRST-CENTURY HEAT

For decades, tourists have headed to the south of France in the summer for warm days on sandy beaches and cool nights sipping wine in sidewalk cafes. That's the kind of thing Debarati Guha-Sapir had in mind when she planned a vacation in the town of Carpentras on the scenic Côte d'Azur in August 2003. Guha-Sapir was envisioning a vacation from her job as director of the Belgium-based Centre for Research on the Epidemiology of Disasters. Little did she know she was heading right into a disaster.

That summer, a heat wave of unprecedented intensity blazed across Europe, killing more than 52,000 people, including as many as 13,400 in France alone. Many of them were elderly left alone as much of the population was enjoying their summer holiday. Public officials were slow to acknowledge the severity of the situation: President Jacques Chirac didn't return from his vacation in Canada until two weeks after the crisis started, and the prime minister was also slow to return from holiday. Eventually the surgeon general resigned over the debacle.

To avoid overheating, Guha-Sapir cut short her visit with friends and her plans to hike, swim, and relax in town. Instead she spent several days driving around, since the car, unlike most French homes, had air conditioning. "The heat was just absolutely total, night and day," she remembers.

. . .

The European heat wave of 2003 didn't just break temperature records—it blew them away. That summer, temperatures were 20–30 percent higher than the seasonal average over much of Europe, from Spain to the Czech Republic. French grape-growing records—which contain historical climate data—indicated that the event had no equal since 1370.

The odds of such a heat wave happening by chance, based on climate trends from 1864 to 2002, were about one in ten million. "The 2003 anomaly can only be accounted for by a combination of global warming and natural variability," wrote Kevin Trenberth of the National Center for Atmospheric Research.

What if a heat wave like the European scorcher of 2003 occurred in the United States? That has become more likely in this changing climate. The 2003 heat wave was an extreme manifestation of a trend toward increased heat waves worldwide, and the probability of such extreme heat has already increased by between two and four times over the past century. By the 2040s, more than half the years could have summers warmer than that of 2003, according to a climate scenario put forth by the IPCC based on business-as-usual greenhouse gas emissions.

In addition to increasing the occurrence of heat waves, climate change also alters their very nature. Climate scientists agree that future heat waves will be hotter and feel hotter. That's because a warmer atmosphere can hold more moisture, raising humidity enough to make conditions unbearable. Nightly minimum temperatures in particular have risen and will continue to rise. Hot nights increase the lethal potential of a heat wave. Normally, falling temperatures at night allow heat-retaining surfaces to cool, offering relief to living beings. But if nighttime minimum temperatures remain significantly elevated for days on end, the compound effects can be devastating.

Heat already kills about 1,500 people per year in the United States, making it more lethal than hurricanes, tornadoes, floods, and earthquakes combined. But unlike those natural disasters, heat waves kill silently, leaving no destruction or physically altered landscapes behind. What's more, their victims are disproportionately elderly, infirm, poor, minority, and living alone. Official figures don't reflect anywhere near the real death toll because there is no uniform definition of heat as a cause of death and because other causes of death triggered by heat, including heart attacks, usually aren't logged as heat-related. But public health officials and researchers know that heat is deadly.

More than seven hundred people died because of the stifling heat wave that slammed Chicago in 1995, for example, when outdoor temperatures hit 41°C (106°F) with high humidity and temperatures inside brick buildings without air conditioning climbed much higher. The city's ambulance fleet was overwhelmed, leaving many suffering victims— and corpses—waiting hours to be picked up. The county morgue was full, and bodies were stored in nine refrigerated semitrucks loaned by a meatpacking company, creating an ominous scene in the morgue's parking lot.

In 1995, Larry Kalkstein was a geography professor at the University of Delaware. He and his colleagues built a computer model that predicted how a heat wave like Chicago's would affect New York City, St. Louis, Philadelphia, Detroit, and Washington, DC. They found there would be more than five thousand excess deaths in those five cities during a summer with such a heat wave. New York—with almost three thousand excess deaths—and St. Louis would have the highest mortality because of the multitude of high rises, brick row houses, and tar roofs that all trap heat. In addition to the direct health risks from the heat and humidity, heat waves indirectly harm human health and the environment in ways that are less quantifiable but no less real. Massive electricity blackouts can stop air conditioning, putting more people at risk, and hinder medical care at hospitals and nursing homes. And heat waves can harm the environment, indirectly harming health. The 2003 European heat wave caused glaciers to melt and incited forest fires that ravaged much of the continent, heavily polluting the air. Heat waves during droughts can wilt crops or spur pest infestations of aphids, whiteflies, or locusts that can slash crop yields. In developing countries, this can lead to food shortages and price increases that affect health and nutrition.

Heat waves can also trigger epidemics of human disease. That's especially true in developing countries where dry, cracked earth won't absorb water from the monsoon rains that follow heat waves, meaning likely contamination of water supplies. People weakened by heat waves are also more susceptible to various diseases. In India, a 2003 heat wave that raised temperatures to 50°C (122°F) killed 1,400 in the state of Andhra Pradesh. When the heat wave yielded to heavy rains, mosquitoes flourished and an outbreak of Japanese B encephalitis killed 110 children.

BEATING THE HEAT

Although dangerous heat waves may be a threat in the years to come, we are far from helpless in dealing with them. Indeed, deaths from heat are actually one of the most avoidable effects of climate change. If people suffering heat exhaustion are found early enough, usually the only things they need to survive without permanent damage are an air-conditioned space, rest, and hydration.

The best way to find victims while there's still time to help them is to predict dangerous heat waves early. Just as Andrew Githeko developed a model to predict malaria epidemics and William Sprigg developed a model to predict dangerous dust storms, Larry Kalkstein has developed a sophisticated warning system for extreme heat. The model is based on data showing that conditions that are either particularly hot and dry or particularly hot and moist have led to increased mortality. The system works by taking National Weather Service data and projections and automatically processing them to calculate the alert level for a specific city.

The poster child for Kalkstein's warning system is Philadelphia, which lost 118 residents to a heat wave in 1993. Kalkstein and his colleagues launched the Heat/Health Watch Warning program in Philadelphia in the spring of 1995, coincidentally just before the city was struck by the same mass of stifling air that settled on Chicago that summer. Philadelphia reported only 18 deaths in that heat wave, compared with Chicago's hundreds.

Of course an alert is nearly meaningless without action. Philadelphia took several steps to help residents stay cool and healthy. Residents can call a twenty-four-hour "heatline" for advice, information, and help; air-conditioned cooling centers are set up around the city, and public outreach plans are in place.

An analysis of the Philadelphia system indicates that it saved an estimated 117 lives between 1995 and 1998. Philadelphia National Weather Service meteorologist Gary Szatkowski thinks it has been a great success and has proved initial doubters wrong. "In this day and age in this society, it's unacceptable someone should die from excessive heat," he said, adding that it's also very possible to create a society in which no one does.

As of late 2008, Kalkstein's warning system was used in forty U.S. municipalities, including New Orleans and Seattle, as well as the nine largest cities in Italy, three Canadian cities, and Seoul. The system is run off servers at Kent State University and the University of Miami, where Kalkstein is now based, and municipalities pay a relatively moderate fee to participate.

Besides early warning systems, other measures can help keep people safe during dangerous heat waves. Cities can push for parks and other green space, buildings with green roofs, and urban gardens. These measures not only keep a city from overheating when the heat is on. They also draw down CO_2. That in turn helps fight the root causes of climate change and will keep the air we breathe cleaner and healthier for generations to come.

Harvest of Trouble

If there were a heaven for biologists, it would probably look a lot like Woods Hole, Massachusetts, a small seaside village situated on a point of land in southernmost Cape Cod.

Not far offshore, the warm Gulf Stream mixes with cooler northern currents, breeding a rare diversity of sea life that attracted the biologists who founded the Marine Biological Laboratory (MBL). Today, the MBL is the oldest, and arguably the most illustrious, private marine laboratory in the western hemisphere. Each summer, biologists from all corners of the globe converge on the MBL to work long hours in its aging laboratories and bandy about ideas. The MBL library, which remains open twenty-four hours a day, is a delight for tireless young investigators. As the noted science writer Lewis Thomas once wrote of the MBL, "If you can think of good questions to ask about the life of the earth, it should be as good a place as any to go for answers."

Back in November 1994, when my colleagues and I were still urging the medical community to recognize the health threat posed by climate change, nine of us in the New Disease Group convened a four-day conference at the MBL to ask some questions of our own. By that point, we had been meeting for more than two years at the Harvard School of Public Health to discuss the source of the plethora of new diseases—the epidemic of epidemics—that had appeared on the scene beginning in the mid-1970s. The year before, the conclusion that numerous infectious diseases were emerging or reemerging had been endorsed by the scientific establishment in a seminal report from the Institute of Medicine of the U.S. National Academy of Sciences.

To gain insights into emerging diseases from other corners of biology, we cast our net wide. We invited more than sixty scientific explorers, some of whom studied human disease by a variety of means, ranging from epidemiology to mathematical modeling to field biology on disease vectors, and others who conducted research on diseases of plants and animals.

By day, we listened raptly to each other's presentations and brainstormed intensely over lunch at the MBL's Swope Center, a gathering place overlooking the green waters of Eel Pond. By night, we strolled to a waterfront saloon, drank beer, and brainstormed some more. On the third day, Pamela Anderson, then a young postdoctoral researcher in Dick Levins's lab, walked to the whiteboard, sketched a diagram, and offered an eye-opening insight.

Anderson was an expert on infectious diseases of plants. As farmers and gardeners well know, plants—like humans—can contract a variety of infectious diseases carried by viruses, fungi, bacteria, or other agents. Anderson told us that, just as there had been an epidemic of human epidemics over the previous two decades, there had also been an epidemic of plant epidemics, many of them caused by viruses. Anderson described the different processes by which viral disease could spread among crop plants and emphasized that the same processes apply to infectious disease in humans.

A host—the organism in which a disease develops—might be introduced to a new area, bringing it in contact with a new virus or a new vector to which it had not previously been exposed. This happened when cacao plants, which are native to South America, were introduced in the late 1800s to west Africa, where they subsequently contracted a viral disease from indigenous trees. This could also happen to a person with no malaria resistance who moved into an area where the disease was endemic.

Alternatively, a virus could be introduced into previously unexposed host populations, as when a citrus virus was accidentally introduced from Africa to South America in the 1920s—or when Europeans deliberately exposed Native Americans to smallpox in the 1600s, causing devastating epidemics.

Or plant populations might become more vulnerable to disease when they are biologically stressed. Anderson described a bean virus in Latin America that infected *Phaseolus vulgaris*, the common bean, which is a staple food for many Central Americans. Strains of bean plants that resist the virus at ordinary temperatures become vulnerable to the virus when they're taxed by high temperatures. Humans can become more vulnerable to infectious disease because of drugs, malnutrition, or emotional or environmental stress, a phenomenon I'd witnessed with AIDS patients I'd attended in the 1980s.

Anderson had synthesized these mechanisms into a unifying framework that helped explain how social and ecological changes could work together to drive waves of infectious disease—whether they were in plants, animals, or humans. One could thus study epidemics in plants and animals and learn a great deal about human epidemics, which was exciting to me as a doctor and public health scientist.

Anderson had adopted a systems theory framework like that of our mutual mentor, Dick Levins. Her cohesive, multidimensional approach helped her account for and probe the multiple causes of real-world ills. The framework she presented depicted an integrated set of pathways by which human activities had unwittingly contributed to the emergence of disease via our social choices and our impacts on ecosystems, including the managed ecosystems we call farms.

If humans were responsible, however, there was hope that we could make different choices and adopt healthier practices. It all clicked into place. Anderson's presentation galvanized our discourse and helped launch the new field of emerging infectious diseases. It was a thrilling, unifying moment.

THE FIELDS OF TOMORROW

At 10 A.M. on a hot midsummer morning in an experimental Illinois farm field, Clare Casteel squatted down, grabbed hold of a healthy deep-green soybean leaf, and attacked it like a hungry beetle.

As her team of undergraduate assistants, two men and one woman, watched closely, Casteel used a small sewing tool to perforate the leaf. "You want to go straight across. Go straight across several veins," said Casteel, a twenty-six-year-old with tousled strawberry-blonde hair, stylish rectangular glasses, and muddy black Converse All-Stars. The result was a dotted line of holes that resemble what the beetle's mouthparts do to a leaf. A few minutes later, her entire team would be mimicking beetle attacks themselves.

Casteel, a graduate student in plant biology at the University of Illinois, Urbana-Champaign, held no ill will toward the plants she was damaging.

FIGURE 17. The Soybean Free Air Concentration Enrichment (SoyFACE) facility near Champaign, Illinois, exposes soybean crops, under realistic field conditions, to the higher levels of carbon dioxide expected by midcentury. Studies here help project how tomorrow's atmosphere will affect production of this staple crop. (Photo by Andrew Leakey, University of Illinois, Urbana-Champaign)

On the contrary, she had devoted three years of her life to understanding them, including many evenings and weekends in the laboratory. Today she was conducting an experiment that would shed light on how soybeans, one of the world's most important crop plants, would fare in the high-CO_2 environment they're likely to experience by midcentury.

Casteel—who hails from Hartsburg, Missouri, population 108—was that day working at a unique experimental station that explores what lies ahead for American agriculture. The station, known as Soybean Free Air Concentration Enrichment (SoyFACE for short), is run by a consortium of scientists at the University of Illinois, Urbana-Champaign. It allows scientists to control CO_2 levels in open air plots, which lets them study how soybeans would fare in the field when there's more CO_2 in the atmosphere (figure 17).

Each of the experimental plots Casteel and her cohorts worked that day sat in a 370-square-meter (4,000-square-foot) octagonal plot whose borders were enclosed by special shin-high, green plastic tubing. By combining wind gauges and other sensors at the center of the plot with sophisticated controls that pump out CO_2 from the plastic tubing on the upwind side of the crop plants, researchers can maintain CO_2 (or, in other experiments, ozone) at steady and elevated levels inside the plot. By sampling the same line of soybeans grown under different conditions, researchers like Casteel can compare, for example, plants grown under the 387 parts per million (ppm) CO_2 that prevailed in 2008 with plants grown under the 550 ppm CO_2 that she and other scientists project may prevail in 2050.

Her experiment that July day in 2008 was designed to elucidate how soybeans in the future might fend off a bane of farmers worldwide: herbivorous insects that eat their crops. Soon Casteel and her cohort of assistants were spread out all over the flat green plot, squatting in the muddy field among the soybean plants. For the next two hours, they worked with a quiet focus. They intentionally injured plants, marked them with colored ribbon, then moved on to do it again at the next plot. On their breaks, they chatted about classes, conferences, plants, and bugs.

Two hours later, they made the same rounds again. They snipped off the injured leaves from the labeled plants, placed them quickly in orange-capped plastic tubes, then walked quickly to freeze them with a hiss in a vat of liquid nitrogen. Later, in the lab, Casteel and her undergraduate assistants would grind up the leaves to extract a cellular compound called ribonucleic acid, or RNA, that can indicate which genes the plants activated to fight off the simulated insect attack. The experiment was one of a long series of studies that would help Casteel and her adviser, plant biologist Evan DeLucia, learn how well soybeans will fend off insect pests as CO_2 levels rise. The answer to that question will help answer a larger question: If the climate changes as scientists project, will we be able to grow enough food to feed humanity?

A GREENING EARTH?

Photosynthesis removes CO_2 from the air with great efficiency. What plants, algae, and blue-green algae (cyanobacteria) gain from photosynthesis are the carbon-based compounds that serve as molecular building blocks for their tissues. They create these compounds via a two-step process. In the first step, known as the light reactions of photosynthesis, plant cells use the energy of sunlight to split molecules of water, releasing molecular oxygen and producing hydrogen ions and energy-rich chemicals such as adenosine triphosphate (ATP) that fuel metabolism. In the dark reactions (so named because they can take place even at night or if the plant is placed in the dark), ATP powers a reaction that combines carbon dioxide and the hydrogen ions to form sugars and starches. Those sugars are then transformed into the molecular building blocks of plant tissue—cellulose, proteins, and more. The most important player in the dark reactions is an enzyme called ribulose bisphosphate carboxylase-oxygenase—Rubisco for short. Rubisco, which is key to photosynthesis, is the most common protein on Earth.

Like all enzymes, Rubisco is a catalyst, meaning that it accelerates a chemical reaction that would otherwise take place far too slowly to be of much value to an organism. Carbon dioxide fixation by Rubisco, like all enzyme-catalyzed reactions, moves faster when there's more reactant, in this case more carbon dioxide. But Rubisco can only work so fast. When CO_2 climbs above a certain level, which varies from species to species, Rubisco is working as fast as it can.

In most plants that evolved in high-latitude temperate regions, the properties of Rubisco and the cells that contain it are such that carbon dioxide above the preindustrial level of 280 ppm is expected to accelerate photosynthesis. This is true for many of the world's most important staple crops, including rice, wheat, and soybeans. These plants use what is called C3 photosynthesis, named this because the first product made with fixed CO_2 is a three-carbon sugar. Higher CO_2 levels also cause C3 plants to close their tiny leaf pores, called stomata, thereby sealing in water and making the plants more drought-tolerant. These two responses to rising CO_2 led early agronomists to predict, beginning in the 1960s, that as CO_2 levels rose in the atmosphere, crop plants like rice, wheat, and soybeans would speed up their photosynthesis and grow faster, thereby increasing crop yields.

This process is different in crops such as corn or sorghum, which—like many plants that evolved in the tropics or subtropics—conduct photosynthesis using a different cellular strategy. Most plants from those regions are C4 plants, so named because the first product of carbon fixation is a four-carbon sugar. C4 plants tend to concentrate CO_2 in their cells prior to photosynthesis, which means that, even at current atmospheric levels, their Rubisco is already working as fast as it can.

By the 1980s, crop scientists had tested the early agronomists' predictions concerning C3 plants. They did this by exposing plants to CO_2 at about 550 parts per million—roughly double preindustrial levels. They exposed plants in greenhouses or in growth chambers with grow lamps in a lab. They also carried out field experiments with transparent plastic enclosures—open-top enclosures surrounded by glorified shower curtains, with CO_2 wafted toward the plants. Rice, wheat, and soybeans inside all these setups photosynthesized faster and produced about 30 percent more biomass. Plant scientists therefore predicted that a warmer, higher-CO_2 world would produce higher yields for farmers, helping to feed the world.

The IPCC initially endorsed the findings, although they added a few caveats. In their third assessment report, released in 2001, the IPCC concluded that although rising temperatures and drier soils would reduce global crop yields through other mechanisms, this CO_2 fertilization would more than make up for it, at least in temperate regions. Total crop yield, the IPCC reported, would rise in temperate regions but fall in the tropics.

Global warming naysayers, including publicists for fossil fuel companies, did not miss the significance of this news, as it was one of the few pieces of seemingly good news on global warming emerging from legitimate scientists. In 1991, the Western Fuels Association, a trade group representing coal-burning utilities, spent $250,000 to produce a video called *The Greening of Planet Earth,* which promised an era of agricultural bounty as carbon dioxide levels climbed. From 1998 through 2006, the Greening Earth Society, which operated out of the offices of the Western Fuels Association, promoted the same view— that rising greenhouse gas emissions were a good thing because they'd promote plant growth and make the environment greener. They claimed that their viewpoint was a "scientifically-sound [sic] perspective on the increasing atmospheric carbon dioxide."

Plant scientists, however, were already learning that reality was a lot more vexing.

PLANTS IN A POT, AND BEYOND

Evan DeLucia is a plant biologist who directs the School of Integrative Biology at the University of Illinois, Urbana-Champaign. He oversees a large research team, including Clare Casteel, that studies how forest and farm ecosystems respond under the atmospheric conditions predicted for the future. That anticipated atmosphere includes elevated carbon dioxide and, in regions like central Illinois that are downwind from major cities, it also includes elevated ground-level ozone from industrial emissions and vehicle exhausts, which stymie plant growth. DeLucia combines laboratory and field studies to understand how global change affects which insects eat which plants, as well as when, where, and why.

On the afternoon following Casteel's field experiment, DeLucia described his research in his spacious, sunlit office in the university's Institute of Genomic Biology, which overlooks a wide patio. A small cornfield enclosed by a hedge sits beyond the patio: the Morrow Plots. Continuously planted since 1876, these are the oldest and longest-running experimental agronomic plots in the United States.

At fifty-two, DeLucia is lanky and youthful looking, with a friendly smile and an air of easygoing efficiency. The native New Yorker had studied forest ecology while earning a doctorate from Duke University in the mid-1980s. Soon after, he headed out to the flatlands of Illinois to take a faculty job. DeLucia was interested in how climate and climate change would affect the physiology and ecology of plants. "It was just clear very early on that this was going to be one of the most important questions that we were going to have to solve as a society," he said. Like other ecologists of his generation, DeLucia had initially investigated pristine ecosystems, flying out regularly to the Sierra Nevada and the Rocky Mountains to examine the effects of climate on forests at different elevations. But after a while, he shifted his focus to the sprawling agricultural landscape he called home.

At that point, DeLucia began to study the effects of climate change on soybeans, which, along with corn, are one of the Midwest's two major cash crops. The early greenhouse and growth chamber studies had produced valuable information about these effects, DeLucia said. "If you're working with a plant in a pot, you can measure how fast it grows, and you can measure its physiology. You can clamp a machine on and measure its photosynthetic rate, and you can measure how many seeds it produces. All that is good stuff, and we want to know that stuff," he explained.

But, he added, a lot of other questions can't be addressed. "It would be like setting out your heart beating in a plate rather than within your body," he said. "You're missing its connection to the greater ecosystem." In the case of soybeans, that meant answering questions like how much nitrogen they are removing from the soil and to what extent elevated levels of CO_2 spur their growth in the real world, where they face sun, clouds, rain, wind, disease, and the bane of farmers everywhere: insects.

To get a better fix on soybean performance in a high-CO_2 world, in 2001 DeLucia and several of his colleagues—plant biologists, entomologists, agronomists, and others—installed a free air CO_2 enrichment (FACE) system in a typically flat, fertile farm field several miles south of Champaign. About two dozen FACE setups exist worldwide, each focused on a different type of plant ecosystem: loblolly pine stands in North Carolina, aspen and birch forests in northern Wisconsin, barley and wheat fields in Germany, and a variety of other plants. "The beauty of the experiment is that it's a large plot and completely open, so anything can come and go," DeLucia says.

THE BEETLE WITH A SWEET TOOTH

One of the first experiments at the FACE plots, performed by DeLucia's colleague Stephen Long, revealed that when carbon dioxide was elevated to levels projected for 2050, the increase in soybean yield—the CO_2 fertilization boon—was half of what researchers typically saw in the growth chamber. The studies were repeated. Time after time, with crop after crop, the results were the same: real-world conditions cut the CO_2 fertilization boon in half. The results, which were published in *Science* in 2006, "go directly to food security," DeLucia said. When climate change kicks in full force, DeLucia asked, "What is going to be the capacity of the human race to feed itself?"

In the summer of 2002, just after they'd published their first papers on their field experiments, DeLucia and his students noticed something curious. "As you walked through the high-CO_2 plots, your jeans got sticky. As you walked through the non-CO_2 plots, they didn't," he recalled. The stickiness in the high-CO_2 plots, it turned out, was produced by soybean aphids, green insects about the size of a lowercase *o*. Soybean aphids, like whiteflies, make a living by puncturing plant leaves and sucking the plant's lifeblood from veins, called phloem, that transport it. Because the lifeblood of soybeans is mostly sugar water, the aphids must suck out proportionally huge amounts to get the nitrogen compounds they need to make protein. Most of that sugar water goes right through them, and they excrete a substance called honeydew that's sweet and sticky.

Around the same time, DeLucia's team observed that Japanese beetles also preferred the high-CO_2 soybean plots to the low-CO_2 plots, by huge majorities. "Being good field biologists, you're saying, 'What's going on?'" DeLucia recalled. They proceeded to spend the next several years finding out.

Japanese beetles chew up soybean leaves in a farm field just as they chew up rose bushes in a suburban backyard. DeLucia first enlisted a squad of undergrads to photograph half-eaten leaves in the high-CO_2 and ambient-CO_2 plots. The undergrads then conducted a computerized image analysis of the photos to quantify the amount the beetles had eaten. The high-CO_2 plots experienced double the damage.

At first, DeLucia believed that the beetles needed to eat more just to survive in a high-CO_2 atmosphere. That's because leaves in almost all plant species, including soybean, get sweeter and less nutritious under elevated CO_2 compared with ambient CO_2. So for a beetle, switching from ambient-air soybeans to high-CO_2 soybeans would be "like going from spinach to Twinkies," DeLucia says. And that meant that they'd have to eat a lot more to get the nutrients they needed. It was also known that Japanese beetles love sugars: if you feed them glass wool dipped in sugar water, they'll eat it until they die. The beetles were drawn to sugar and would need to eat more under high-CO_2 levels to survive; DeLucia thought perhaps that was why they chowed down on the leaves of the future (figure 18).

At that point, DeLucia turned to his longtime collaborator, May Berenbaum, the head of the university's entomology department, who focused her research on the interactions between herbivorous insects and their host plants. A graduate student of Berenbaum's named Bridget O'Neill decided to investigate just what the beetles gained by voraciously consuming the high-CO_2 plants.

Japanese beetles lay eggs in soil underneath turf grass—lawns—and spend their early life stages underground before emerging as adults. In one of her first experiments, O'Neill trapped virgin adult Japanese beetles as they emerged for the first time into sunlight, took them back to the lab, and mixed males and females in ice-cream containers with some soil at the bottom for the females to lay their eggs in.

Each morning, O'Neill would drive to the experimental field, cut fresh soybean leaves from ambient and high-CO_2 plots, return to the lab, and feed the leaves to her beetles. Each day, she'd count beetles in the containers to see how many were still alive and sift the soil in the containers for eggs to see how much the survivors had reproduced. After a few months, it was clear that beetles fed high-CO_2 leaves were faring better, since they lived longer and laid more eggs.

The results meant that "the opportunity for crop losses might go up in the future," DeLucia explained. But why were the beetles living longer and reproducing more? Was it because they were eating sweeter leaves? DeLucia didn't think so. For humans and beetles both, subsisting mostly on sugar was usually a bad thing. So what was happening?

DeLucia pulled out a pencil and sketched a vase containing a soybean

FIGURE 18. Japanese beetles devouring soybean leaves. Studies at the SoyFACE facility by Evan DeLucia, May Berenbaum, and colleagues have shown that when carbon dioxide levels rise, the soybean leaves are less nutritious. Plant-eating beetles consume more, resulting in lower crop yields. (Photo by Bridget O'Neill, University of Illinois, Urbana-Champaign)

leaf. He explained that, to find out what was happening, O'Neill took leaves from the ambient-CO_2 plots and artificially sweetened each by placing its stem in a vase of sugar water. Artificially sweetened soybean leaves—unlike the leaves from the actual high-CO_2 plots—did not extend the beetles' lives or cause them to lay more eggs. That meant something besides sugar was extending the beetles' lives. What was it? The answer turned out to be less obvious and more interesting, and its backstory reaches deep into evolutionary history.

AN ARMS RACE FOR THE AGES

Plants initially spread on land more than four hundred million years ago. At first, they had an easy go of it. But they were just too good a food source for evolution to pass up, and it didn't take long for animals to start eating them. Among the first to do so were insects.

A simple thought experiment reveals what might have ensued. As herbivorous (plant-eating) insects bred, multiplied, and evolved, they would have had a defenseless food supply to devour. According to simple ecological theory, insects would have eaten plants into extinction. Then,

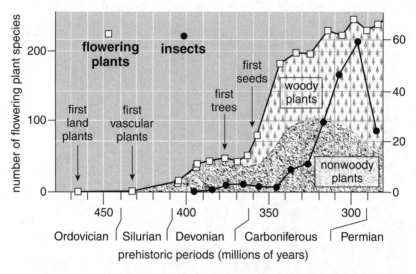

FIGURE 19. Locked in battle. Insects and land plants dueled for supremacy from the start, as plants developed chemical defenses and insects developed ways around them. Their evolutionary arms race has continued ever since. (From G. J. Retallack, et al., "Early Forest Soils and Their Role in Devonian Global Change," *Science* 276 [1997]: 583)

lacking food, insect populations would have crashed, driving themselves into extinction as well, DeLucia explained. "But if you look outside, it's a green world, so obviously that theory is not working."

DeLucia gave his informed take on what occurred. "Insects chew on plants, plants die, and the ones that don't die had something special about them that prevented them from dying or from suffering damage, or that prevented the insects from eating them," he said. The specially adapted plants survived and thrived, and the insects next turned to other plants for a meal. "This is natural selection," DeLucia said. But what was special about the plants that made it through?

Some of them had evolved to produce chemicals that poison, or at least repel, herbivores. Such strategies are ubiquitous among today's plants: cassava plants produce cyanide, and willow plants produce salicylic acid (from which we get aspirin). Other plants employ more sophisticated strategies. Some plants, when they're chewed on, emit chemical signals called pheromones that lure friendly insects to their rescue. For example, some of these pheromones draw parasitic wasps, which lay their eggs inside the caterpillars that threaten the plant. The wasps then hatch and eat the caterpillars from the inside. Plant defenses such as these allowed ancient plants to survive the onslaught of herbivores and to grow into mighty forests and colonize vast prairies.

Meanwhile, insects did not sit idly by. Many evolved ways to get around plant defenses, such as producing enzymes that detoxify the plant's poisons. Natural selection thus set up an evolutionary arms race between insects and plants that continues to this day (figure 19).

This arms race is part of the reason why plants make thousands of defense compounds, since "if one is no longer useful, they keep building new ones," DeLucia says. We humans, in turn, have benefited from this epic battle, as many of our medicines are derived from plant defense compounds. For example, the active ingredient in Tamiflu, which combats bird flu, comes from a defense compound of the star anise plant, a relative of the plant that's used to make licorice. And in the case of agriculture, we've benefited nutritionally from the skills that crops like soybeans display at the not-so-gentle art of self-defense.

. . .

By the time O'Neill was feeding fresh soybean leaves to her captive beetles, scientists had learned that one of the soybean plant's defenses against being eaten was to produce a compound called a protease inhibitor that prevents beetles that eat it from digesting proteins. "I'm not a beetle," DeLucia said, "but this has got to be like the Thanksgiving dinner from hell. You have just gone way over the top, and you are lying on the couch watching football, and it is not going anywhere." Japanese beetles respond to such a nightmare by flying off, if they can, or wandering off to die.

What's more, a soybean plant can actually sense when it's being chewed on, and it responds by producing an alarm chemical called jasmonic acid. Jasmonic acid functions as adrenaline does in humans— it puts the whole body on high alert. The plant responds by ramping up production of its insect-killing poisons. The researchers wondered if high-CO_2 conditions were weakening the soybean plant's ability to defend itself.

Clare Casteel did the experiments that helped answer this question. With undergrads in tow, she collected Japanese beetles and starved them for a day in the lab. The next day, at the experimental plots, she released them on soybean plants under ambient and high-CO_2 conditions. Then she cut the leaves, quick-froze them, and ran the aforementioned molecular test to see which genes were turned on. According to Casteel's studies, which were published in *Plant, Cell and Environment* in 2008, and subsequent studies done by Argentine postdoc Jorge Zavala in the laboratory, the high-CO_2 world diminished jasmonic acid production, which cut protease inhibitor production, which allowed beetles to eat freely on the poorly defended plants. "The startling thing we discovered," DeLucia explains, "is that plants grown under high CO_2 have trouble ringing that alarm bell."

More research is needed to determine whether tomorrow's CO_2 levels will do the same to other crop plants, or to plants in general, DeLucia said. But if they do, it would mean that more CO_2 and climate change are tipping the plant–insect balance of power in favor of the insects. And that could make it harder to grow enough food to feed humanity.

Scientists don't know yet if that will happen, but a look back in evolutionary time by another scientific team does not augur well. Geological evidence indicates that 55.8 million years ago, several thousand billion tons of CO_2 was pumped fairly quickly into the atmosphere by an enormous burst of volcanic activity—roughly the same amount that will be pumped into the atmosphere over this century if we keep burning fossil fuels and removing forests from the Earth. During that time, known as the Paleocene–Eocene Thermal Maximum (PETM), temperatures spiked by 5°C (9°F).

A team led by Ellen Currano of the Smithsonian Institution has examined more than five thousand fossil leaves from the era found in Wyoming. According to an article by DeLucia and his colleagues that commented on the study, the leaves contained "a dazzling array of damage types, including gaping holes inflicted by chewing insects, galls formed by wasps, delicate mines formed by larval moths and flies, and holes caused by aphids and mites." The fossil evidence indicated that as CO_2 and global temperature spiked, so did herbivory, the consumption of live plant tissue by animals.

Currano and her colleagues concluded that "the dramatic rise in diversity and frequency of herbivore attack on all abundant plant species during the PETM suggests that anthropogenic influence on atmosphere and climate could eventually have similar consequences," which would threaten not just farms but the forests that clean the air, take up CO_2, and produce oxygen.

A modern-day experiment called Biosphere II, located just outside Tucson, Arizona, provided more cause for concern. Beginning in 1991, groups of up to eight people lived there in what was the largest closed ecological system ever created. The experimental environment was nearly as large as three football fields; under its clear roof was a rain forest, a small ocean with a small coral reef, and several plots of planted cropland. The designers meant for the forests, cropland, and ocean to keep CO_2 levels relatively constant, but instead they rose steadily, and insects flourished. There were many explanations offered for the rise of CO_2 in the Biosphere II atmosphere. The fact that it happened, however, suggests that we vastly underestimate just how much plant matter and how big an ocean is needed to sustain animal life in Biosphere I—also known as Earth.

THE SCOURGE OF SOYBEAN RUST

Plants succumb to more than hungry insects, as farmers and gardeners know well. The mere mention of one plant disease—soybean rust—strikes dread in the hearts of farmers. The fungal disease causes crumply brown and yellow sores on the plant's leaves, which spread quickly and cause the leaves to die and drop off. Whole fields can be defoliated within two weeks, turning almost overnight from a lush green to dry brown. The fungus is considered so dangerous, it is labeled a bioterrorism "select agent" by the federal government and is allowed to be grown for research only at Fort Detrick in Maryland, the same facility where anthrax is studied.

Soybean rust was first recorded in Japan in 1902. From the 1930s through the 1960s, it was also seen in Australia, India, China, and other parts of Asia, and by the 1990s, it was infecting crops in Africa. More than 80 percent of the world's soybeans come from the Americas. Soy is key to the economies of Brazil and several other South American countries, and it is also key to the economy of the United States, where it is an $18 billion annual industry.

In the 1980s, projections were made about the economic impacts if soybean rust were to take hold in the Americas, and the results weren't pretty. Farmers and agronomists were shaken in 2001 to see the disease pop up in Brazil and Paraguay, where it didn't take long to reach epidemic proportions. In 2003, the disease cost Brazil $1.3 billion in crop losses, despite massive fungicide applications that cost $750 million. The disease hit Brazil even harder in 2004, costing more than $2.3 billion. As if that weren't bad enough, a drought in southern Brazil and Argentina in 2004 facilitated an outbreak of another fungal disease, "charcoal rot," which also decimated soybean crops.

. . .

In the fall of 2004, Louisiana State University plant pathologist Ray Schneider was giving a tour to a visiting soybean farmer when the two found signs of the dreaded disease. Samples sent to the U.S. Department of Agriculture lab in Bethesda, Maryland, confirmed their fears: soybean rust had made it to the United States. Scientists later figured spores had probably been transported up from South America in September on the winds of Hurricane Ivan. Within days of the hurricane, the USDA sent scientists throughout Louisiana and then other states looking for signs of the disease. It was found in Arkansas, Mississippi, Alabama, and Florida.

Scientists were ready for the invasion. Even before soybean rust blew in, they had been working on modeling to evaluate its spread, tracking it by satellite or airplane, and developing genetically modified soybean varieties resistant to the disease. Widespread testing and fungicide have so far kept soybean rust losses in the United States relatively low. In addition, the fungus dies during cold winters, making the United States less susceptible than hotter countries. But it will clearly be an ongoing battle.

From its start in the Southeast, soybean rust spread northward into the Midwest, and by 2006, it had struck crops in 274 counties in fifteen states, including Illinois, Indiana, and Missouri. In 2007, the disease had spread to nineteen states. It grows on many plants other than soybeans, including the ubiquitous kudzu vine, making it extremely hard to eradicate once it has alighted. Climate change will likely open new frontiers for the disease; warmer, wetter winters and longer growing seasons could help the disease spread farther. Soybean rust losses in the United States could run from $240 million to $2 billion a year, which could prove devastating to farmers.

Other pests and pathogens have also progressed northward with a warming climate. In the past, cold temperatures at night and during the winter killed off pests in the northern United States, but now climate change is raising daily and yearly minimum temperatures, which means fewer pest-killing frosts. In many areas, climate change is also bringing increased humidity, more heavy rains, and flooding, all of which can facilitate the growth of plant pests and diseases, including those caused by bacteria, nematodes, and fungi. After the historic 1993 floods in the Mississippi River basin, midwestern crop fields were extensively damaged by fungal epidemics. Climate change is also expected to bring more drought, which favors other kinds of pests, including aphids, whiteflies, and locusts, and more extreme weather (hailstorms, heat waves) could also bring significantly more disease and damage.

Even in years when the weather holds relatively steady, an altered climate will make crops more susceptible to weeds. "In most fields, eight to ten weed species compete with crops," says Lewis Ziska, the plant physiologist with the U.S. Department of Agriculture who worked on ragweed pollen. Ziska, who's an expert on weeds, says that agronomists and farmers alike have sought "uniformity, uniformity, uniformity" when breeding crop lines. But when the environment changes, it's the fast-growing, cross-breeding, genetically diverse denizens of a farm field—the weeds—that adapt readily.

Already, weeds claim about a tenth of global crop yields annually, and the combination of pests, pathogens, and weeds already reduces global yields and stored grains by some 42 percent—an annual loss of about $300 billion. Losses will only rise with climate change. Combating these outbreaks will require increased spending on pesticides, herbicides, fungicides, and insecticides—pollutants that contaminate surface and underground water supplies and food and threaten the health of farmworkers and consumers.

If crops aren't eaten away by pests or choked out by weeds that thrive with elevated CO_2 and climate change, the heat itself may kill them. Corn is very sensitive to drought and heat, and its pollen loses vitality when temperatures exceed 36°C (97°F). Such temperatures kill soybean seedlings, so spring heat waves could be particularly damaging. Scientists have projected that an average global temperature increase of just 1°C would decrease annual wheat, rice, and corn yields by 10 percent.

PARCHED

A huge electronic map in swirling colors loomed above the frantic trading on the floor of the Chicago Mercantile Exchange in April 2008, giving traders in agricultural goods up-to-the-minute information on the weather worldwide. Global climate is key to determining future predictions of agricultural yields and to setting prices at the Chicago exchange, which essentially determine agricultural prices worldwide.

Traders that spring day were seeing record grain prices nearly across the board. Those record prices were leading to stockpiling of grain, government controls on crop exports, and food shortages around the world, which in turn sparked riots. One of the main reasons cited for the high grain prices and attendant strife was an ongoing multiyear drought in Australia that Aussies call "the Big Dry."

Drought can wipe out crops and decimate local economies, and climate change, with its warming temperatures and changing weather patterns, is bringing more drought to the center of many continents. It is estimated that 35–50 percent of the world's wheat crops, including those in Australia, the United States, India, and China, are at risk from drought.

Many climatologists attribute Australia's severe and recurrent drought to changing patterns of westerly winds that whip around Antarctica, which, thanks in part to global warming and in part to the southern hemisphere ozone hole, now seem to be circling closer to the southern continent than before, starving Australia of moisture.

That drought and the resulting crop failures have brought despair to the Australian countryside. "[Farmers] just have absolutely nothing left," a Salvation Army drought support coordinator told an Australian newspaper. "They have sold everything they could, reduced their stock, and borrowed heavily to plant crops at the start of the season, which they now face losing." As a result, depression and domestic violence are on the rise, and as of 2007, more than one hundred farmers in the bush had committed suicide as their livelihoods slipped away.

. . .

As hard as drought has hit Down Under, at least Australians have enough to eat. In the developing world, the toll of drought can be even harsher. The nearly five hundred million people who live in rural areas in dry and semiarid lands—mostly in Asia and Africa but also in parts of Mexico and Brazil—tend to have the lowest levels of health and well-being. Many such communities live directly off the land, subsisting via agricultural, pastoral, and nomadic herding. Droughts exacerbate environmental degradation and make such lifestyles much harder to sustain. This makes these communities highly vulnerable to changes in climate. For example, higher-than-average levels of rainfall in the 1950s and 1960s drew people to the Sahel, but when typically dry conditions resumed in the 1970s, the ecosystem could no longer sustain them, and the resulting famine killed 250,000 people.

There's solid evidence that drought, induced by climate change, is already fostering famine and malnutrition in Ethiopia. In this spare and beautiful country, farmers depend on rain to water their crops, and too little rain can mean too little food. To help evaluate Ethiopia's ability to grow food for its citizens, geographer Chris Funk of the University of California, Santa Barbara, and his colleagues studied Ethiopia's rainfall from 1960 to 2004. They found that the crucial rains, which fall from March to May, have diminished in the southeast and northeast of the country since 1980 and 1996, respectively, suggesting a lasting change in climate.

They also linked the changing regional climate to global climate change. Until recently, prevailing winds typically swept across the central Indian Ocean, picking up moisture and depositing it in East Africa as rain. But the surface waters of the central Indian Ocean have been steadily warming, as have surface waters in other oceans. This has changed wind and rain patterns in the region. Winds now tend to blow dry air into Ethiopia from inland areas of Africa, rather than moist air from offshore, and more rain falls over the ocean itself and less over land. This in turn produces drought in southern and eastern Africa.

Challenges to crop yields from droughts and climate change couldn't come at a worse time. Populations in these two regions have been growing rapidly, just as food and water have become scarce. Modeling done by Funk and his colleagues have predicted that this trend will continue. This will bring an "explosive combination of drought and declining per capita agricultural capacity," they've written.

Climate change may foster food shortages in many other parts of the world. According to IPCC predictions, global warming is projected to melt mountain glaciers in Asia, Africa, Europe, and the Americas within several decades. The huge populations that today depend on glacial melt to feed rivers, irrigate farms, and produce hydropower will be forced to depend on rain rather than glacial melt. This includes Southeast Asia, where rice production is already insufficient to meet the needs of a growing population. And just as glacial melt slows to a trickle, changing weather patterns will spread and intensify drought.

In many parts of the world, disappearing mountain glaciers and droughts will make fresh, clean water for drinking, bathing, and other necessary human (and livestock) uses a scarce and valuable commodity. Already one billion people worldwide lack access to safe drinking water. Each year, there are 1.7 million deaths and fifty-four million healthy life years lost because of lack of adequate water, sanitation, and hygiene. If present consumption patterns continue, the United Nations Environment Programme estimates, two out of every three people on Earth will live under water-stressed conditions by the year 2025. And these dire projections do not even take into account a changing climate. Climate change will affect the quality of available drinking water, as well as the quantity. Some pathogens that infect drinking water sources thrive when temperatures warm. And intense storms (also linked to climate change) will more often pollute drinking water sources with runoff from industry, confined livestock operations, agriculture, and sewage overflows.

In the United States, water conflicts have already begun to play out in courtrooms and legislative chambers: Georgia, Mississippi, and Alabama have been locked in a bitter conflict since the early 1990s over the water resources of the Apalachicola, Chattahoochee, and Flint river basins. A drought that started in 2006 worsened the situation. It left thousands of square miles of parched and cracked land in the Southeast, and it left the water level in Lake Lanier, the reservoir that provides Atlanta's water, almost twenty feet below normal levels. The Southwest could also run dry. At least one study predicts that as the region dries further, its reservoirs could be seriously depleted by the 2020s, depriving millions in San Diego, Arizona, and other areas of drinking water. And in parts of the developing world, water conflicts could have even more serious consequences, resulting in physical violence, economic retaliation, and mass migrations.

THE ROAD TO SUSTAINABILITY

The risk of starvation, malnourishment, and water shortages should serve as an international call to action on climate change and on agricultural development. Several steps should be taken to preserve health as water grows scarcer. To ease shortages, water delivery systems should be improved, and this essential, precious resource should be used more efficiently. To limit the spread of disease, sanitation systems should be improved. Appropriate technologies could greatly improve the lives of rural farmers toiling away with hoes and planting sticks. Farmers may also need to adapt to climate change by changing when they sow their seed or by switching crops. A move from corn to sorghum, for example, would reduce water needs and help farmers adapt to the higher temperatures. To implement such changes, nongovernmental organizations and other civil society institutions can play crucial roles by assisting, investing, and working to change international trading rules. What the planet does not need, however, is a rush in the developing world toward more industrialized agriculture, which has contributed so much to deforestation and to soil and water depletion.

As a plant ecologist, Evan DeLucia, the scientist working on the soybean leaves, has seen firsthand some of the pitfalls of modern industrialized agriculture, which include the never-ending swaths of corn and soybean that cover Illinois. He's seen how nitrogen from fertilizer used on midwestern fields runs downstream, contributing to devastating blooms of algae that kill shellfish in the Gulf of Mexico. And he knows how industrialized agriculture has contributed to the loss from rich soils of fully half the carbon they once held, cycling it back to the atmosphere and fueling climate change.

"We have treated this ecosystem as a machine for producing food, fuel, and fiber. What we have ignored are its effects on the nitrogen cycle, the hydrologic cycle, the carbon cycle, and its role in global ecology," DeLucia explained.

For DeLucia, living amid the vast corn and soybean farms of central Illinois gradually kindled an epiphany. "It does not take long living in this landscape to have a couple of things happen to you," DeLucia says. "One is to really embrace its sublime beauty. The other thing is to understand its ecological role, and this is actually something that's not widely appreciated.

"You still need the food, you need the fuel, you need the fiber," DeLucia continued, "but instead of maximizing the production of those things, can we optimize the production of those things against other ecosystem goods and services—storing carbon, holding nitrogen in the landscape, wildlife habitat? Even though it is a completely human-dominated landscape, it is still an ecosystem. And if we took that bigger view of this landscape, we might be doing agriculture in a more sustainable way."

6

Sea Change

In the salt air of Woods Hole, it's natural to ponder marine life, and at our 1993 conference at the Marine Biological Laboratory, I paid close attention when it was Ken Sherman's turn to present. Sherman is an innovative ecologist and oceanographer who works for the National Oceanographic and Atmospheric Administration's National Marine Fisheries Service in Narragansett, Rhode Island. In the mid-1980s, Sherman and his colleague, Lewis Alexander of the University of Rhode Island, came to the conclusion that the world's oceans were made up of dozens of large zones that function as semi-independent ecosystems. These large marine ecosystems, as they became known to scientists and policy makers, straddle national boundaries and ring almost every continent, as well as large island systems. Most cover more than seventy-five thousand square miles of sea.

Sherman and Alexander realized that large marine ecosystems operated by the laws of nature and didn't give a hoot about politically determined fishing quotas or state or national boundaries. They also realized that the best way to manage these large marine ecosystems sustainably—that is, without overfishing or polluting them to death—was to treat each of them as the ecosystem it was, with its own biologically determined limits. To make that concept work in practice, fisheries scientists, marine ecologists, and other scientists had to share data and consult with each other before making management decisions. Ocean managers immediately saw the utility of the concept, and it was quickly adopted by a variety of regional, national, and international agencies concerned with preserving the health of the oceans. This had never occurred before, which is why oceanographers and ocean policy experts regard the large marine ecosystem concept as a major advance.

The year of the Woods Hole conference, *The Lancet* had published the series of journal articles I had coedited. Those papers included the one I'd coauthored with Rita Colwell, who uncovered the plankton–cholera connection, and Tim Ford, whose specialty was water quality and disease. In that paper, we'd written of climate change and cholera, and we'd raised the question of other health impacts to come. We'd also addressed the global epidemic of red tides, brown tides, and other harmful algal blooms, explosive growths of certain types of algae that can cover thousands of square miles of ocean, lakes, and ponds. The toxins produced by some of these algal blooms can be nasty: some damage the human nervous system, and some cause diarrhea. These examples illustrated how our health is inextricably linked to the health of the ocean. I was drawn to explore this link more deeply.

In Ken Sherman, I recognized a kindred spirit—someone who could help me achieve the broader goal of looking at whole ecosystems and our role in them. Sherman had also thought deeply about how to integrate different fields of science to derive a more accurate picture of the natural world, and he too was applying that cross-disciplinary science to address real-world challenges. We began to collaborate professionally, and we became friends.

Sherman and I explored how best to assess the combined influences of human interference, via pollution, habitat destruction, overharvesting, and climate change, on our coastal oceans. How were these factors affecting marine ecosystems, and how were the resulting changes to those ecosystems affecting human health?

Typically, scientists interested in addressing such broad questions hold a conference and invite specialists in each of the relevant fields. In our case, this would have included experts in red tides; fisheries; chemical pollution affecting whales, seals, and other marine mammals; sea urchins; shore birds; and more. Such conferences are convened in the sometimes amorphous hope that intellectual cross-pollination will be enough to shed light on the larger questions.

We knew that such an approach wouldn't be enough for us. We did need to assemble scientists with a variety of specialties, but to adequately address the questions we were asking, we also needed to think creatively about how to integrate results from our various fields. Instead of focusing on one species or a group of related species at a time, we would use systems science to achieve a more comprehensive picture about what we humans were doing to our large marine ecosystems.

We would start by conceiving of the world—including the ocean and its many inhabitants—as one inextricably linked system, divided into subsystems. In this broader view, a disease of sea urchins or dolphins becomes a symptom of ill health in the larger system, with wide-ranging effects that might not be appreciated for years.

But first we needed a reality check. By the early 1990s, reports of fish kills, red tides, and shellfish-induced food poisoning were rolling in, and the rise in harmful algal blooms was receiving international attention. But marine ecosystems and organisms had always been plagued by disease and disturbances that periodically flared up and settled down. Was disease in the oceans really on the rise? And if so, was climate change playing a significant role? Sherman and I began a project to tackle these questions.

DISEASES OF THE SEA

In March 1994, more than fifty scientists from diverse fields converged in Cambridge, Massachusetts, for a conference to scope out the project. We assembled at the Harvard Divinity School, not far from Harvard Square, and the group of scientists present—ornithologists, ichthyologists, medical doctors, ecologists, climatologists, marine biologists, and more—had rarely or never worked with one another. Normally, the bird people don't talk to the fish people even though seabirds eat fish; the whale people don't talk to the algal researchers, even though baleen whales feed on algae. But the scientists visiting Harvard that day realized they had something in common: an interest in the interconnections between global environmental change and the health of both humans and marine life. They began realizing the possibilities of such a collaboration, seeing that they could combine their seemingly isolated data sets into a mosaic that would yield surprising and enlightening results for all. It would take dozens of scientists several years to complete, but conducting the first systematic assessment of coastal seas would be worth the effort. As the ambitious scope of our new project became clear, Sherman exclaimed, "This is huge. This is wonderful. This is impossible. Let's do it!"

Researchers interested in the connection between the environment and health have a tough time getting funding because the environment is the purview of the National Science Foundation, while health belongs to the National Institutes of Health; the two agencies are beginning to work together now, but they rarely did back when we were launching our study. We did, however, secure funding from the National Oceanographic and Atmospheric Administration's Office of Global Programs and from the National Aeronautics and Space Administration (NASA), funders who were starting investigations into climate's role in agriculture, human health, and fisheries. We called our project the Health, Ecological and Economic Dimensions of Global Change Program, or HEED Global Change for short.

In our study, we sought the answer to three questions. First, had there been a real long-term rise in diseases in the coastal marine environment? Second, were there common links among the diseases that occurred in different species? (If they happened at the same time or in the same place, then one might cause the other, or, alternately, they might share a common cause.) And third, was there a discernible link between climate change and marine diseases? As we dove with relish into this group exercise, we used systems science as a framework. In other words, we examined the entire ecosystem without focusing, as specialists did, on one particular type of organism or phenomenon that occurred within it.

The three broad questions we asked broke down into dozens of smaller ones, each of them informative. For example, did harmful algal blooms happen in the same place as seabird mortalities? (If so, the toxins they produced might be killing the birds.) Did harmful algal blooms precede human disease from shellfish poisoning? (If so, it suggested that algal population explosions could make people sick.) Did warmer sea surface temperatures occur in the same places as harmful algal blooms? (If so, it suggested that climate change, which is warming ocean surface waters, was raising the risk of diseases in humans and many other species.)

To address questions such as these, we first had to systematically assemble all the data we could find on diseases in many species. In that, we faced a big challenge. Although the ocean covers more than two-thirds of our planet's surface and harbors exponentially more biodiversity than the land, our understanding of these life-forms was—and still is—incomplete at best. Scientific attention to the ocean, and particularly to marine disease outbreaks, had been sporadic. There was no long-term trove of data akin to a century of daily and monthly weather records, and nothing at all like the continual measurements of atmospheric CO_2 from the monitoring station that Charles Keeling had built on Mauna Loa.

To deal with the gaps in the historical and epidemiological data, we scaled up our study. Working with records from 1972 to 1996, we searched over a large area—the North American Eastern Seaboard from Labrador down to the Caribbean—and collected data on diseases in ten taxonomic groups, including turtles, fish, marine mammals, sea grasses, seabirds, invertebrates, and humans. We knew that if diseases appeared in the same place in different groups of organisms, that would suggest a common cause. Likewise, if one group of organisms was infected with the same disease at the same time at distant locations, that could indicate that at the root of the outbreaks lay large-scale environmental change—perhaps climate change.

We had no shortage of material. By the early 1990s, many scientists were concerned about the numerous diseases that were being reported with increasing frequency in marine organisms from all parts of the world. For example, during a large El Niño event in 1983, the Caribbean Sea warmed more than usual, and an infectious disease decimated the dominant species of sea urchin, which clean algae from coral reefs, thereby maintaining their health. The repercussions were huge: Jamaica's renowned ring of reefs was snuffed out from mats of algae, and they have still not recovered. The 1983 El Niño, another El Niño in 1987, and gradually warming seas all contributed to other losses in the Caribbean: four thousand hectares of turtle grass perished in Florida Bay, several species of coral disappeared in multiple locales, and dolphins and other marine mammals experienced mass die-offs. In the North Atlantic, mass die-offs of marine mammals had become increasingly common, with our pollution appearing to play a major role.

Overall, we collected reports from state and federal agencies, more than 2,100 peer-reviewed journal articles, and more than 2,000 news reports, all describing such disease outbreaks and ecological disturbances. We were fortunate that Sherman's son Ben, a crack computer programmer, was available to build us a large interactive database that allowed us to enter data on such outbreaks, including time, place, disease, and the associated microbe. The system he built allowed us to layer these data sets, as well as satellite-generated images, on a map using an early geographic information system (GIS). (Today, one can do this sort of mapping using Google Earth, but in those days the process was much more laborious, and we employed four junior researchers to search the literature and enter reams of data.) The mapping process enabled us to step back, see the whole, and look for large-scale trends.

The results were startling.

Our combined data sets revealed many more outbreaks than we had expected, of both new diseases and old ones that had reemerged. We saw clusters of diseases affecting many types of marine life, grouped around geographical hot spots such as the Chesapeake Bay. We saw a clear upward trend in disease reports over the span of two and a half decades, and we found that disease outbreaks in many life-forms occurred roughly in the same place and at the same time as harmful algal blooms.

What's more, we saw that diseases in many species peaked during El Niño events. This indicated that warming and extreme weather were influencing the incidence of marine disease. (Colwell had used similar reasoning to help link climate to cholera.) The association of diseases with El Niño demonstrated the link to climate and weather patterns, in general. More specifically, the results gave us a preview of how rising ocean temperatures and increased storms may affect coastal marine ecosystems, and how that may affect our health and the health of the other life-forms that rely on the sea.

We first reported our findings on emerging marine diseases in 1998, and in 1999 they were included in a multiauthored article in *Science* on the decline of marine ecosystems. In both reports, we concluded that coastal marine ecosystems had been beset in recent decades with new and more severe outbreaks of disease and that a series of human-induced environmental insults, including pollution and climate change, had been partly responsible. It was clear that projected changes in the climate threatened to exact an enormous toll.

Our findings received significant media coverage, and we conducted briefings on the findings for legislative aides on Capitol Hill. Science may have a lot yet to learn about ocean ecosystems, but we know for sure that humans are drastically changing those systems, and not for the better.

AMNESIA FROM SEAFOOD

In late November 1987, more than 150 people on Canada's Prince Edward Island were struck by a mysterious illness. All of those sickened developed nausea, vomiting, diarrhea, and abdominal cramps. For some, though, it was worse. The more severely afflicted developed headaches, became confused, and lost their short-term memory, sometimes permanently, which led health officials to name the new disease amnesic shellfish poisoning. Some had difficulty breathing, or seizures.

Nineteen people were hospitalized, and twelve required intensive care. Four people died.

Canadian health officials began investigating. It turned out that everyone who had gotten sick had eaten cultivated mussels, and those mussels had all come from an estuary known as Cardigan Bay. At the time the shellfish were harvested, there had been a bloom of an alga known as *Pseudonitzschia pungens,* which is a diatom, a type of single-celled alga with an intricate silica-based skeleton. Diatoms are common and almost universally beneficial; they account for 23 percent of the world's photosynthesis and form the base of many marine food webs. The Prince Edward Island cases were the first ever in which a diatom had been known to cause disease.

Health officials identified a toxin called domoic acid that had caused the damage. The chemical structure of domoic acid closely resembles a compound called glutamic acid that's a neurotransmitter—a chemical messenger that brain cells use to communicate. Glutamic acid binds to receptors on brain cells in the hippocampus, a part of the brain that helps store short-term memories. Because of this close resemblance to glutamic acid, domoic acid binds to the same receptors. Unlike glutamic acid, however, domoic acid doesn't let go once it's bound. The unlucky brain cells to which domoic acid has bound fill with calcium, then burst and die. The unlucky owner of the brain loses attention span and short-term memory, sometimes forever.

At the time of the initial outbreak of amnesic shellfish poisoning on Prince Edward Island in 1987, unusual conditions were stirring in the ocean. The outbreak occurred during an El Niño event, which brought warmer-than-usual waters from the Gulf Stream near the shore. El Niño conditions also included heavy rains that flushed out to the estuary large amounts of the nitrogen and phosphorus compounds that algae use as nutrients to fuel their growth.

Following the 1987 Prince Edward Island outbreak, domoic acid poisoning next appeared in the Pacific off the West Coast of the United States, where it was blamed for causing strandings of sick and dying birds on beaches and strange behavior in sea lions. Levels of the toxin in West Coast waters skyrocketed for a decade beginning in the mid-1990s. At one point, domoic acid levels were more than ten times the established safe limits near beaches where Native American and Canadian First Nations tribes harvest shellfish for subsistence. A study of Native Americans on the West Coast who have diets rich in mussels and razor clams found that 23 out of 653 surveyed had suffered poisoning. Harmful algal blooms threaten hundreds of millions of people around the world—especially those in isolated islands and coastal regions in developing countries who, like the Native Americans poisoned by domoic acid, depend on seafood for protein.

All over the world, our practices have created conditions like these in nearshore ocean water. We inadvertently add nutrients to our water sources via fertilizer runoff from farms, sewage runoff from towns and cities, and atmospheric nitrogen fallout from fossil fuel burning, and, via climate change, we are inadvertently warming the seas. We're making a dangerous brew of our seas, and we're still learning how much it may harm us.

. . .

Although the amnesic shellfish poisoning was caused by a diatom, it is typically other types of algae that bloom to dangerous levels. The term *algae* refers to both multicellular plants and microscopic single-celled organisms called protists; it's the latter that bloom in coastal oceans. Protists include diatoms, which are coated with silica like that in sand, and dinoflagellates, mobile microbes with tails that wriggle. Another class of organisms, blue-green algae (which are actually a type of bacteria), can also bloom and produce biological toxins. All float on the ocean's surface, and some migrate to the depths at night. Most are harmless, but dozens of species are not.

Among the dangerous species are the dinoflagellates that cause red tides. In biblical times, a red tide was thought to be the blood of whales or menstrual blood, but now we know it is a huge bloom of pigmented dinoflagellates, some of which produce toxins. One of the most common and foreboding ailments from red tides is paralytic shellfish poisoning, whose symptoms range from mild numbness and tingling of the lips, nausea, and headache to severe paralysis, respiratory failure, and death. Beachgoers can also breathe in the toxins in sea spray, sometimes leading to severe respiratory problems. In Florida, lifeguards often wear protective masks during red tide outbreaks, and regional emergency room visits have been found to increase during red tides. Red tides also cause ciguatera fish poisoning, the most common seafood-borne illness in the tropics, which occurs when humans eat reef-dwelling fish that ate small fish that ate toxic dinoflagellates. Ingesting the resulting toxins can disrupt the sense of taste and cause weeks of nausea and vomiting.

In our study on ocean-borne disease, we'd documented a dramatic rise in poisonings from harmful algal blooms, the scientific name for red tides, from the early 1970s through the mid-1990s in eastern North American coastal waters, the Gulf of Mexico, and the Caribbean Sea. Although toxins from red tides (which can actually be brown, pink, orange, or other hues) have occurred through the ages, their dramatic rise in recent decades illustrates how human-induced damage to marine ecosystems, caused in part by climate change, is breeding illness.

The rise in harmful algal blooms calls for a public health response. For decades, health officials have monitored shellfish beds and closed them when they became dangerous. Predicting the algal blooms before they contaminate shellfish would be even better—something that can be accomplished by using satellite sensors to detect changes in the ocean surface indicative of algal blooms.

In New England, Don Anderson and his staff at the Woods Hole Oceanographic Institution closely monitor prevailing winds and sea surface temperatures by satellite, dispatching ships to scoop up samples when the satellite images detect signs of dangerous algal blooms. They've spotted incipient blooms of *Alexandrium tamarense,* a dinoflagellate that causes paralytic shellfish poisoning, in time to close shellfish beds in areas bordering Massachusetts Bay. Harmful algal blooms are yet another clarion call for us to attend to the deteriorating health of coastal marine ecosystems.

THE TALE OF THE OYSTER

Oysters live in brackish water where the sea and freshwater mix—in river mouths, salt marshes, and estuaries. For millennia, Native Americans feasted on oysters that bred abundantly in the rivers and salt marshes of the Atlantic coast. They steamed the shellfish over fires to open them, made necklaces out of the pearls they often found inside, and left the shells in huge piles called middens.

When Captain John Smith first sailed into the Chesapeake Bay early in the seventeenth century, the oyster beds were so abundant that he had a hard time navigating. Dutch and English settlers were delighted to find the mouth of the Hudson River, Boston's rivers, and other northeastern waterways ringed with oyster beds that would supply them with food during the harsh winters. They'd collect hundreds of pounds of oysters each time they foraged, from beds so abundant that the harvests would quickly be replenished by new growth. Throughout the 1800s and 1900s, oysters remained a significant part of the North American economy; the briny, succulent shellfish were shipped throughout the

United States for consumption in cozy bars and fancy restaurants. The State of Virginia's constitution, written in the closing decades of the eighteenth century, specifically protected oyster beds as a public good.

Along with their culinary virtues, oysters perform a crucial function in estuarine ecosystems. A single oyster filters thirty to fifty gallons of water each day, removing excess nutrients, chemical pollutants, and algae. This essential filtering role makes oysters what ecologists call a keystone species. A keystone is the wedge-shaped rock placed at the top of a stone arch; when the keystone is removed, the arch collapses. Keystone species are similarly crucial to the integrity of an ecosystem, influencing it far out of proportion to their biomass or numbers. By cleaning the water, oysters make bays and estuaries habitable, not only for themselves, but also for fish, sandpipers, herons, turtles, whales, and virtually every other species that dwells in the area either temporarily or permanently.

Unfortunately, oysters in the eastern United States have had a tough go of it in recent decades. Pollution has evidently weakened the oysters' defense systems, for they have become infected with two opportunistic parasites known as Dermo and MSX. The infections make oysters shrivel and become unpalatable, though they are not thought to harm humans.

Historically, Dermo and MSX were found in the Gulf of Mexico and off the southeast U.S. coast, but they were killed off farther north when frigid winters cooled the water. As coastal waters warmed due to climate change, however, oysters were unable to shake the invaders as they might have in years past, and Dermo and MSX marched northward. By the 1980s, the parasites infected oyster beds in the Chesapeake Bay— beds that are now all but gone. In the 1990s, Dermo became epidemic in Delaware Bay, and in 1995 it appeared in Maine. By 2003, it was killing oysters all the way up to Newfoundland.

When oyster populations fail, inlets accumulate nutrients, sediments, and algae. The results are plainly seen, for the water becomes cloudy with particles and, often, jellyfish. This is what happened in the Chesapeake Bay. When oysters were abundant there, the bay's entire volume of water flowed through bivalves every three days and was filtered in the process. Now, with so few oysters, the bay's waters are filtered but once or twice a year. Nutrients build up, which leads to harmful algal blooms and dead zones; shellfish poisoning has increased, as have the populations of a nasty cousin of the cholera bacteria called *Vibrio parahaemolyticus* that also causes gastrointestinal disease. The oyster fishery has crashed, disrupting life for the bay's storied watermen. Despite ongoing attempts to restore the bay, it has not yet recovered.

TREASURES FROM THE DEEP

The tale of the oyster is a parable for the precarious state of our marine ecosystems, which support human health in countless ways.

Marine life constitutes perhaps the world's largest repository of biodiversity that, among other gifts, could provide us with pharmaceuticals to treat cancer, AIDS, and other ills. For example, there is a painkiller derived from a marine cone snail that is one thousand times more powerful than morphine and doesn't cause drowsiness or lead to addiction. But this cone snail and its cousins live in the coral reefs, and they have suffered mightily from climate change and other environmental insults. The ocean is also considered a source for a variety of new antibiotics, which are especially important since so many bacteria have become resistant to today's antibiotics. A 2008 interagency report by the federal government supports this theory, stating, "Because of its unparalleled biodiversity, the ocean holds huge possibilities for new products that could improve and perhaps even revolutionize health care."

Marine organisms also are used as biomedical models to better understand human health and the effects of contamination and pathogens. For example, the heart of the zebra fish bears significant similarities to the human heart; because these fish are easy to keep, grow fast, and have transparent bodies, they are ideal surrogates for scientists to investigate. Scientists have treated zebra fish with polycyclic aromatic hydrocarbons, one of the contaminants in the 1989 *Exxon Valdez* oil spill in Alaska. They've learned that the hydrocarbons cause irregular heart development, which suggests that the chemicals might also cause heart problems in humans. Since so many marine creatures are unknown to science or poorly understood, damaged marine ecosystems may cost us lifesaving medicines and medical testing models we'll never know existed.

But marine life offers us much more than advances in medicine. The biggest gifts come from some of the smallest life-forms: the phytoplankton, or floating "vegetables of the sea," as Rachel Carson called them. Phytoplankton include diatoms, blue-green algae, and other tiny algae, and they grow in unfathomable numbers in the ocean. Phytoplankton are known as primary producers because they, like land plants, conduct photosynthesis, using the sun's energy to turn CO_2 into the compounds that compose their tissues. So abundant that they conduct about 70 percent of the photosynthesis that occurs on the planet, phytoplankton absorb huge amounts of carbon dioxide and produce oxygen. Once aloft in the atmosphere, that oxygen supports animal and human life and generates the protective ozone layer, which absorbs the ultraviolet light that would otherwise sear us. Phytoplankton also serve as food for organisms as diverse as mussels and baleen whales, both of which are filter feeders that consume them directly. Almost every organism in the sea feeds on something that counts plankton in its food web.

Plankton also play a little-appreciated but critical role in regulating global temperature and climate. When they die, they emit a gas called dimethyl sulfide that seeds clouds, thereby deflecting the sun's rays and generating precipitation, which cools the Earth's surface. Warmer temperatures favor phytoplankton growth, which means that phytoplankton growth functions as a rheostat: more heat leads to more dimethyl sulfide, more clouds, cooler temperatures, then less phytoplankton growth. This negative feedback loop works to keep the Earth's temperature and climate steady (although today it is being pushed beyond its limit by humanity's greenhouse gas pollution).

Coral reefs, too, play crucial roles in the health of the ocean. A healthy coral reef is one of nature's wonders. It's a symbiosis of two types of organisms, an alga and an animal. The alga grows on sunlight and CO_2 and supplies the animal (the coral polyp) with food; the animal in turn provides the alga with protection and nutrients. A coral reef consists of many thousands of live polyps perched on a foundation of dead and hardened skeletons that they collectively produce. Reefs provide habitat for one-fourth of the ocean's fish species and as many as nine million species of organisms, many of which are found nowhere else.

The coral reef ecosystem serves coastal and island dwellers in numerous ways. It serves as a nursery to tropical fish of every hue, spawning fisheries where there otherwise would be little. It's a sturdy port in a tropical storm, providing a crucial physical barrier that protects coastlines from storm surges. It prevents saltwater intrusion into the fresh groundwater lenses underneath islands that residents rely on for drinking water, cooking, bathing, and growing crops. And it lures tourists, who spend money locally. Economists have estimated that these services of the coral reef ecosystem are worth $20,000–$151,000 per square kilometer in Southeast Asia. In the Caribbean region, economists have calculated that just the reefs' ability to protect shorelines from storm damage is worth $740 million to $2.2 billion per year. (Of course, such numbers give no indication of the reefs' true value . . . should we lose them.)

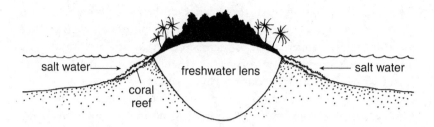

FIGURE 20. Creeping threats from dying reefs. Salt water from rising seas infiltrates damaged coral reefs and intrudes upon groundwater. The saltier groundwater impedes crop growth, fostering malnutrition among the poor. It also leads to high blood pressure among those who use it for drinking water.

Despite their value, coral reefs face so many threats that some coral experts say the ecosystem itself, which has survived hundreds of millions of years, is at risk of extinction. Pollution, bottom trawling, dynamite fishing, coastal development, and sediment from erosion can all damage coral's skeleton. Storms can flush chemical pollutants, fertilizers, and sediments from farm fields and factories out onto reefs, weakening them. Weakened reefs are then vulnerable to disease agents, including bacteria and fungi. For example, coral reefs in the Indian and Pacific oceans have suffered heavily from fungal diseases and lethal "orange disease," in which a slimy orange alga spreads over the coral and kills it. Even seemingly minor changes can disrupt the delicate ecological balance of a coral reef, with disastrous consequences, as when fish and sea urchins declined in Jamaica in the 1980s, causing mats of macroalgae to suffocate reefs, which then collapsed.

Climate change also jeopardizes coral reefs. Increasingly intense storms and associated storm surges can swamp and damage reefs. Warming surface waters can become hotter than the 25°C–28°C (77°F–82°F) range in which reefs thrive. If they stay just one or two degrees above that for several days, corals can bleach—that is, lose their symbiotic algae, becoming colorless and in danger of dying. Currently, warm season temperatures hover around 30°C (86°F) in tropical coastal zones, and reefs are struggling. When reefs bleach, bacteria and fungi can gain the upper hand and cause disease. Salt water can intrude into the fresh groundwater lenses that lie inland (figure 20). Saltier groundwater means less fertile soil, which can lead to lower crop yields, rising food prices, and malnutrition. It also means saltier drinking and cooking water, which can lead to high blood pressure. Damaged reefs may also harbor microbial pathogens that can contaminate the food supply or pose health risks for bathers.

Under assault by so many environmental threats, about 27 percent of the world's reefs have already bleached, and 60 percent are considered highly vulnerable to bleaching. Coral reef degradation is one of the clearest and earliest signs that global environmental change, including climate change, is profoundly disrupting marine ecosystems. At current rates of warming, pollution, and overfishing, coral reefs could collapse entirely within decades, and recovery would be an extremely slow process, if it occurred at all. Such a loss would have enormous consequences for our health and the health of the global environment.

FISHING IN A WARMER WORLD

As climate change hammers agriculture in poor areas, people will turn to the sea for food. But it may not be there, which could make for a hungrier and more malnourished world. Fish and shellfish provide 20 percent of animal protein for more than 2.8 billion people, mostly in developing countries, and they provide more than 50 percent of the animal protein consumed in many nations. As fisheries decline, these vulnerable populations will suffer from food shortages and shortages of the essential protein, minerals, and omega-3 fatty acids they get from fish.

What's more, approximately forty-two million people worldwide make their living in fisheries, with hundreds of millions more earning a living in related sectors. Fishing pulls in more revenue for developing countries than rice, coffee, sugar, rubber, bananas, and tea combined, putting a lot of bread, not just fish, on tables in poor parts of the world. As fisheries dwindle, economic stress may lead to mass migration and fights over scarce aquatic resources. Especially vulnerable areas include South Asia and Southeast Asia, where about 80 percent of the world's fishers live and where low-lying coastlines are likely to be hit hard by storms, floods, and saltwater incursion from rising sea levels.

Climate change will also affect fish and shellfish farming, which produces almost half the seafood the world eats. Aquaculture along river basins in Bangladesh and the Mekong Delta is expected to suffer heavily, as intense rains flood coastal ponds, drop salinity levels, wash in pollutants, and flush the young out of aquaculture nurseries. To be sustainable, aquaculture and mariculture industries will need to invest, at the very least, in dams, barriers, and other infrastructure to protect their ponds from deluges and flooding.

The combined damage to conventional fisheries and aquaculture could fray a critical economic lifeline for communities across the globe. By harming global oceans, climate change threatens to impoverish and undernourish hundreds of millions of people.

OCEAN ACIDIFICATION

Pop open a bottle of soda, and the fizz you hear represents the release of dissolved carbon dioxide. Like that unopened soda, oceans store carbon dioxide as well. The amount the oceans store is determined by a simple chemical equilibrium: the more carbon dioxide in the atmosphere, the more the oceans must store.

We've added almost two trillion tons of CO_2 to the atmosphere since the dawn of the industrial revolution, and all that carbon dioxide has had to go somewhere. It has gone to one of three types of "sinks" that absorb excess atmospheric CO_2: oceans, trees, and soil. Approximately 20 percent has gone into the oceans and 30 percent into trees and soils, leaving 50 percent remaining in the atmosphere.

As we've added CO_2 to the ocean, we've made it more acidic. The ocean is naturally alkaline, with a pH of about 7.8 to 8.5. (The pH of pure water, 7.0, is defined as neutral.) But when CO_2 dissolves in water, it produces carbonic acid, a mild acid, which neutralizes the alkalinity of ocean water. The pH of the oceans has already declined by 0.1 pH unit since the preindustrial age. Because the pH scale, like the Richter scale for earthquakes, is logarithmic, that means the world's oceans are already 30 percent more acidic than they've been for eons. Models based on current emission levels forecast pH declines of at least 0.3 pH units by midcentury, making the ocean two and a half times as acidic as it used to be.

This extra acid will make it harder for many marine organisms—including coral and mollusks like oysters, clams, and mussels—to make their shells. As the ocean becomes more acidic, the calcium carbonate that forms the scaffold of coral reefs and the shells of shellfish becomes more likely to dissolve, and the organisms have to put more energy into building and maintaining their shells. Add enough CO_2, and it's even possible that the extra acid could make their shells dissolve, just as drinking cola can dissolve tooth enamel.

Marine biologists have begun testing how acidic oceans might affect shell-building creatures by placing them in more acidic seawater and seeing how they grow. They've shown that slight increases in acidity slow shell building in snails (which, like oysters, clean seawater), clams, and sea urchins and deform tiny plankton called coccolithophores, which are normally covered by multiple hard, oval-shaped plates, like miniature light-reflecting balls in a discotheque.

In medicine, we call such decalcification (loss of calcium) osteoporosis. This oceanic osteoporosis could alter coral reefs and plankton enough to decimate entire species and dramatically restructure marine ecosystems, marine biologists say. This has led some of them to issue ominous predictions for the world's oceans. Thomas Lovejoy, president of the H. John Heinz III Center for Science, Economics and the Environment, predicted that acidification and reduced calcification could lead to a "reign of jellyfish." German marine biologist Ulf Riebesell put it a bit more vividly. The future of the oceans, he said, will be marked by "the rise of slime."

7

Forests in Trouble

As the 1990s progressed, it became clear to those of us in public health that a wave of infectious diseases was striking humans and many other forms of life. Humans faced multi-drug-resistant tuberculosis, Ebola, HIV, and dozens of other new pathogens. Crops were becoming infested with insects and infected with emerging viruses. Dolphins, whales, and seals were suffering from measles-like viruses, while fish were going belly up en masse with increasing frequency. Even trees were in trouble.

Throughout the 1990s, I'd been trying to forge a new synthesis that would explain how a changing world could breed a wave of epidemics. I had begun my intellectual exploration into health and global change by gazing at a diagram that hangs in frames on the walls of public health offices nationwide. The diagram consists of three interlocking circles, emblazoned with the letters A, H, and E. It is known as a Venn diagram, and it holds the key to epidemiology, the study of epidemics.

The circle labeled A represents the agent—meaning the bacterium, virus, parasite, or fungus that can, if conditions are right, infect a person, plant, or animal. The second circle, H, represents the host—the organism that becomes infected with the agent. The third circle, E, represents the environment—the external conditions that determine whether the agent will invade a host. The same three factors—agent, host, and environment—control whether people develop other diseases as well. The take-home lesson is that there are most often multiple causes for any one person's sickness.

The agent causing tuberculosis, for example, is the bacterium *Mycobacterium tuberculosis*. But even if *M. tuberculosis* is present, the disease won't always develop. It will be more likely to occur if the host is weakened, perhaps by malnutrition or an HIV infection, and if conditions are ripe for transmission. Ideal tuberculosis-transmitting conditions occur in close quarters, such as those in the gold mines of South Africa or 1980s-era crack houses or prisons in New York City, where infected people cough profusely into common airspace. A host with a strong immune system living in a healthy environment can usually fight off the infection by surrounding the slow-growing bacteria with immune cells, effectively quarantining them and preventing the disease. But a weak host in close quarters is more likely to be infected and have trouble fighting the disease.

This framework of agent, host, and environment, I'd realized, could be adapted to assess the impacts of global change. When global change is considered, ecosystems are the host. This analogy works on several levels.

First, like the human immune system, both land- and ocean-based ecosystems have components that fight disease. In the immune system, antibodies stun invading pathogens, and white blood cells devour them. In terrestrial (land-based) ecosystems, birds of prey, like the spotted owls of the U.S. Pacific Northwest, eat rodents that can carry Lyme-disease-infected ticks, hantavirus, and bubonic plague. In marine (ocean-based) ecosystems, baleen whales and oysters filter-feed on algae and animal plankton, preventing the plankton from overgrowing into harmful algal blooms.

Second, just as a host is influenced by its environment, every ecosystem is influenced by the global environment. This includes the conditions in the lower atmosphere (troposphere), the upper atmosphere (stratosphere), the biosphere, the ice cover (cryosphere), and the world ocean.

Even disregarding climate change, humans have made huge changes to the global environment. By using chlorofluorocarbons and related chemicals in our air conditioners and antiperspirants, we've damaged the ozone layer in the stratosphere that protects all land-based life from the sun's damaging ultraviolet rays. By overfishing, we've decimated once-abundant populations of cod and many other species. By unwittingly releasing dangerous synthetic chemicals that act like hormones in animal bodies, we've altered the fate of countless species. The list of disturbances goes on.

Climate change portends larger changes by affecting the viability of

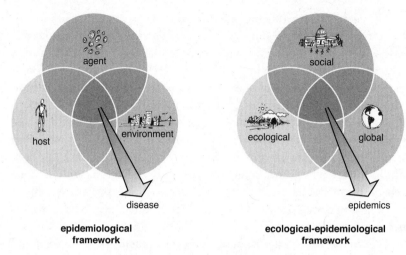

FIGURE 21. The framework. Disease occurs when an agent (bacteria, virus, or parasite) infects a host organism weakened by a poor environment (left). Epidemics occur when human society damages ecosystems already weakened by global ecological and climate change (right).

entire ecosystems. Persistent warming can kill off vegetation, turning grasslands into deserts, as it did when a changing climate transformed the Sudanese Sahara 5,500 years ago from a semiarid grassland suitable for grazing sheep to the bone-dry desert it is today. Warming seas make it harder for coral to reproduce, contributing to the coral bleaching events that are destroying reefs worldwide. Warmer and more variable weather can enable insects, including crop pests, to thrive.

Finally, the analogy works because human civilization is disrupting the functioning of the ecosystems that supply us with healthy food, clean air, and pure water, just as pathogenic microbes disrupt the functioning of the host's body, upon whom they rely for life support. Human civilizations are also disrupting the global environment. The overlapping circles of this Venn diagram represent these interactions. All of our social structures—our economy, our legal system, our energy system—influence both ecosystems and the global environment. That means that the agent in this analogy is us (figure 21).

This Venn diagram is useful in part because it offers an easy-to-grasp framework that illustrates how complex real-world systems work. Conditions in a host, whether human or ecosystem, must be conducive for an agent, be it microbe or human society, to flourish. And when conditions are such that the agent, host, and environment are all disturbed, a small problem can turn into a big one. A cold virus invading the throat of an overstressed person can cause days of sickness; a wildlife disease invading a disturbed and weakened ecosystem can spread and become an epidemic.

ROOFTOP OF THE ROCKIES

On a cold, clear June morning high in Wyoming's Wind River Mountains, Jesse Logan stopped on a snow-covered hillside and pointed with his ski pole to a large pine tree. A few of its needles were turning red, a sign of trouble. About a dozen people gathered around him on the snow to listen.

"We got a live one?" a man asked hopefully, pointing out the green needles at the top of a tree with a foot-wide trunk.

"It's dead and gone, John," Logan, a trim man with dark brown eyes, a neat gray goatee, and the self-contained manner of a traditional rural westerner, replied. "It ain't gonna be green long."

Until he retired in 2006, Logan ran the beetle research unit for the U.S. Forest Service's Rocky Mountain Laboratory in Utah. For thirty years, Logan had investigated the relationship between trees and the pest insects that plague them, combining sophisticated mathematical modeling of insect–plant warfare and boots-on-the-snow field observations.

On this morning, he was teaching a group of amateur wilderness enthusiasts and conservationists how to size up the health of a long-lived conifer called the whitebark pine, as part of an effort to scientifically assess the health of the tree's namesake ecosystem. Whitebark pines, which can grow up to sixty feet tall and have whitish trunks, anchor the high-altitude forest of the Rocky Mountains and the Cascades, growing in cold, windswept locations too harsh for other evergreens. Whitebark pine provides essential support to this ecosystem, which is located just below the timberline, creating soil where there was rock, shade where there was sun, and cover for wildlife where there was none. The deep-green carpet of whitebark pine that covers the West's high country is so important it's been called the "rooftop of the Rockies."

Today, however, the whitebark pine is in trouble, and the rooftop of the Rockies is crumbling.

. . .

On a cool, windswept mountainside, a lone whitebark pine seed sprouts in sparse soil. Over one year, two years, ten years, it sinks its roots and soaks up sun with its distinct bundles of five needles. Above, toward the mountain's summit, there is only rust-colored and dark-green lichen, clinging to bare rock. The only trees nearby are a handful of other white-bark pines, as it is too cold and rugged for any other tree. Each summer, the sapling soaks up sun, and each winter it hunkers down, buried under many feet of snow.

Slowly, over the decades, the tree grows taller and broader. After half a century, it produces pinecones of its own. It doesn't produce a huge number of seeds, also called pine nuts, but it invests well in those it does produce, making them rich with fat. Over many centuries, it grows sturdy and develops a broad crown. Then it grows old, seriously old, hollowed but very much alive. "Perhaps no other tree embodies the sense of the American West to quite the extent of a gnarled, 1,500-year-old whitebark pine clinging to life under some of the harshest conditions on the continent," Logan has written.

As it grows, the whitebark pine's roots pry apart the rock to help form soil, creating habitat for other trees, such as Engelmann spruce. After half a century, it finally develops its broad crown, which provides wildlife with vital shade and cover. Elk climb mountains to calve in stands of whitebark pines. A noisy gray-and-white bird called Clark's nutcracker relies on the whitebark pine's nuts for food, eating some and storing others, thereby planting new trees. The whitebark's rich pine nuts provide crucial nourishment for red squirrels, which store the nuts in large middens. All these crucial ecological roles make whitebark pine a keystone species, as oysters are in estuarine ecosystems.

While many species depend on the whitebark, no species feels its influence more strongly than the endangered grizzly bear, particularly in Yellowstone National Park, home of one of the two last wild populations of grizzlies in the continental United States. In the late summer and fall, when nutritious food is in short supply, the grizzlies raid the red squirrel middens and chow down on pine nuts. The bears need the fatty nuts to sock away enough fat of their own. If a female grizzly has too little fat, then a fertilized bear ova won't implant, and, the next year, the female won't have a cub, according to studies by brave grizzly bear biologists.

Grizzlies who can't get a good meal near the treeline tend to move downhill to find food, occasionally encountering humans. Such encoun-ters almost always end badly for the bears—and they occasionally end badly for humans, Logan says.

The whitebark pine ecosystem also plays a crucial role in protecting western watersheds, helping furnish water to the increasingly parched western foothills and plains. The Rocky Mountains are the headwaters for several major rivers that western states rely on, including the Missouri, the Snake, and the Colorado.

In the Rockies, most of the winter precipitation falls as snow, and healthy whitebark pine forests trap snow the way a fence or building does, causing it to form deep drifts. In early summer, when the sun warms the mountains, snow melts first in open areas; only later will the snow melt beneath the shaded whitebark stands. The snowpack serves as a reservoir, dispensing an even supply of meltwater to streams and brooks, supplying water to populated downstream regions through the dry summer.

The whitebark pine ecosystem exhibits the traits of all healthy ecosystems, traits Ken Sherman taught me during our coastal oceans study. Ecosystems are composed of communities of organisms and the nonliving factors they rely on, such as soil, water, and sunlight. While different in appearance, the subalpine whitebark pine ecosystem shares common traits with tallgrass prairies and coastal salt marshes.

Healthy whitebark pines live for centuries, and the grizzlies that feed in whitebark stands and the elk that calve there return year after year. Healthy ecosystems are resilient. When disturbed, the stand tends to recover. When trees are taken out in an isolated avalanche or small lightning-ignited fires, enough healthy trees and Clark's nutcrackers remain in the vicinity to replace them with new seedlings.

The species in a healthy ecosystem can play multiple roles. The whitebark pine tree, for example, provides food for squirrels and grizzlies, provides cover for elk and deer, and helps create new soil that opens up space for subalpine fir and other conifers. The numbers of different species fluctuate as a biological community moves from infancy through adolescence and on to maturity. But the system itself remains intact, year after year, century after century.

Healthy ecosystems have a diversity of components—a mix of species, with each part playing many starring roles. This provides the system with flexibility and resilience—the ability to bounce back after stress, such as fire, avalanche, severe weather, or invasive species. And in healthy ecosystems, there's a place for both generalists, like the elk that wander from high-altitude forests to grasslands in the foothills, and specialists, like the Clark's nutcracker and red squirrels, which have focused their lifestyle around the fruits of the whitebark pine.

BATTLING THE BEETLES

The mountain pine beetle is an indomitable, fingernail-size black beetle that burrows under the bark of pine trees by the thousands, eating the tree's phloem (nutrient-transporting tissue beneath the bark) from within. Mountain pine beetles are native to the West, and they have always been able to invade and eat just about any pine species. But they could never gain a foothold in the high-elevation forests of the Rockies and Cascades, where temperatures during winter cold snaps would occasionally plunge to −40°C (−40°F), frigid enough to kill the burrowing beetles. Instead, they lived primarily on the lodgepole and ponderosa pines that predominate at lower elevations.

To lay their eggs, female mountain pine beetles burrow into the bark. If the tree seems like a promising host, the pioneering beetles send out pheromones to lure their comrades. The trees try to fend off this influx by secreting sap that drowns the beetles and their eggs. "Trees are pretty vicious; they go after the beetle and its children," says bark beetle expert Kenneth Raffa, an entomologist at the University of Wisconsin. He adds that trees can change their chemical makeup to make themselves less attractive to the invader or even choke off nutrition to a section of their own tissue in order to strand the beetle in dead wood. The lodgepole's formidable chemical defenses can ward off small numbers of beetles, keeping them from reproducing.

But the beetles have ways of fighting back. They harbor spores of a symbiotic blue stain fungus, which they inject into the tree, and the fungus inhibits sap production and grows to supply the beetle larvae with an essential nutrient. Male and female beetles work together to carve out troughlike chambers, called galleries, for their eggs under the bark. The galleries run perpendicular to the wood's grain, with little niches or troughs off a main channel. When the eggs hatch, the wrinkled white larvae, each about a quarter inch long, will feast on the tree's phloem and pupate under its bark. To overcome a tree's defenses, beetles gang up on it by the thousands, and an infested tree will have thousands of beetle larvae developing in its tissues. Then, when the time is right, they emerge en masse as adults and head off to attack other trees.

Mountain pine beetles battled like this with lodgepoles and other low-elevation evergreens for millennia, occasionally wiping out stands—particularly those with diseased or dying trees—but basically fighting them to a draw. These days, though, as the western mountains get warmer and drier, the balance of power has shifted, just as it has between crops

FIGURE 22. Mountain pine beetles like this, shown here on bark from a recently infested tree in Togwotee Pass, Wyoming, are devouring centuries-old whitebark pine trees at high elevations throughout the Rockies. (Photo by Dan Ferber)

and crop pests. Drought and warmer weather dry trees' sap, making it harder to drown the beetles and nail them with defensive chemicals, and bark beetles easily overcome such weakened defenses. The situation is a perfect example of the environment weakening a host organism, making it more vulnerable to a disease-causing agent.

What's more, it no longer gets cold enough in the Rockies to freeze the burrowing beetles, which are now attacking whitebark pines in droves. And whitebark pines, unlike pines from lower elevations, evolved without having to face beetle attack and lack adequate defenses to fend off the ferocious beetles.

. . .

At Togwotee Pass, Jesse Logan took out a small hatchet and hacked at the foot-wide trunk of a tree, removing a piece of bark. Holding it, he pointed out a white grub half the size of a grain of rice—an early-stage pine beetle larva—and two black fingernail-size beetles (figure 22). He pointed out the J-shaped galleries on the bark and on the stripped trunk, spaced an inch or two apart. These were the chambers that adult beetles carve and that young beetles enlarge as they develop from eggs to larvae to adults. As Logan was explaining, the budding scientists peered in for a closer look.

Logan described how the female beetle burrows into the tree, followed by the male; how they mate and carve out the J-shaped chamber; how the female lays eggs, and how the larvae feast on the tree's phloem, the nutrient-carrying tissue inside the bark. He explained how the larvae pupate, transform into adults, emerge from the tree, and, when the time and temperature are right, fly off to attack new trees.

John, who runs wilderness trips for a living, interrupted with a question. "I train a lot of college students, so I need to know how far from the J is the love chamber."

"That's a research project you guys can start," Logan quipped. "I'm sure that the paper would be well received by the *Journal of Irreproducible Results*."

ROCKY MOUNTAIN MAN

High in the Rockies, Logan was in his element. His love for the outdoors started early, while fishing high-country trout streams with his dad and brother near his family's Pueblo, Colorado, home. A family vacation at age twelve introduced him to the wild lands of northern Yellowstone National Park, where the spectacular landscape and the trout streams in particular hooked him for life.

In college at Colorado State University, he abandoned his lifelong plan to run a ranch or farm in order to major in biology. He wanted to work outdoors but enjoyed the sense of order and structure that mathematics provided, and so he became fascinated for the first time with the intellectual side of life. The high country always beckoned, however, and Logan, a natural athlete, spent as much time as he could in it. In the winter, he'd ski into the backcountry, sometimes for days at a time, and in the summer he'd fly-fish for cutthroat trout in the region's cool mountain streams. "If you're fishing for trout, then by definition you're in great country," Logan said.

While he was a graduate student in entomology at Washington State University in the mid-1970s, Logan did a seminal piece of science that would one day help predict the dangers climate change poses to forests across the North American West.

At Washington State, Logan's adviser was Alan Berryman, a theoretical biologist who built mathematical models that predicted how insects would infest forests and agricultural crops. At the time, a pest insect called the McDaniel spider mite was devastating Washington state apple orchards. An even smaller predatory mite could eat the McDaniel spider mite, and the two organisms maintained a balance of power that protected the apple crop enough to ensure a reasonable harvest. But when summer temperatures were particularly hot, the predatory mites failed to control the spider mites, and apple yields fell. To help control the mites, Logan set out to develop a theoretical model that explained what was going on in the field.

Logan knew that, as cold-blooded animals, insects and mites develop more quickly when weather is warmer. If he could predict how quickly the spider mite and its predator developed, he realized, he'd be halfway home in his effort to predict field populations. At the time, ecologists believed that as temperatures rose, insects developed proportionally faster, in what mathematicians call a linear relationship. But in the field, they knew, populations would explode as temperatures rose, then crash as it got too hot—clearly a nonlinear relationship. Undaunted by the sophisticated mathematics needed to tackle the problem, Logan dove in deep.

Logan was taking an applied mathematics class, and he had an epiphany one day when the professor, an applied mathematician named David Wollkind, was discussing the mathematical methods used to model the air pressure very near an airplane wing, which determines the plane's lift. The behavior of the air in a thin layer right next to the wing differed dramatically from its behavior farther away—that is, the relationship between distance from the wing and air pressure was nonlinear. The exact same mathematics, Logan realized, could also predict the relationship between temperature and rates of insect development. "I thought, wow, this would really work," he recalled.

Logan then combined his model of developmental rates with a mathematical model developed from the Lotka–Volterra equations, which describe how predator and prey populations interact and fluctuate over time. The combined model was able to predict the impacts of rising temperatures on predator–prey interactions, including the spider mite and its minute but deadly predator. The model predicted a balance of power between the two insects at cooler temperatures. But as temperatures rose, it predicted, the prey would win out—just as they were doing in the field.

Because insects everywhere respond to even minimal temperature changes, other entomologists saw the potential of Logan's modeling method to predict how the insects they studied would behave in the field. Logan and his colleagues reported the work in the journal

Environmental Entomology. It was very well received, and for the next two and a half decades, Logan used the tools of mathematics and biology to predict how insect populations would respond to weather and climate. By the late 1980s, he'd become one of the top experts in that field. Then a colleague from the U.S. Forest Service suggested that the Forest Service fund a graduate student to work with Logan, who was by then a professor at Colorado State University, to model the effects of temperature on the mountain pine beetle.

GHOST FORESTS OF THE NORTH

At the time, global warming was not yet on Logan's radar, and Logan and his Forest Service colleagues were trying to solve an intriguing biological problem with major consequences for the health of western pine forests. Some insects go into a sort of hibernation called diapause, emerging when some physiological cycle in their bodies comes to fruition. This allows those insects to emerge en masse almost simultaneously. Mountain pine beetles use this strategy, which helps them swarm the next pine tree and kill it. But they had no mechanism for diapause. So how did all the beetles manage to emerge from trees at once?

Logan and the graduate student who took on the project, Barbara Bentz, suspected that temperature was the cue. Over several years, Bentz adapted Logan's modeling methods to mountain pine beetles. Like other insects, beetles develop through several distinct life stages—egg, four stages of larvae, pupa, then the adult beetle. Each stage responds to temperature differently. The model Bentz and Logan developed accounted for the effects of small temperature changes on the complexities of the beetles' larval development, and it predicted that temperature cues alone could cause the beetles to emerge from pine trees in synchrony.

Earlier Forest Service studies had shown that changes in temperature did in fact make beetles emerge in synchrony in a controlled-temperature growth chamber in the laboratory. Bentz wanted to see if the model predicted pine beetle behavior in the field. By then, she had earned her doctorate and was working at the U.S. Forest Service Laboratory in Logan, Utah. She set up a state-of-the-art field experiment in the Stanley Basin of central Idaho. There, she inserted tiny thermometers called thermocouples just underneath the bark of lodgepole pine trees and linked them to a computer, which allowed her to measure temperatures inside the bark, where beetles lived, in real time. She counted the numbers of mountain pine beetle eggs, larvae, pupae, and adults infecting the trees and cross-checked those numbers with weather data. The model accurately predicted abundance of beetles at each stage of their life cycle. Having validated their model, Logan and Bentz then used the model to predict whether mountain pine beetles could successfully establish themselves in whitebark pine stands. At then-current temperatures, the model predicted that mountain pine beetles couldn't establish themselves, and in fact they hadn't.

By the time they finished that work in the early 1990s, the IPCC had issued its first report, which predicted warming of about 2.5°C (4.5°F) in the American West. Since extreme cold kept mountain pine beetle infections at bay at high elevations and high latitudes, Logan and Bentz wondered how those rising temperatures would affect where mountain pine beetles could live.

Their model predicted that just 2.5°C (4.5°F) of warming was enough to allow the beetle to thrive in high-elevation whitebark pine stands in the White Cloud Mountains of central Idaho, where it had never before been able to live, and to expand its range several hundred miles northward into British Columbia. The beetles would then be able to spread even farther north, the models predicted, deep into the lodgepole forests of British Columbia and potentially into the vast and pristine boreal forests of northern Canada. This was an important prediction, Logan realized. He decided they'd better test it in the field.

In 1994, Logan and some other colleagues set up several state-of-the-art weather stations in a pristine whitebark pine ecosystem on a high-elevation site near the Stanley Basin. There were high-quality weather records for the site, known as Railroad Ridge, dating back to the 1940s. He got called on the carpet by his bosses at U.S. Forest Service headquarters in Washington, DC, for spending money to study a nonexistent problem. But, by then, the money was spent and the weather stations were a done deal, so there was not much his bosses could do.

Also at Railroad Ridge was an eerie and unusual whitebark pine forest. A warm stretch in the 1930s had led to mountain pine beetle outbreaks in the area. When cold temperatures returned, the beetles were killed off, but it was too late for the trees. Large numbers of centuries-old whitebark pine died, leaving a collection of standing skeletons—a "ghost forest," as Logan and a colleague described it in a 2001 publication—that only hinted at the forest's past grandeur. Now global warming threatened to turn much larger expanses of whitebark

FIGURE 23. Former Forest Service entomologist Jesse Logan sizing up a centuries-old whitebark pine that was killed in recent years by mountain pine beetles. (Photo by Dan Ferber)

pine into ghost forests, perhaps endangering the entire whitebark pine ecosystem (figure 23).

In the late 1990s, Logan made several trips to central and western Idaho to check out what was happening with the whitebarks. Mostly they were fine. But then in 2003, he returned to Railroad Ridge. As he looked out across the ridge, he saw one whitebark pine after another turning red. "Oh, shit," he thought. "It's happening."

BATTERED ECOSYSTEMS

Ecosystems have an inherent resilience, and the healthy ones can withstand a great deal of abuse, although they may change. For example, when chestnut blight was introduced to North America from Asia early in the twentieth century, it wiped out the dominant tree in large swaths of eastern forest. Other deciduous tree species, such as red maples and oaks, wound up taking over as dominant species and the forests survived.

But ecosystems, like all systems, have their limits. As humans disturb ecosystems—by cutting down forests, acidifying rain, filling in wetlands, hunting or fishing species to near extinction, dynamiting coral reefs, and introducing invasive species, to name just a few of our destructive habits—each successive disturbance builds on the previous ones, making the ecosystems sicker, weaker, and more vulnerable to additional stresses. These additional stresses may come from habitat fragmentation, in which we carve up large blocks of wildland that animals like bears need to roam, or they may take the form of infections by pathogens such as the chestnut blight, parasites like bark beetles, or invasive species like kudzu, which has overgrown native vegetation in many parts of the U.S. South. They may also come from chemical toxins, overharvesting, overdrawing groundwater, and clear-cutting large patches of essential habitats.

Global climate change adds insult to injury. Climate change has already been shown to alter a number of ecological systems. In temperate climate zones, the comma butterfly and nine others have extended their ranges northward in England since 1982. This change has been detected by careful field surveys, repeated over a period of years. Sometimes the damage done has been more dramatic, as when Costa Rica's high-altitude cloud forests warmed and dried, which led to the rapid extinction of the golden toad.

As humans alter ecosystems, the illnesses and infestations present in those systems can boomerang and make us sick. The previous chapter described epidemics of harmful algal blooms, which sicken humans and marine life. Bark beetles spread and destroy forests we depend on, with uncertain consequences. According to the Millennium Ecosystem Assessment, a landmark assessment of ecological conditions published in 2005 by the World Resources Institute, the links between environmental change and human health are "complex because often they're indirect, displaced in space and time, and dependent on a number of modifying forces." But, the authors add, "ecosystem services are indispensable to the well-being and health of people everywhere."

How can we diagnose sick ecosystems? The best diagnostic tools we have are biological indicators. We can see if long-lived and specialist creatures are declining and being replaced by versatile generalists like rats, jellyfish, or bark beetles. We can notice a shift in the balance of power that favors prey over predators (such as mosquitoes over darning needles and lacewings); fast-growing, weedy plants over slower-growing plants; and microbes over larger, long-lived organisms. We can monitor species diversity, to observe if it declines. We can watch for the rise in emerging infectious diseases of animals and plants, which may be viewed as symptoms of yet greater ills (figure 24).

FIGURE 24. Climate change has allowed mountain pine beetles to survive at high altitudes where they once would have died of cold, and now they're destroying large tracts of high-altitude whitebark pine forest, an ecosystem known as the rooftop of the Rockies. (Photo by Dan Ferber)

The signs and symptoms of ecological distress are all around us. The havoc we're wreaking is profound, and climate change has only just begun.

BARK BEETLES TAKE BRITISH COLUMBIA

To the north of Idaho, in majestic, green lodgepole pine forests of British Columbia, mountain pine beetles have long been a natural presence, and, as in the U.S. lodgepole forests, the insects and trees had long battled to a draw. Starting around 1997, however, this uneasy equilibrium was broken and the bark beetles seized the upper hand. The main reasons were drought and warming temperatures, which allow beetles to move north and to higher elevations—up to at least 1,600 meters elevation (almost one mile) from the prevailing 1,100 meters. Warmer weather allows beetles to cram a life cycle into a single year instead of two and to attack trees with thinner bark that previously wouldn't have kept them insulated enough to survive.

British Columbia foresters watched in alarm as pines began turning a rusty red and then a dull dead gray in increasing numbers, threatening to convert formerly deep-green expanses into the ghost forests described by Logan. Canadian scientists and foresters leapt into action to curb the epidemic. To try to root out infected trees, the government doubled the permissible timber harvest and allowed logging in parks where it had been banned. This pleased timber companies and created logging and milling jobs, but the foresters still couldn't fell trees quickly enough.

The foresters tried injecting trees—sixty-eight thousand in all—with an arsenic-based pesticide. But they lost track of which trees had been treated, and the trees were logged and burned, inadvertently exposing wildlife and possibly humans to the poison. They baited "trap trees" with pheromones to attract as many beetles as possible to isolated stands of trees, to prevent them from spreading, but there were just too many beetles reproducing too quickly to contain them. Each winter, they enlisted workers to ride to the front lines of the infestation on snowmobiles or drop in by helicopter. Foresters burned thousands of infected trees, one at a time, in a procedure known as a "fall and burn on spot." They sprayed pesticides rather than injecting them. The effort was costing British Columbia almost $800,000 a year.

After several years, scientists and foresters realized the battle was lost. They could only wait for the beetles to "literally eat themselves out of house and home," in the words of Alaskan bark beetle expert Ed Berg. Beetles prey only on mature trees, and once they've been eaten, beetles starve, their populations drop, and young trees can regrow. It's an extreme conclusion to an epidemic, but it's what's happening in British Columbia. Already, thirty-three million acres of lodgepole pine forests and 3.5 billion mature trees have been killed by the mountain pine beetle. By 2013, a whopping 80 percent of British Columbia's mature trees, mostly lodgepole pine, could be destroyed.

"Where once a green carpet of trees laid, now it's a brown-red horror show," bemoaned a crestfallen *Toronto Sun* writer. Another journalist called British Columbia a "rust-coloured monument to the consequences of failure."

Such failure has ramifications for the planet. Models used by Canadian Forest Service researchers predicted that between 2000 and 2020, the beetle outbreak in south-central British Columbia would increase atmospheric carbon by 270 million tons, which is more CO_2 than the entire country of Canada reported as emissions in 2005. Still, an even more frightening prospect loomed. Bark beetles emerge from their host trees when storms and wind are brewing, and they can travel long distances on air currents. They could potentially cross the Continental Divide of the Rockies, invading Alberta's expansive, fragile forests. These forests are the gateway to the vast boreal forest of the north, known as the "lungs of the Earth" for its role producing oxygen and socking away carbon, that stretches all the way to the northeastern United States. In the worst-case scenario, they'd provide a transcontinental corridor of bark beetle destruction.

. . .

In the summer of 2006, a large swarm of adult mountain pine beetles emerged as a strong westerly wind was blowing, and they were air-lifted across the Continental Divide into Alberta, where they became entrenched and began to spread. Today, the beetles have infested even the hybridized lodgepole pine and jack pine forests that continue unbroken to the northern jack pine boreal forests, the lungs of the Earth.

Experts debate whether mountain pine beetles could colonize the boreal forest, as jack pines are scragglier than lodgepole pines, with a thinner nutrient-carrying phloem layer in their bark that may offer less sustenance to the beetles. But if they do kill large tracts of northern jack pines, fires sweeping through the crowns of the trees could release billions of tons of carbon, making the boreal forest a culprit in global warming instead of a stabilizing carbon sink, just as the bark beetle infestation had in British Columbia.

The Albertan government managed to hold the outbreak at bay through 2008, at a cost of nearly $200 million and aided by two unusually frigid winters in 2007 and 2008. Nonetheless, by the summer of 2008, Alberta had lost about five million acres to the beetles, and the fight against them continues.

In British Columbia, it was a different story. By 2008, the infestation was waning, as there were relatively few trees left for the beetles to kill.

THE WEST AFLAME

Should Canadian foresters want an image of what may be in store for their ravaged forests, they can just look south to the U.S. West.

Global warming has hit the mountains of the U.S. West harder than anywhere else in the continental United States, raising average temperatures in the eleven-state region a full 1.1°C (2°F) above the twentieth-century average. The IPCC predicts more warming and longer, more intense droughts in the future. These changes have fostered bark beetle infestations throughout the region.

The piñon pines of the Southwest are falling to their own type of bark beetles, known as piñon ips, and disappearing from the beloved piñon-juniper ecosystem, thanks to climate change–related drought, warming temperatures, and water-sapping human development. And new seedlings are not springing up to take their place.

About 4.8 million of Colorado's lodgepole pines had fallen prey to mountain pine beetles by the summer of 2007, and up to 90 percent of them are expected to die by 2012. More than four million acres of once-lush spruce forest on Alaska's Kenai Peninsula—as much as 95 percent of the mature trees—have been consumed by spruce bark beetle infestations.

In a beetle-ravaged forest, dead logs can be ignited by lightning and smolder for weeks or even months until warm dry winds kindle the embers into a full-blown blaze. Trees that have perished in the previous year or two are still full of highly combustible resin, and they can serve as "fire ladders" that let the licking flames climb up into the treetops, or crowns. One recent study does suggest that beetle infestations may actually reduce fire risk in ponderosa pines by thinning their crowns, and more research is needed to clarify whether and under what circumstances this is true. But most evidence suggests that a beetle-ravaged forest is like a field of roman candles, ready to burst into flame. And burst into flame they have.

Fires have grown larger and more frequent from Mexico to Alaska. In 1996, the Lone Fire burned 60,000 acres, three times more than any previous recorded blaze in Arizona. In 2004, a lightning-sparked fire in the state's Mazatzal Wilderness consumed 119,000 acres. In 2005, another lightning fire in the state consumed 248,000 acres. Giant fires burned farther north as well. In Oregon, the infamous Biscuit Fire burned almost 500,000 acres in 2002. In Alaska, 6.38 million acres went up in flames in 2004 alone.

California, with its large populations and semiarid climate, has been particularly hard hit. A series of blazes in October 2007, for example, swept across Southern California, from Santa Barbara down to the Mexican border, burning half a million acres and forcing a quarter million people to evacuate. The Cedar Fire, which scorched San Diego County in 2003, sent walls of fire taller than the Statue of Liberty to devour evergreens almost as tall. Solar panels melted and charred cows lay by roadsides, as quaint tourist towns were transformed into moonscapes.

WHENCE THE WILDFIRES?

What was causing these increasingly menacing conflagrations? Although it was convenient to blame climate change, most forestry experts blamed it instead on overzealous fire suppression, which had allowed forests in the region to become overgrown with densely packed, water-starved trees that burn hotter, longer, and over a wider area. Fire suppression also allowed thick layers of flammable underbrush to accumulate, and as people built homes and moved into wooded, fire-prone areas, chances for accidental ignition increased.

Tony Westerling, a scientist with the Sierra Nevada Research Institute at the University of California, Merced, didn't doubt any of this. But he suspected that climate change might be even more critical.

Documenting climate change's role in wildfires would not be easy. There was no centralized, uniform historical data set on American wildfires, and parsing out the effects of climate change from other factors would be a challenge.

In 2002, Westerling visited the offices of one federal or state agency after another in search of fire data, spending hours squinting at handwritten logs and delving into often spotty and disorganized documents. "Before 1990, they didn't keep very good records. Once the fires went out, they just went home and had a beer," remembers Westerling's colleague Steven Running, a forest ecologist at the University of Montana. Westerling ultimately mapped reams of data representing more than 1,100 large fires between 1970 and 2003, creating the most comprehensive historical fire record ever for the western United States. He divided the study into two time periods, 1970–1986 and 1987–2003, and cross-checked fire data with weather data, including temperature and dates of spring snowmelt. Between 1987 and 2003, the fire season had become a full two and a half months longer; the frequency of large fires had increased, and the average burn time of large fires had jumped from one week to five.

One fact strongly implicated climate change. Wildfires in mid-elevation forests—around 2,100 meters (6,800 feet) in elevation—were burning earlier, longer, and significantly more often. Since fires in these high, sparse zones had previously occurred only about once a century, fire suppression alone would not have worsened fires at these altitudes. Other factors must have been responsible.

Westerling's data analysis pointed to warmer ground temperatures and earlier snowmelt. Fire is highly unlikely to catch on and spread in

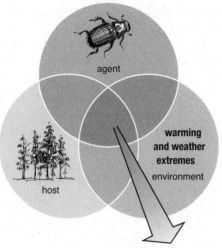

agent

host

warming
and weather
extremes

environment

forest diebacks and wildfires

FIGURE 25. Perils of the pines. Warming and climate change favor bark beetle infestations of spruce and other pine forests, which cause forests to die back, creating large tracts of tinder. Drought and hot weather make it more likely that the tinder will fuel devastating wildfires.

a forest blanketed by snow. But ground temperatures had increased an average of almost 1°C (1.8°F) in the 1987–2003 years compared with 1970–1986. Now snow was melting weeks earlier, vegetation was drier by summer, and more potential fire days occurred each year, bolstering Westerling's theory that climate change was behind the increase in wildfires (figure 25).

He expects the trend to continue. Models from a subsequent study he worked on predict that climate change will increase burned area by 54 percent in the western United States by the 2050s, with the incidence of fire nearly doubling in the Pacific Northwest and nearly tripling in the Rocky Mountains.

None of this bodes well for human health because where there's fire, there's smoke. Smoke from wildfires can affect air quality hundreds or even thousands of miles away, bringing multiple health problems, especially in the young, the elderly, and people with existing heart or respiratory problems. By consuming trees, houses, and electrical wires, wildfires release compounds known to be toxic and, in some cases, carcinogenic, such as carbon monoxide, polycyclic aromatic hydrocarbons, and volatile organic compounds like formaldehyde.

Smoke from forest fires also contains an assortment of tiny but dangerous particles. The tiniest of them, each less than one-twentieth the width of a human hair, cause the greatest health damage because they penetrate the tiniest airways and can lodge in the alveoli, the gas-exchange areas of the lungs.

The toxic smoke from the 2002 Hayman Fire in Colorado left people suffering for months afterward. Following the 1998 fires in Florida, complaints of chest pain increased by 37 percent, asthma by 91 percent, and bronchitis by 132 percent. In California's Central Valley, the smoke from frequent wildfires can remain trapped for days on end, combining in the heat with existing smog to make the outside air nearly unbreathable.

"There are absolutely significant health effects," said atmospheric and environmental chemistry scientist Dan Jaffe of the University of Washington–Bothell. "Fires are the dominant air pollution problem in the West, whether you're talking about Denver, Salt Lake City, or L.A."

RACING THE FLAMES

It was about 1 A.M. on October 26, 2003, when Bob Younger, a photographer who lived with his journalist wife, Sandra, in the chaparral canyons northeast of San Diego, woke up to the acrid smell of smoke. He roused Sandra, and the two of them, bleary-eyed but increasingly nervous, ventured outside into a blasting hot Santa Ana wind. Ashes were flying through the air, and, from their home in Wildcat Canyon, they saw an ominous reddish glow in the distance. Sandra got on the Internet but found no relevant news about the blaze. She called two fire stations and was told a fire was spreading, but not to worry; winds would move it in a direction opposite from their home, the dispatcher said.

Not fully convinced, she and Bob decided to get a few hours of sleep and then reassess the situation. The next thing Sandra remembers was being jolted awake by a yell from her husband. The air outside was thick with smoke. The canyon across from their home was engulfed in flames, and they could see a fiery glow in the bottom of the canyon between the burning ridge and their property. They knew they had to get out of there, and fast. They spent five minutes sweeping framed photos off tabletops and walls into a laundry basket, and Bob grabbed a few boxes of his best negatives. Then they piled into Sandra's white Acura, along with their two large Newfoundland dogs and their cockatiel.

The smoke was so thick that it was impossible to see the winding dirt road, which hugged the side of the mountain from their home to the main highway half a mile away. "It was like looking out an airplane window when you're in a cloud and can't see a thing," Sandra recalls. Terrified of driving off the road and tumbling down the hill, but knowing the fire was advancing, Sandra inched the car forward. All of a sudden, an animal jumped out of the smoke into their headlights. It seemed to be a bobcat and, after realizing it could see the road, they followed the animal closely through the haze.

Finally, she felt the car's wheels meet the pavement of the highway, where they continued to drive between two curtains of flame. "There was fire blowing across the road," said Sandra. "The car was scorched. I kept thinking I had to get out of there as fast as I could because I didn't want the gas tank to explode."

They sped for a mile or so and at last left the blaze behind.

Twelve of their Wildcat Canyon neighbors weren't so lucky. The victims of the Cedar Fire included a couple, ages sixty-three and fifty-eight, and their thirty-two-year-old son, who were overcome by flames as they tried to reach the waters of the local reservoir; two of their bodies were found in their truck, one outside on the road. A fifty-five-year-old man died after he came up against a wall of fire while fleeing in his RV with his four Irish wolfhounds. Steve Rucker, a thirty-eight-year-old firefighter, died after driving four hundred miles from his home to help battle the blaze.

Dizzy with relief and exhaustion, Sandra and Bob stopped at a gas station for water. Their mouths were so dry they could barely speak. They realized their cockatiel had lost most of its feathers and was bleeding from frantically flapping around its cage in the trunk, so they brought it to an emergency twenty-four-hour veterinary clinic where others escaping the fire were also pouring in with animals. "We figured the bird is about all we have left; we better take care of it," Sandra said. Finally they made it to a downtown San Diego hotel that was welcoming fire refugees with pets, where they would stay for nine days. "By the end of the day it looked like Noah's Ark," she recalled.

The Cedar Fire, which they had just fled, was just one of several fires scorching San Diego County. The Santa Ana winds blew smoke toward the coast; the air downtown was thick with smoke, the sky was brown, and the Sun blood red. Schools and stores closed, and people stayed locked inside. It was hard for Sandra and Bob to find an open store to buy some extra clothes and provisions.

Five years after the blaze, the Youngers had rebuilt their home and their lives, but it was a long, costly, and draining process. Now Sandra looks at the chaparral that quickly grew up on the fire-scorched land around their home with a mix of admiration and suspicion. She feels lucky to have survived the Cedar Fire and to have had decent insurance, but the trauma still hangs like a smoky cloud in their past. Among the various health effects of wildfires, she notes, acute and lasting mental health impacts like the ones she and her neighbors suffered cannot be underestimated.

"It draws a line through your life," she said. "Everything becomes 'before the fire' and 'after the fire.'"

8

Storms and Sickness

To scientists who were paying close attention in the spring of 1997, conditions in the Pacific Ocean portended trouble. The Pacific's easterly trade winds had piled up warm water in the western Pacific, and sea levels near Indonesia were a full foot and a half higher than those in the eastern Pacific off Peru. Warm air was steaming off the warm water. Climate modelers had taken reams of data from a network of scientific buoys and satellites, fed them into their models, and run the models on supercomputers. The models revealed that an El Niño was brewing.

In September 1997, I flew from Boston to Maputo, Mozambique, to advise officials from key government ministries from Mozambique and several other southern African countries about how they might ease the impacts of epidemics we feared would soon be upon them. The El Niño was expected to begin in earnest during the last few months of that year. Climatologists knew that El Niños typically increased the extreme weather occurring in southern Africa, which meant that health and relief workers in the region could face enormous challenges.

I'd returned to Mozambique several times since leaving in 1980, a year that had seen the beginning of a long and hellish war. The people of Mozambique had benefited greatly from the health care and educational reforms that Samora Machel and FRELIMO had introduced, some of which we *cooperantes* had helped to put in place. The prospects of similar reforms held strong appeal for the disenfranchised residents of neighboring South Africa. But South Africa's apartheid leaders, who had already witnessed the Soweto uprising in 1976, a year after Mozambique's independence, were threatened by the prospects of an empowered black populace next door, and they responded by funding rightist guerrillas to destabilize Mozambique's government and cow its population.

To aid that struggling populace, my wife, Andy, and I made a return visit in 1985 to the country we had grown to love. Working as volunteers for an aid organization called American Jewish World Service, we helped airlift medicines into Mozambique's war zones. We returned again in 1990 for a conference of health personnel from throughout southern Africa as they envisioned a postapartheid health system in newly liberated South Africa. Apartheid in that country officially crumbled in 1994, and, not coincidentally, the war in Mozambique ended the same year. In 1995, we returned again with other volunteers from Physicians for Human Rights to investigate the continuing danger and harm from hundreds of thousands of land mines scattered about the nation.

By September 1997, I was serving as an informal adviser to the U.S. National Oceanic and Atmospheric Administration's Office of Global Programs, focusing on public health in the developing world. In that capacity, I'd participated in workshops in Peru, Uruguay, and Brazil, explaining to health, agricultural, water, and meteorological ministers how to use El Niño forecasting to anticipate, prevent, and treat health problems.

Now I was back in Mozambique again and, on a shaded campus on the outskirts of Maputo, I made my way to the site of the workshop, a one-story building that housed the country's agriculture ministry. Earlier El Niño events had almost invariably brought drought to southern Africa, so colleagues from Columbia University's International Institute for Climate Prediction and I planned to instruct the assembled officials how to prevent drought-related diseases and crop pests and how to ensure adequate food and water during the expected drought.

The day before the workshop, though, I got an inkling that things might not turn out as we expected. I'd dropped by the Ministry of Agriculture to receive a briefing, and, on the way into the building, I spotted a wall map of southern Africa that incorporated satellite imagery and the greening index, which reflected precipitation during the unusually large El Niño of 1982–1983. The map showed a regionwide drought, with the exception of Mozambique's 2,735-kilometer (1,700-mile) long coastal plains, which were bathed in green, meaning that plenty of rain had fallen during that large El Niño. I hesitated. Perhaps not all El Niños were alike. Did weather patterns during large El Niños differ from those of run-of-the-mill El Niños?

These thoughts lingered in my mind as I entered the workshop the next day, but I put them aside when it came my turn to speak. I intended to help get the health and meteorology ministries to collaborate to protect the nation's health. The meeting was chaired by the head of Calamidades, the Mozambican intragovernmental agency charged with preparing for natural calamities and disasters. About forty of us, including those from South Africa and Zimbabwe, gathered in a large classroom. Then I gave the presentation I'd prepared.

My presentation described how drought led to poor hygiene, raising the risk for waterborne diseases, including cholera. Then, Godfrey Chikwenhere, a colleague from Zimbabwe, and I recounted our joint work on how the drought in 1983 had sparked a rodent population boom in his country, which led to a devastating drop in maize production and, near the Mozambique–Malawi border, an outbreak of bubonic plague. Following our presentation, Dr. Julie Cliff, an Australian physician, recounted how a severe and prolonged drought in 1981 in the northern province of Nampula had led to an outbreak of konzo—cyanide-induced neurological poisoning among children who had eaten cassava.

We advised health officials to educate the public about the need to adequately process cassava, and to boil water and stock treatment centers with the salt-and-sugar solution packets that, when dissolved in clean water, could save lives in case of a cholera outbreak. We suggested measures that water and energy officials could take to ensure sufficient hydropower and supplies of fresh water. Others from the agriculture ministry suggested that rice farmers along the coast might have to switch to wheat, a far less water-intensive crop. It was a well-thought-through scenario planning exercise.

Then, outside the classroom window, drops of rain began to fall. It rained for the rest of the conference, and for many weeks after. The El Niño had begun right on schedule that September, but it had brought rains instead of drought. While my group of scientists had been careful to say that drought was highly probable rather than certain, the Mozambican newspapers did not couch their headlines in such nuances: "El Niño Will Bring Drought, Experts Warn"—or something of that sort.

Our failure to predict rains in Mozambique made officials there slow to accept climate-related predictions and preparations for several years. But climate models have improved since then, making them more useful for generating early warning systems for human disease and for crops. While their forecasts are not foolproof, they're an important way to prepare for uncertainty and thereby protect public health. This will be increasingly true in the years to come, when extreme weather becomes more common than it is today.

A PRISON CRUMBLES, AND DANGER MOUNTS

Almost a year after our attempt to predict the effects of extreme weather in Mozambique, nature served up a powerful reminder of the devastation extreme weather can cause. The day was October 30, 1998, the setting was Honduras, and the reminder that came ashore was a monster storm called Hurricane Mitch, one of the largest and most powerful Atlantic hurricanes on record.

On that day, skies darkened to a sinister gray over the Honduran capital of Tegucigalpa. Wind flattened large trees. Rain pounded. And, when he got to the city's central prison that day, so did Dr. Juan Almendares's heart.

Visiting the prison to treat prisoners was part of Almendares's normal routine. Almendares ran a nongovernmental organization devoted to combating torture and helping heal its victims. To try to prevent abuses at the city's central prison, he and his staff would visit the prison two to three times a week, making it clear to guards that they were paying attention and bearing witness if prisoners had been abused. Almendares's nonjudgmental approach and his national reputation as a humanitarian and human rights activist earned him the respect of both prisoners and guards.

On the day of the hurricane, however, Rio Choluteca, the main river bisecting Tegucigalpa, was a raging torrent dozens of feet above flood stage, and the rain was quickly destroying the adobe and stone walls of the century-old prison, which sat just a few yards from the river. As the building began to crumble, a contingent of gun-bearing, uniformed police unfamiliar with Almendares arrived to help transfer the prisoners to higher ground, a busload at a time. Guards yelled at the prisoners, tied their hands behind their backs, and beat them to control them. Guards tried to move the prisoners out, but it was taking too long. Some prisoners ran, and some jumped into the river to escape. The guards responded by shooting at them.

The fast-rising river swelled toward Puente San Rafael, the bridge that Almendares and two coworkers had crossed on their way to the prison, threatening to cut off their route home. As they watched from outside the prison, Almendares and the two women with him, the assistant director and a nurse from his torture-prevention center, heard the rat-a-tat-tat of machine-gun fire. "It was horrible," Almendares recalled.

THE MAKING OF A HUMAN RIGHTS ACTIVIST

By the time Hurricane Mitch descended on Honduras, Almendares had been promoting human rights for his fellow Hondurans for several decades. Today, he is known internationally for his courageous and visionary work as a medical doctor, healer, human rights activist, and environmental leader. He runs a free health clinic for poor Hondurans and has established three nongovernmental organizations: his torture-prevention center; the Honduran Peace Action Committee, a grassroots women's organization that works to improve the living conditions of low-income families in marginalized communities; and Movimiento Madre Tierra, the Honduran arm of Friends of the Earth International. He has been recognized by the World Health Organization for his efforts to combat tobacco use in Honduras, he has spoken before the United Nations, and he has run for president of Honduras.

Born in 1939, Almendares grew up poor in San Pedro Sula, in the banana-growing north of Honduras. His family lived in the town's red light district, a part of town known by locals as *abajo de la linea*—literally, "under the line," meaning on the other side of the tracks—and as a boy he witnessed the destructive behavior common in such neighborhoods, including rampant alcohol use, prostitution, and violence. One night when he was eight, his mother informed him that his father had been murdered. "People hired by a landowner did it," he says.

Influenced by his mother—a midwife and healer who promoted religious tolerance—Almendares shunned violence, drugs, and alcohol and pursued his studies avidly, gaining admission at age eighteen to the National Autonomous University of Honduras Medical School. As a medical student, he was so poor he couldn't afford to buy his own textbooks. "I had to wait until my *compañeros* were asleep; then I would study from their books," he said. Nevertheless, Almendares finished medical school determined to serve the poor, become an educator, and do scientific research.

A stint as a postdoctoral researcher at the Cardiovascular Research Institute in San Francisco in the late 1960s provided lessons in physiology and clinical medicine, and other lasting lessons as well. "I came to

FIGURE 26. Dr. Juan Almendares comforts a woman inside her home in Choluteca, Honduras. (Photo by Jason Lindsey/Perceptive Visions)

know the conquests of the civil rights movement," he recalled, as well as "the spirit for peace that existed among the youth."

As a professor at the National Autonomous University of Honduras Medical School a few years later, Almendares put his experience and beliefs to work. After Hurricane Fifi slammed Honduras in 1974, killing about eight thousand people, he organized thousands of students, including medical students, into medical brigades and had them fan out across the nation to help the sick and injured. He pushed his medical school into providing free help to poor rural people, thereby changing the tone of the health profession in Honduras (figure 26).

Such courageous, humanitarian leadership made Almendares a well-known and well-loved figure, and by 1978, he'd been elected rector (chancellor) of the National Autonomous University of Honduras. As an outspoken critic of U.S. military intervention in Honduras and its support for right-wing Contras in neighboring Nicaragua and El Salvador, Almendares gained some powerful enemies. Military leaders engineered Almendares's dismissal as university rector, with the tacit support of the Reagan administration, and for three dark years beginning in 1982, he was harassed, isolated from his family and his medical practice, then captured and tortured by right-wing military squads. Today, he said, "I am compelled to speak out against every kind of torture and terrorism."

To fight violence, in 1996 Almendares founded his torture prevention and treatment organization, the Center for Prevention, Treatment and Rehabilitation of Torture Victims and Their Families (CPTRT). It was his torture prevention and medical work through this center that led him to Tegucigalpa's central prison on the momentous day when Hurricane Mitch roared into town.

. . .

During the hurricane, Almendares and his two companions remained at the central prison the entire day without food or drink in order to prevent human rights abuses. Inside and outside the prison, the pandemonium continued. Some prisoners started fires, and some were burnt. Others tried to escape. Almendares kept talking to police, telling them to stop shooting. He's convinced he saved some lives, but later he and his staff at his torture-prevention center investigated, estimated that fifty prisoners were killed during the storm, and denounced the killings, although the authorities never admitted them. As evening fell, Almendares remained at the prison despite the lack of electricity and a citywide curfew that gave authorities the right to shoot violators. "We had this responsibility. We were afraid they would kill more prisoners," he said.

By that evening, it was bedlam all over Tegucigalpa, and it remained that way for days as Mitch's rains continued unabated. Roads turned into rivers, with gaping holes and deep gullies, and people abandoned cars. Bridges collapsed, as did buildings. In the *colonias* (poor neighborhoods) of Tegucigalpa and its sister city, Comayagüela, where poor people dwell on steep, denuded land, the results were worse. Colonia Soto, a neighborhood built on a steep hillside overlooking Rio Choluteca, collapsed in a mud slide one morning, taking three hundred homes with it.

For two rainy days following that long day outside the central prison, Almendares would minister to more than one hundred prisoners at the local soccer stadium, giving them medicine, treating them for skin ulcers from floodwaters, even defusing a potential riot. But, at about 10 o'clock that first evening, the rain had eased and the bridge had held, so Almendares headed home for the night.

KNOCKOUT BLOW

When we hear about hurricanes on the news, we typically hear about events like these: howling winds, raging floods, homes destroyed, families torn apart. When the storm is as devastating as Hurricane Mitch or

Hurricane Katrina, we hear heartrending follow-up stories for weeks about efforts to rescue the stranded and help those who lost all. But when the news of short-term impacts slips from page one, the damage continues.

Hurricane Mitch generated many short-term horror stories. It hit Honduras so hard that some believed it might send the country back to the preindustrial age. Nationwide, 189 bridges and more than a hundred highways were destroyed, at a cost of $417 million, and two hundred thousand kilometers of electrical lines, transformers, and towers were flattened, according to Godofredo Andino, who runs the program of emergencies and disasters for Honduras's secretary of health. The storm ruined 70 percent of the agricultural production of Honduras that year, including $800 million worth of bananas. The people of Morolica, a village in southern Honduras not far from Nicaragua, lost all their roads, phone lines, and means of communication with the outside world. After waiting a week for help that never came, the townspeople, led by the mayor, walked sixty miles to the capital city of Tegucigalpa to make sure they got some assistance.

In terms of public health, the short-term aftermath of Mitch was nothing short of disastrous. More than 6,500 people died in Honduras, and many more in neighboring Nicaragua. Hundreds of neighborhoods in Tegucigalpa lost access to public water. More than three-quarters of the hospitals in Honduras had major problems with water supplies and sanitation because infrastructure was so damaged, as did numerous small clinics. The two months that followed saw spikes in diarrhea, respiratory infections, rodent- and waterborne leptospirosis, and dengue fever, outbreaks of which public health officials had just subdued when the storm hit.

Such outbreaks typically follow major hurricanes and flooding, particularly in the developing world. Epidemics of bacterial and viral disease have occurred in Brazil, Russia, Argentina, Philippines, and India following heavy flooding events during the last two decades. The heavy rains of large storms can increase the spread of vector-borne infectious diseases by creating conditions that enable vector species like rats, mosquitoes, and snails to breed.

The developing world will continue to be hit hardest by disasters linked to climate change and the resulting clusters of epidemics. That's because many developing countries are in geographically vulnerable areas with low-lying coastlines, and because they are less structurally, politically, and financially equipped to prepare for disasters and deal with their aftermath. But no nation is immune from such devastation.

That point was seared into the collective consciousness by images of

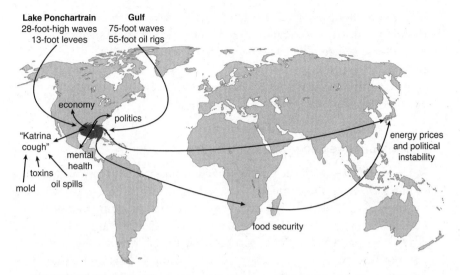

Lake Ponchartrain
28-foot-high waves
13-foot levees

Gulf
75-foot waves
55-foot oil rigs

economy

politics

"Katrina cough"

mental health

toxins

mold

oil spills

energy prices and political instability

food security

FIGURE 27. Health and the hurricane. In Hurricane Katrina's wake, floods spread toxins, and damaged buildings grew mold, which fueled the "Katrina cough"; post-traumatic stress disorder also spiked. Hurricane Katrina also caused an oil spill the size of the *Exxon Valdez* spill, poisoning wetland ecosystems, and stalled food shipments to southern Africa, which desperately needed the food.

Hurricane Katrina: desperate and dying people stranded on rooftops in sweltering heat as bloated bodies floated by, houses carried for blocks by surging waters near the levee breaks, the Superdome packed with refugees huddled amid filth and chaos, and a national administration that, for a while, seemed not to care.

Hurricane Katrina killed more than 1,450, and 2,000 more were still missing six months later. And victims suffered a barrage of health problems during and after the disaster. The storm and ensuing flood brought physical injury as people fled flooded homes or suffered from violence in the suddenly lawless city. Survivors, many already destitute, disenfranchised, and thus in poor health, found themselves homeless. Some suffered heatstroke, severe dehydration, and stress, which brought on heart attacks or strokes.

Other health problems emerged later. Toxic chemicals spread widely. Katrina victims contracted waterborne infections from norovirus, *Salmonella, Vibrio* bacteria (which killed five people), and other pathogens. Shortly after people returned to their destroyed and sodden homes, rumors started flying about "Katrina cough." New Orleans residents reported suffering a persistent cough, inflamed nasal passages, infected eyes, and the like, which many blamed on rampant mold left behind by receding floodwaters. Though mold and dust are known to cause respiratory distress, several studies found rates of such symptoms in New Orleans were no higher than in other areas. The reports were so widespread, however, that in 2008, Tulane Medical School launched a comprehensive five-year study of one thousand New Orleans workers to examine the suspected phenomenon (figure 27).

CLIMATE CHANGE AND STRONGER HURRICANES

We're all familiar with the god's-eye view from satellites that can warn of hurricanes spinning ominously toward land. Such storms begin when the tropical sea surface builds up more heat than it can dump into ocean currents flowing poleward from the equator. This heat causes water to evaporate from the ocean's surface, creating upward drafts, which carry the moisture aloft, where it condenses into clouds. This condensation releases energy back into the air, creating more wind, which then causes more evaporation, and more wind. The result is a gigantic upside-down whirlpool that sucks heat out of the ocean—a vast convulsion of the atmosphere. And as they swirl and pass over more warm waters, hurricanes can grow in size and strength.

When hurricanes move over land, they cut themselves off from their source of energy and eventually run down. Even over the ocean, there's a natural brake that prevents hurricanes from becoming too powerful. When a hurricane strengthens, it stirs up the ocean's layers, mixing the warm surface water with cooler deep water, thereby powering down the warm water driving the hurricane.

As the globe warms, warmer oceans store more energy to feed hurricanes. And as warming penetrates deeper into the oceans, they may be letting off on the natural brake, allowing hurricanes to grow stronger. These effects may have contributed to the devastating power of hurricanes Mitch and Katrina, as well as recent clusters of hurricanes that follow one another in close sequence, as Rita, Katrina, and Wilma did in 2005 and as four hurricanes did in Florida in 2004.

Studies over the last five years have shown that hurricanes are getting stronger. In a 2005 paper in *Nature,* meteorologist Kerry Emanuel of the Massachusetts Institute of Technology described a new measure, the power dissipation index, that captures the amount of energy that flows through a hurricane over its lifetime. He found that the power dissipated by hurricanes in the North Atlantic and North Pacific oceans has nearly doubled over the thirty-five years since accurate satellite measurements

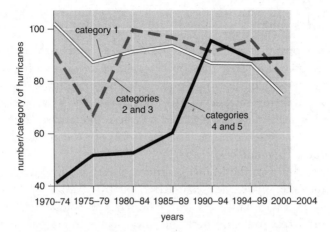

FIGURE 28. Killer storms on the rise. Since 1970, devastating category 4 and 5 hurricanes have become more than twice as common worldwide. Milder hurricanes (categories 1–3) have not. (From P. J. Webster, et al., "Changes in Tropical Cyclone Number, Duration, and Intensity in a Warming Environment," *Science* 309 [2005]: 1844)

became possible. The same year, Peter Webster's team at Georgia Tech University reported in *Science* that while the total number of tropical cyclones per year stayed roughly constant worldwide, the strongest storms—category 4 or 5 storms, whose maximum winds topped 210 kilometers per hour (131 miles per hour) or 259 kilometers per hour (155 miles per hour)—nearly doubled over thirty-five years (figure 28).

Some researchers have questioned the evidence behind these studies, but Emanuel said the link is most likely real. "The weight of the evidence favors the idea that global warming is increasing the incidence of intense hurricanes globally, but the case is not yet airtight," he said.

Top hurricane researchers do agree that as the globe warms, hurricanes will intensify, though the total number of hurricanes will remain relatively constant, or perhaps even drop. In 2010, *Science* published a study by Morris Bender, Tom Knutson, and Robert Tuleya of the National Oceanic and Atmospheric Administration—three prominent, previously skeptical modelers. They and their colleagues concluded from modeling studies that category 4 and category 5 hurricanes would double in frequency in the Atlantic this century because of ocean warming. "It is likely that hurricanes in the future will be more intense on average and have higher rainfall rates than present-day hurricanes," Knutson has written on his Web site.

AFTER THE FLOOD

When a monster storm like Hurricane Mitch or Hurricane Katrina strikes another locale, it tugs on heart strings, but it's often all too easy to convince ourselves that we're safe and sound in regions of the globe that are temperate, developed, and not prone to hurricanes. But as climate becomes more unstable and weather more erratic, it can bring dangers much closer to home, as a lot of folks in Cedar Rapids, Iowa, found out in the summer of 2008.

That summer, as the Midwest suffered its second five-hundred-year flood since the 1990s, the Cedar River burst its banks and inundated Cedar Rapids with up to ten feet of water. The National Guard cordoned off access to floodwaters and stood sentry to prevent distraught residents and curious onlookers from coming into contact with waters that had become a stew of toxic chemicals and pathogens from sewage.

People regularly die in floods like the Cedar Rapids flood, sometimes trapped in vehicles crossing swollen rivers, sometimes drowned inside their homes. Downed power lines and soaked car batteries can cause electrocution, and damaged heaters and stoves pose the risk of carbon monoxide poisoning. There is even a risk from wild, stray, or lost animals confused and displaced by floods. In rural areas, snake bites mount.

But the most widespread health risks from flooding arise as the waters ebb. When the waters subsided in Cedar Rapids, recovery efforts were put on hold after utility workers fell through a sodden second-story floor. When residents donned heavy rubber boots and gloves and crept back into their damaged homes, they found furniture tossed about, sagging waterlogged ceilings and staircases, ruined belongings, and thick layers of gritty, stinking mud. The whole city smelled, a stench halfway between an open sewer and a wet dog, a stink that could make one's eyes water and stomach clench. And the stinking muck blanketing the city posed a potent health hazard. Mold grows quickly on damp surfaces, especially when it's warm and dank, and it can cause serious respiratory illness.

The risks don't stop there. More than seven hundred U.S. cities and towns have "combined sewer overflow systems," in which stormwater runoff and sewage combine in the same outflow pipes and pollute streams and rivers with raw sewage, which carries about a hundred pathogens, including hepatitis A and E viruses, *E. coli, Campylobacter,* and norovirus. These pathogens, which sicken hundreds in the United

States each year, can then infect people while they fish, swim, or wade. They can contaminate drinking water sources. The most infamous of such outbreaks was a 1993 epidemic of cryptosporidiosis in Milwaukee that was caused by a protozoan parasite transmitted in fecal matter. It killed more than 100 people, most of them elderly or with weakened immune systems, or both, and it sickened 403,000. To prevent such outbreaks in the stormier, flood-prone years to come, it's essential to repair the country's crumbling water infrastructure.

MONSTER STORM, LASTING DAMAGE

On a sunny Thursday morning in January 2008, Juan Almendares gathered with about two dozen woman and one man from the Honduran Peace Action Committee, the grassroots organization he runs that helps poor Honduran communities obtain social services and restore their local environment.

The women had made their way by foot and bus from poor neighborhoods across Tegucigalpa and Comayagüela to the headquarters of CPTRT, Almendares's torture-prevention center, to meet with Almendares and to tell their stories to a visiting American journalist and photographer. Posters around the office offered reminders of the critical human rights work done here. *Los detenidos tienen derechos.* (Prisoners have rights.) *Para prevenir el violencia, predique con el ejemplo.* (To prevent violence, lead by example.) Gradually the group filtered in and gathered on plastic chairs on a shaded stone patio at CPTRT's headquarters.

As a twenty-something Canadian intern served snacks—flatbread with papaya and banana—on paper plates, the local women sat in a circle on white plastic chairs and chatted. But when Almendares walked in, their voices quieted in respect for the man they simply call *El Doctor*. Almendares leaned on a table, his body relaxed, his dark, seen-it-all eyes welcoming each visitor. *"Buenos dias,"* he said warmly. He introduced the American visitors and asked the local women to speak about their experiences during Hurricane Mitch *con confianza*—with trust and familiarity. He nodded over to the visitors and said in Spanish, "They are our great friends now."

One woman after another from the *colonias* got up and described how, over a few days in 1998, Hurricane Mitch had upended their neighborhoods, their families, their health, their lives. A woman named Carmen said the houses in her hilly neighborhood sank into the ground: "We were living with contaminated water but, thank God, it's now much better." The cars on the narrow city streets outside rumbled by, and the air smelled of car exhaust.

A woman from Colonia Ramon Amaya Amador described the flooded streets, the four hundred families who'd lost everything, the residents who had worked so hard to rebuild their own homes. Irma Yolanda Vargas, a community leader from a *colonia* called Campo Cielo, lost her home in the floods, as well as the land it sat on. "For twenty-three days [after the storm], families were surviving on very little water. They had very little help," she said. A woman named Maria described how, in the weeks following the storm, respiratory and skin diseases were widespread in her neighborhood. Although the particulars of each woman's story varied, the common threads were dirty water, lost shelter, cold, and sickness.

All of the people at the meeting had put their lives back together. But for some the damage had been long-lasting, and, a decade later, they still lived in conditions that were far from healthy (figure 29). A woman named Yolanda Lucas Rodriguez obtained a government loan after Mitch to rebuild her house, but it didn't go far enough, and soon after, she lost her factory job. Today she's still scraping together what money and materials she can to rebuild it little by little. So far, there's just a wall in the front and a wall in the back, and still no kitchen. "I need a new roof. Even now when it rains, I get wet. I'm hoping somebody listens. I'm not a whiner. I really need it," she said. The crowd applauded.

In some neighborhoods, there was still no drainage, no sewer system. Of the hundreds of neighborhoods in Tegucigalpa that had lost access to public drinking water supplies during Mitch, three hundred still didn't have it in January 2008, almost a decade after the storm.

The situation in those neighborhoods illustrates how health hazards can linger after rescue crews have left. This is particularly true in poor nations, such as Honduras, or vulnerable cities, like New Orleans, that failed to receive the resources they needed to recover. And it could be true of the more intense storms expected in coming decades.

To measure the lasting damage of Hurricane Mitch, Almendares conducted a pilot study to document whether Hurricane Mitch, in combination with social and economic changes, left Honduras more vulnerable to the health effects of climate change. Almendares and his assistants, who were Peace Action Committee members and community leaders, surveyed the residents of nine *colonias* on the outskirts of Tegucigalpa.

Supervised by Almendares, each community leader conducted a survey in her home community, and each participated in a focus group.

FIGURE 29. Residents of Colonia Soto in front of their apartment building, which was damaged in Hurricane Mitch. The storm's torrential rains swamped the neighborhood; the dirt road in front of their apartment turned into a torrent, and not far down the road, a steep hill collapsed in a mud slide, taking three hundred homes with it. A decade later, the building still hadn't been repaired, residents still struggled, and help was still in short supply. (Photo by Jason Lindsey/ Perceptive Visions)

Eighty-five percent of the people surveyed perceived an increase in new diseases after Mitch; 71 percent believed people were recovering more slowly from common illnesses. Almendares's team also collected people's perceptions of the local environment. At least 75 percent of them reported that water was more expensive following Mitch and that the incidence of landslides, fire, and dust—all of which can cause injuries and disease—had increased (figure 30).

Almendares and his coworkers also documented social problems that can affect mental health. Eighty-eight percent of the people surveyed had their homes damaged or destroyed; 77 percent had lost jobs. The average

FIGURE 30. Residents get their daily supply of water in Colonia Soto, a poor neighborhood in Comayagüela, Honduras. As climate change brings increases in both flooding and drought and as populations rise, supplies of clean water may grow scarcer. (Photo by Jason Lindsey/Perceptive Visions)

income in these communities in 2006 was $75 per month. Perhaps as a result of these challenges, 60 percent had had a family member emigrate to the United States or another country to earn money to send back to the family.

Almendares told the North American visitors that the problems continue. Although he and the women of Peace Action Committee had asked the government for help, many families in the *colonias* still lived in dangerous places where two to three days of rain could once again destroy their houses. They still often lacked water and adequate sanitation.

As they spoke, the Peace Action Committee members praised Almendares for his help in rebuilding their lives. Yolanda Vargas described how, after her family lost their home and the land it sat on, Almendares helped them find new land on higher ground where she could grow a garden to supply her family with food. "Thanks to God for the help of the doctor. He has helped us very much."

Almendares's calm demeanor did not change while the women heaped praise on him. Instead he deflected the credit back to the group. "It's important to recognize that they have worked together like a family, like a community," he said.

KATRINA'S LEGACY

On a hot day in July 2008, twenty-six-year-old Teddy Thomas unloaded groceries and his baby son from his car outside the FEMA trailer his family had been living in since the hurricane.

After Katrina's floodwaters subsided, the Federal Emergency Management Agency (FEMA) had distributed tens of thousands of the trailers to house people "temporarily," for what turned into years. Shortly after moving into the trailers, many residents reported bloody noses, rashes, respiratory problems, and other ailments. Two years later, the federal government admitted that the trailers contained dangerously high levels of formaldehyde, which can cause all of these problems, and cancer as well. By 2008, the government had ordered residents to vacate all FEMA trailers, but Thomas's neighborhood in the Lower Ninth Ward was still dotted with them.

"Of course we're worried, especially since we have an infant," Thomas said. "But if you have nowhere else to go, what are you going to do?"

The trailer crisis illustrates how pernicious and often unanticipated public health hazards arise after hurricanes, even in developed countries, when people are displaced and cast on the mercy of governments or charities unable to deal with the burden.

Almost five years after the storm, many Katrina survivors still suffer from the trauma, which is reawakened each fall by threats of new hurricanes. Several months after Katrina, rates of both serious and moderate mental illness had more than doubled from prestorm rates. Children were especially affected, according to a 2008 study by a Tulane psychiatrist: one in ten kids had a friend or relative killed during Katrina, and one in four was separated from his or her caregivers.

Nor did Katrina survivors necessarily bounce back, according to a follow-up study by Ronald Kessler of Harvard Medical School that was done two years after the storm hit. For example, shortly after Katrina, people had low rates of suicidal thoughts and plans despite their depression and despair. But in the two years that followed, suicidal thoughts and plans more than doubled, especially among those living outside the New Orleans metropolitan area. "Usually after two years, the vast majority of people have recovered, and that's not the case here," Kessler said.

Sitting in a slice of shade next to an abandoned yellow house in the Lower Ninth Ward in July 2008, Marguerite Burke, a stoic woman with a leathery face creased beyond her sixty-four years, said she was just trying to take it day by day and keep the events of the previous three years from killing her. She managed to move back into her nearby home, but most of her neighbors were gone.

"I didn't have insurance; I lost everything," she said. "But I have emphysema [and] I can't get my nerves too upset, so I try not to think about it."

The Ailing Earth

The 1997–98 El Niño that prompted our workshop in Mozambique turned out to be the most powerful of the twentieth century. By the time it drew mercifully to a close in May 1998, the extreme weather that came with it had caused at least $33 billion in property damage and killed thousands. The Horn of Africa had received forty times its usual rainfall, destroying roads, isolating villages, and triggering a series of epidemics. Reports of malaria spiked. More than eighty-five thousand people contracted cholera in East Africa, far more than previous years, and more than four thousand people died. An epidemic of Rift Valley fever infected eighty-nine thousand people and killed almost one thousand.

In the spring of 1998, I was coteaching a class called Global Environmental Change and Human Health with James J. McCarthy, a biological oceanographer and Harvard colleague. As part of the class, we launched a project. We wanted to understand the extent to which extreme weather caused disease outbreaks and other health problems. Several scientific reports had demonstrated that heavy rainfall and drought triggered outbreaks of a variety of infectious diseases. Nothing breeds extreme weather like an El Niño, so we focused on the pattern of disease outbreaks during the one that was still unfolding.

The dozen or so students in the class combed through bulletins of disease outbreaks put out by health agencies to discover when and where outbreaks had occurred. They tapped meteorological data worldwide to determine where heavy rainfall or drought had occurred. As with our study of coastal oceans, we used geographic information systems to portray outbreaks on a world map, then overlaid the map with different shades representing wet and dry areas. Our center published the report, with the map, in 1999.

The map revealed outbreaks of cholera, Rift Valley fever, malaria, dengue fever, and hantavirus pulmonary syndrome clustered in areas of flooding and drought, but few in regions of the world where weather had not been extreme. Hard-hit areas included South and Central America, eastern and southern Africa, China, and Southeast Asia. What's more, drought in Southeast Asia had led to massive forest fires whose plumes of smoke led to respiratory ills in hundreds of thousands of people. The results showed clearly that extreme weather events were often followed by clusters of health problems (figure 31).

As public health professionals, we seek to prevent health problems from arising. If we know a health hazard exists, we try to put in place preventive measures to keep people healthy. Understanding the links between extreme weather and disease was just a first step toward prevention. In various professional and public forums, I began to urge my colleagues to test early health warning systems.

Such systems would encompass the sort of warning system Ken Linthicum had developed to predict the extreme rains that can lead to Rift Valley fever. But to advance this approach, we would need to go further.

We'd need to integrate surveillance of diseases and disease vectors and store the data in a centralized and searchable repository, rather than rely on the current system of isolated reports. Such a central database would allow scientists to map outbreaks across the landscape to identify geographic clusters of disease. The database would need to include information on weather and climate, as well as social conditions, such as population distribution and the level of trained health personnel in an area.

We'd also need to incorporate programs that monitor the health of ecosystems on land and at sea, including watersheds and coastal zones. We'd need to track levels of disease-carrying organisms, including mosquitoes and rodents on land and algae at sea, for their populations respond rapidly to changing environmental conditions. These organisms could serve as biological indicators (bioindicators), whose numbers and distribution indicate the odds of disease outbreaks. This could be done, for example, by satellite monitoring of oceans to detect algal blooms, and research vessels could sample coastal oceans to detect noxious species and the pathogens that accompany them.

LEGEND

||| abnormally wet

⫴ abnormally dry

□ hantavirus pulmonary syndrome

▨ respiratory illness due to fire and smoke

△ cholera

■ malaria

○ dengue fever

◉ encephalitis

◉ Rift Valley fever

FIGURE 31. El Niño and disease. Outbreaks of infectious diseases carried by mosquitoes, rodents, and water often follow storms and floods. Droughts also lead to waterborne diseases, dengue fever, and the harmful health effects of fires. The events here occurred in 1997–98, during the twentieth century's largest El Niño. [Redrawn from map by Bryan Christie for August 2000 issue of *Scientific American* and used with permission]

We'd also need to monitor the health of animals that harbor diseases that can jump to humans, such as the pigs and birds that harbor swine flu and the crows that harbor West Nile virus. In this way we'd be more prepared for newly emerging diseases, since most of them are caused by microbes that make the jump from animals to humans. By combining these forms of disease surveillance, we could generate warning systems to spot potential epidemic trouble spots months early and mobilize medics and medicine to the area.

Despite the tremendous promise it holds to prevent disease, such a fully integrated health monitoring system has yet to be built.

. . .

By the time we wrapped up our study of the 1997–98 El Niño, the scientific tussle about climate change's effects on infectious disease had reached a fever pitch. We had begun to alert the medical community of the infectious disease risk, with a chapter in the landmark 1996 IPCC report.

At the same time, however, another group of scientists was denying the connection between climate change and disease. They made some valid points. Outbreaks putatively caused by climate change could be linked to other factors. Travelers can carry an infection from one country to the next, and a poor community is more likely to experience an outbreak of that virus than a richer one. A region with ill-prepared health officials could suffer more frequent and more severe outbreaks than a region that employs the preventive tools of modern public health. Newly drug-resistant microbes could spread despite public health measures that previously controlled their drug-susceptible cousins.

But this group took their argument one critical step further. They argued that because these factors could lead to a particular disease outbreak, it meant that we could not pin any outbreak on climate change. This was incorrect. Just as heart attacks can be simultaneously caused by high cholesterol, genetic disposition, and excess stress, disease outbreaks too have multiple causes. A declining public health infrastructure may raise the odds of an outbreak, and climate change can raise the odds further. And the more climate changes, the more conditions favor the spread of infectious diseases. By focusing on what we can't prove and what we don't know, this group has delayed action on what we do know: that climate change means more extreme weather, and extreme weather raises the risk of disease outbreaks.

In the summer of 1999, I accepted an invitation to speak before an expert panel at the Institute of Medicine, an influential organization created to provide unbiased scientific advice to the U.S. government and the general public. Policy makers all over Washington use the findings in the institute's reports to craft policy, and courts use them to decide important cases.

I was heartened at first by the invitation because it suggested that the Institute of Medicine was committed to undertaking its first serious examination of the health threats presented by climate change. I hoped to persuade my colleagues to accept that climate change was contributing to the spread of vectors and, with them, infectious disease.

I had critical new data to bolster my case. Just a few days before I was scheduled to speak to the expert panel, *Science* had published our map linking downpours and droughts to outbreaks of more than half a dozen infectious diseases. In the same issue of *Science,* Ken Linthicum's team described an example of such an outbreak, linking the year-to-year climate changes of El Niño to outbreaks of Rift Valley fever in Kenya, as described in chapter 3.

On a sweltering July day in 1999—part of a three-week heat wave that was helping to rev up West Nile virus in Queens, New York—I presented our group's results to my colleagues. I made my case, flew home to Boston, and hoped for the best.

Over the next year or so, my colleagues heard from other experts and deliberated on the evidence. Their report, titled *Under the Weather,* came out in 2001. Unfortunately, I'd been overruled, and the panel, while acknowledging the connections between climate and disease, had taken what I viewed as an overly cautious tone. "The potential disease impacts of global climate change remain highly uncertain," it concluded. On it went in a similar vein, highlighting what scientists could not prove and what we did not know. Since the report came from such an influential panel, it meant that those who opposed laws or rules to lower CO_2 emissions would be able to cite the uncertainties highlighted by the report to justify inaction. The outcome was hugely disappointing, perhaps to some of the authors themselves.

BACKSLIDING WITH BUSH

As the twenty-first century began, the international community of nations, especially those in Europe, was forging ahead to address the forces driving climate change. The United States, however, had just elected George W. Bush as president.

As a young man, George W. Bush had followed the footsteps of his father, George H. W. Bush, into the oil business, and by the time he ran for president, his financial gains in that business and as owner of the Texas Rangers baseball team had made him a millionaire many times over.

During the presidential campaign, George W. Bush had tacked to the center on climate change, promising "mandatory reduction targets" for carbon dioxide from power plants. But when Bush selected another oilman, former congressman Dick Cheney, as his running mate, the die was cast regarding his administration's policies on climate change and energy.

Cheney was an old Washington hand who had served as a congressman from Wyoming and had run the Pentagon under President George H. W. Bush. When George W. Bush selected him, Cheney was serving as chairman and chief executive officer of Halliburton, a multinational oilfield services company, where he had earned tens of millions of dollars. Under Bush, he would become one of the most powerful vice presidents in United States history.

By the time Bush and Cheney took office, there was no longer any doubt among mainstream scientists that climate was changing in response to human activities. The debate had become focused on how much, how fast, where, and to what extent climate change was taking place. But less than two months after taking office, as moderate Republicans in the Senate were preparing legislation designed to limit carbon dioxide emissions from power plants, Bush reversed his campaign pledge and opposed it at the urging of Vice President Cheney.

Meanwhile, Bush charged Cheney with developing a new energy policy for the nation, and Cheney wasted no time convening a task force to do so. Although Cheney fought tenaciously for years to keep the list of his advisers secret, a remarkable series of lawsuits and investigations gradually revealed who they were. They included executives from Enron, ExxonMobil, more than a dozen oil and drilling companies, major utilities, and three dozen industrial trade associations. The task force met just once with representatives from environmental groups. At that point, the report had already been drafted.

Cheney's energy task force recommended the construction of new oil refineries, several hundred thousand miles of oil pipelines, tens of thousands of miles of new gas pipelines, and a slew of new power plants, most of which were to be powered by coal or oil. The task force warned that if their recommendations were not followed, there would be an energy shortage that would "inevitably undermine our economy, our standard of living, and our national security." Cheney told the Associated Press that year, "Conservation may be a sign of personal virtue, but it is not a sufficient basis for a sound, comprehensive energy policy."

Bush and Cheney remained intimately tied with the fossil fuel industries, actively belittling and quashing reports of the likely effects of climate change. In 2002, the Bush administration, at the urging of ExxonMobil, ousted the chairman of the IPCC, climate scientist Robert Watson, who had spoken very strongly about the need for action to combat global warming. The same year, President Bush dismissed a carefully researched report from his own Environmental Protection Agency that contained dire predictions of harm from unabated climate change. Bush said it had been "put out by the bureaucracy."

CLIMATE CHANGE FUTURES

The next year, 2003, our center launched a large two-and-a-half-year study of climate change unlike any that had preceded it. The study was somewhat akin to our earlier studies on the oceans and extreme weather in that it relied on systems thinking and integrated data from a variety of disciplines. This time, however, we brought together scientific experts and representatives from businesses, nongovernmental organizations, and UN agencies. Everyone involved accepted that climate change was occurring and that it would affect human health, ecosystem health (via pest infestations and in other ways), and economies, and we sought to examine how the impacts might play out.

The study was sponsored by the reinsurance company Swiss Re (an insurance company that insures other insurers) and the United Nations Development Programme. The goal was to project how climate change and loss of biodiversity would affect human health worldwide. We kicked off the study with a two-day conference at the United Nations headquarters in Manhattan.

In projecting the effects of climate change, the IPCC and other scientific bodies have used scenarios of climate trajectories, primarily based on the pace of carbon emissions. Scenarios are more than simple projections. When done well, scenarios use the best information available to create plausible alternative stories of how the future might unfold. Scenarios are therefore helpful for dealing with uncertainty and making policy.

IPCC emissions scenarios are useful but limited in that none of the currently relevant scenarios diverge in average global temperatures until after 2035. To plan for the coming decades, scenarios of the impacts of climate change can complement scenarios of emissions. In the realm of public policy, impact scenarios can be especially helpful for short- and medium-term planning and for the assessment of losses, both insured and uninsured.

We developed two scenarios that highlight possibilities not adequately considered in past assessments. Both envision a climate with gradual warming and growing variability of weather, with more weather extremes. Our report, which I coedited with Evan Mills of the Lawrence Berkeley National Laboratory, was released in November 2005. The two scenarios tell different stories of how biological systems may respond to ongoing climate change.

Our first scenario specified that gradual climate change would lead to escalating impacts. It projected a world with extreme weather events that are more frequent and also tend to cluster in the same place or occur at the same time (something insurers are very concerned with). Under this scenario, most glaciers would retreat and the water cycle would accelerate. Together, this combination would undermine water supplies in some regions and land integrity in others. Permafrost in the Arctic would melt much more than it does today, threatening native peoples and northern ecosystems and releasing stored methane. Gradually rising seas, compounded by more destructive storms, would cascade over deteriorating barrier reefs, threatening low-lying coastal regions worldwide.

Our second scenario was worse. It explored a future in which warming and variable weather pushes ecosystems into accelerated change and past tipping points, with surprisingly destructive consequences.

Such changes would severely compromise resources and ecological functions in land-based and marine systems. Drought and pests would cause widespread diebacks of temperate and northern forests. Coral reefs would collapse from warming, ocean acidification, and disease. Severe storms would occur at the same time, or in succession, across the globe, overwhelming the adaptive capacities of even developed nations. Some developed nations would be humbled into conditions resembling those in developing countries for extended periods.

Our second scenario, in particular, contains changes that would rattle the sense of stability we've developed as a civilization, and for the insurance sector, it means imagining unmanageable risks. Even so, our two scenarios were not the worst possible outcome.

THINKING THE UNTHINKABLE

A few years ago, the director of a little-known but influential division in the Pentagon released an eye-opening report. Andrew Marshall directed the Office of Net Assessment, which functions like a strategic think tank within the Pentagon. Marshall had served every president since Richard Nixon, and he was legendary in Beltway circles for conducting one clear-eyed, penetrating analysis after another, some of which changed important aspects of U.S. military policy. President George W. Bush had drawn on Marshall in 2001 to lead a review of the Pentagon's capacities. As Stephen P. Rosen, a professor of national security and military affairs at Harvard University, has said, "Every Secretary of Defense for twenty-five years, regardless of party, has kept Andrew Marshall close to him, because Marshall spoke truth to power."

Just as President George W. Bush was launching his campaign for reelection, Marshall took the unusual step of sending an unclassified report to reporters at *Fortune* magazine. He had commissioned the report, published in October 2003, from the Global Business Network, a consulting firm that has pioneered the use of scenarios to help the military, corporations, and global organizations plan for an uncertain future.

"The purpose of this report is to imagine the unthinkable," wrote the report's authors, Global Business Network analysts Peter Schwartz and Doug Randall. The title of the report, "An Abrupt Climate Change Scenario and Its Implications for United States National Security," barely hinted at the frightening scenario its authors had described within.

The choice of title made clear that the authors were not talking about the climate change of conventional wisdom. For decades, scientists had assumed that climate would change gradually and steadily, giving humanity time to adapt, an idea reflected in the benign-sounding term *global warming*. And there is indeed evidence for steady change. Measurements of CO_2 from atop the dormant volcano Mauna Loa in Hawaii show a steady rise in the gas's concentration since a monitoring station was installed there in 1958. Average temperatures rose fairly steadily through the twentieth century, hiking the global average temperature about 0.75°C (1.4°F). Sea levels have been creeping higher, now by three centimeters (a bit more than an inch) each decade. As late as 2000, the IPCC issued a report with scenarios for projected climate change that assumed the world would warm steadily.

Schwartz and Randall's report, however, had news for the Pentagon: "This view of climate change may be a dangerous act of self-deception."

Evidence came from a growing body of science. Two years earlier, the National Academy of Sciences had issued an extremely sobering report called *Abrupt Climate Change: Inevitable Surprises* that extensively reviewed research on ancient climates and concluded that "climate had changed much more rapidly—sometimes abruptly—in the past and therefore could do so again in the future." Schwartz and Randall drew from that report, interviewed a series of leading climate scientists, and developed a new scenario that they honed by bouncing repeated versions off the scientists. To be sure, it was a worst-case scenario meant to help the Pentagon plan for contingencies. But the scientists told Schwartz and Randall that every element of it was plausible.

The scenario began in 2004, as the global warming of the twentieth century continued. But it really kicked in starting in 2010, when the world's warming climate hurtled over a precipice.

In 2010, large currents in the Atlantic Ocean would change abruptly, Schwartz and Randall wrote in their scenario. As the Gulf Stream moves along the surface of the Atlantic, it carries warm water northward. Some of it evaporates along the way, warming the nearby land, including the East Coast of North America as far north as Newfoundland. Meanwhile, the remaining water in the current cools and becomes saltier, which makes it heavy. The current splits into two arms, and each plunges to the bottom of the ocean, one near Iceland and one near Greenland's southern tip. Like a factory conveyor belt, the overturning cold, salty water flows south in the ocean depths, pulling the warmer surface water—the Gulf Stream—north to replace it. Warm water heading north warms the U.S. Northeast and allows residents of Dublin, London, and Oslo to bask in temperate conditions, even though they're at the same latitude as Anchorage.

Melting and shedding of Arctic and Greenland ice, however, has spread a layer of buoyant cold freshwater on the surface of the North Atlantic. In the Pentagon planning scenario, that fresher surface layer slowed the conveyor's usual swift dive, which in turn slowed the flow of warm water north. This could happen as more ice in the Arctic melts and more rain continues to fall in northern latitudes because of the evaporation of heated tropical seas. The far North Atlantic is already becoming measurably less salty, and recent studies suggest that the ocean conveyor, while variable, may already be slowing down.

In the Pentagon scenario, the conveyor crosses a tipping point and abruptly shuts down. This actually happened about 12,700 years ago, according to ice core records, when the planet was emerging from the last ice age. At the time, the thick ice sheet covering North America was dwindling, and climate researchers suspect that a vast lake of meltwater from that ice sheet, covering most of the present-day Canadian province of Manitoba, suddenly broke free and poured into the Atlantic. That freshened the water so much that the conveyor shot straight across to France, as indicated by marine algae called foraminifora found in ancient seabed cores. This reversed the world's warming, suddenly throwing it into a 1,300-year-long period of icy climates known as the Younger Dryas.

The scenario's authors presume a similar sequence of climatic events—warming, this time powered by the human-made changes to the greenhouse, is followed by rapid cooling in just a few years. The sudden chilling of the climate in 2010, according to the scenario, is catastrophic.

In the years following 2010, the climate cools markedly through much of North America, Europe, and Asia. In the United States, agriculture is hit hard, as growing seasons become shorter in the Northeast, rains grow scarce in the Southwest, and the South becomes drier. Europe is pounded by winter storms, and its winds intensify. "By the end of the decade, Europe's climate is more like Siberia's," the authors wrote. People begin to migrate south from Scandinavia and other parts of northern Europe in search of warmth.

Freshwater supplies worldwide become scarcer in many regions, and decades-long droughts grip key farming areas in eastern North America, Europe, and southern China. Crop yields fall 10–25 percent worldwide. China struggles to feed its huge population and suffers widespread famine.

Although its capacity is markedly diminished, the United States has the resources to remain self-sufficient, but it must secure its borders, using the military to keep out starving immigrants from Caribbean islands, Mexico, and South America. The United States keeps all the water from the Colorado River, provoking tensions with Mexico. By 2025, the European Union nears collapse, and dire conditions in China lead to civil war.

The authors dispassionately describe the Earth's carrying capacity for humans—the ability of its ecosystems to support human life. "As abrupt climate change lowers the world's carrying capacity, aggressive wars are likely to be fought over food, water, and energy. Deaths from war as well as starvation and disease will decrease population size, which over time will re-balance with carrying capacity."

In other words, civilization as we've known it would change forever. In its place would develop a world of resource haves and have-nots, a poorer world, a world of massive human suffering.

FLIPPING THE SWITCH

So far, man-made global warming hasn't pushed the climate to shift suddenly, as described in the Pentagon scenario or the National Academy's 2002 report on abrupt climate change. Instead, our climate has warmed gradually and developed altered patterns of precipitation. But clues unearthed by paleoclimatologists, researchers who study ancient climates, suggest that we'd be very unwise to push our luck.

Paleoclimatologists use a variety of indirect measures, or proxies, to determine what past climates were like. Trees grow more during years when it's warmer and wetter, generating thicker annual rings. Scientists can drill cores in a tree and count rings from the outside in, to determine a tree's age. By cross-checking dates with a ring's thickness, scientists can generate a record of past climate.

Similarly, scientists can drill cores in the ice sheets of Greenland and Antarctica, which are up to three kilometers (two miles) thick, to obtain snapshots of past atmospheres and past climates. These ice sheets build as year upon year of snow accumulates without melting, trapping air from that year's atmosphere and preserving it in annual layers as snowfall hardens into ice.

Researchers determine a temperature for a particular year by testing the year's air sample for two isotopes of oxygen. Isotopes are variants of a chemical element that have slightly different weights; for example, oxygen-18 is slightly heavier than oxygen-16. The ratio of these two oxygen isotopes varies according to the temperature at which the ice formed, so measuring this ratio creates what scientists call a paleothermometer. By cross-checking dates of layers with the temperatures of each layer, scientists have charted how a region's temperature has changed over the hundreds of thousands of years of the ice sheet's existence.

Paleoclimatologists have learned a great deal about the behavior of climate from such ice cores. In the early 1990s, an international team of researchers drilled a three-kilometer-long (1.9-mile-long) core of Greenland's ice sheet and took readings of each layer. The temperature was amazingly constant for the past 10,000 years—the Holocene era, in which human civilization arose. In contrast, temperature during most of the previous 250,000 years had been marked by a series of sharp dips and rises (with the exception of one 10,000-year warm, interglacial period from 130,000 to 120,000 years ago). The stability of climate in our era—the Holocene—is an exception, not the rule.

Such studies have also demonstrated that the climate can change surprisingly fast. In the ancient past, some regions have warmed as much as 16°C (29°F) in a decade. Also, weather patterns can flip into a new state in as little as three years, according to a study of Greenland's ancient climate. Such climate flip-flops happened as our planet emerged from the last ice age, and temperatures flickered up and down rapidly. Most of the warming that brought the Earth out of the last ice age happened in just two quick bursts: one 12,800 years ago and another 11,500 years ago. Indeed, this sort of behavior seems to be a property of systems in general: stretches of stability, then bursts of rapid change when systems are pushed past a tipping point.

There are at least two ways climate could change very rapidly. One is through calving and melting of icebergs and other increases of meltwaters from the Greenland ice sheet, which could cause a sudden change in ocean currents, as in the Pentagon planning scenario. A second way is by a catastrophic release of methane, a greenhouse gas that is shorter-lived than CO_2 in the atmosphere but more powerful in its warming effects. (Unfortunately, both could happen; we are approaching both tipping points at the same time.)

Methane is the main component of natural gas, which today comes largely from rotting landfills and the belches of cattle and sheep. But fifty-five million years ago, a huge release of underwater methane caused temperatures to jump 8°C (14°F), leading to a mass extinction. By looking at different isotopes of carbon atoms preserved in seafloor sediments, scientists discovered a massive release of carbon-12, the lighter of the element's two isotopes. The only possible culprit, most researchers agree, would have been a sudden melting of methane hydrate, a solid form of methane that had combined with water at high pressure, low temperatures, or both. Methane hydrates can accumulate in pockets under sediment, in the deep sea, and under frozen seabeds in the Arctic. Today there are already ominous signs that Arctic Ocean stores are being released as warming waters cause the sea-bottom sediment to thaw.

Fifty-five million years ago, an increase in volcanic activity or change in ocean currents or temperatures may have pushed these methane stores past a tipping point. As a bit of the gas escaped, it presumably warmed the planet, causing further release of the gas and more warming. This positive, amplifying feedback mechanism could kick in again in the not-too-distant future: Carbon dioxide levels of 450 parts per million, which we could reach in less than three decades, could warm the Earth enough to unlock large stores of methane in tundra and under seabeds. This could release enough methane from permafrost and the thawing seabed to create the equivalent of 1,000 ppm of carbon dioxide in the atmosphere, according to physicist Joseph Romm of the Center for American Progress. That would lead to much more warming than we'd expect from our CO_2 emissions alone.

A LIVING EARTH

Are we pushing our planet toward a tipping point? Just as scientists can measure oxygen in trapped ice-core air bubbles, they can also measure carbon dioxide. By analyzing ancient Antarctic ice cores using this method, we've learned that the current CO_2 level, about 390 parts per million and rising, is higher than it's been for at least eight hundred thousand years.

By adding CO_2, methane, and other greenhouse gases from human activities to the atmosphere, we have created a heat-trapping blanket. You could say that the planet has developed a fever.

This idea seems at first only a metaphor. How could the planet and its parts be akin to a human being and his or her organs and limbs, fingernails and hair? But the idea of a living Earth was more than a metaphor for an unassuming British chemist named James Lovelock.

Beginning in the mid-1960s, Lovelock developed a scientific hypothesis that presumed that the Earth was a giant superorganism—that it was alive. His neighbor in Bowerchalke, England, the Nobel Prize–winning novelist William Golding, author of *Lord of the Flies*, told Lovelock that the concept reminded him of the supreme goddess of Earth in Greek mythology. And so Lovelock's idea acquired its name: the Gaia hypothesis.

Early in his career, Lovelock demonstrated a knack for building precision instruments. The electron capture detector he built revealed widespread pesticide residues, paving the way for the wake-up call in Rachel Carson's 1962 environmental classic, *Silent Spring*. On a cruise to the Antarctic, he used the same type of detector to spot chlorofluorocarbons (CFCs)—gases used in refrigerators and air conditioners—in the air above remote reaches of the ocean. This discovery enabled other scientists to show that these chemicals were eating a hole in the ozone layer.

These accomplishments cemented Lovelock's scientific reputation and his role in the burgeoning environmental movement. This would have been enough for many scientists. And it seems unlikely that a designer of precision instruments—someone you'd expect to be wedded to the details, who might miss the forest for the trees—would also be a big thinker who floats controversial ideas. Yet Lovelock played both roles well.

Lovelock arrived at the idea of Gaia in a roundabout way, via a mental journey through the cold, dry, and apparently lifeless planet of Mars. In the early 1960s, NASA was assembling the Viking mission that would search for signs of life on the Red Planet. They hired Lovelock to help design precision detectors to look for residues from living creatures such as microbes, past or present. Other scientists involved thought to look for life by sifting through Martian dirt for molecules such as DNA, which stores genetic information, at least in Earthly life-forms. But, Lovelock wondered, what if Martian life didn't use DNA? Were there more general signs of life that the Mars missions could try to detect?

By considering the properties of life on Earth, Lovelock hit upon the idea of looking for signs of life, not in the dirt, but in the atmosphere. Earth's atmosphere is rich in oxygen, which reacts readily with other chemicals. This is why iron left outside soon rusts. According to basic chemistry and thermodynamics, almost all that oxygen should have reacted with iron and other minerals and disappeared long ago, creating a state called chemical equilibrium, in which the levels of the reactants no longer change. Lovelock knew it was living organisms—specifically the algae and green plants that carry out photosynthesis—that continually supply oxygen to the atmosphere, thereby pushing Earth's atmosphere away from chemical equilibrium. He generalized from that insight: other forms of life, even life very different from Earth's, would also push a planet's atmosphere away from chemical equilibrium.

When they examined Mars using images from satellites orbiting Earth, Lovelock and a colleague argued that its atmosphere was close to chemical equilibrium, which suggested there was no life on the Red Planet. The Viking mission went ahead in 1975, and in the end, the other scientists agreed with him: Mars then seemed devoid of living organisms.

Lovelock next turned his sights back to Earth. Our planet's oxy-

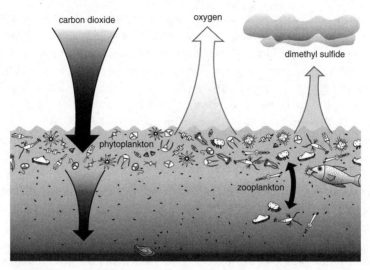

carbon dioxide

oxygen

dimethyl sulfide

phytoplankton

zooplankton

FIGURE 32. Plankton and climate control. When seas warm, populations of floating algae—phytoplankton—bloom, followed by zooplankton, the minute animals that eat them. When both types of creatures die and decompose, they release dimethyl sulfide, which seeds clouds. Clouds in turn reflect sunlight and generate rain, cooling the planet. Phytoplankton also draw down CO_2 and supply the air with oxygen in vast amounts.

gen levels not only had remained elevated but, according to geological evidence then available, had remained remarkably stable at about 20 percent of the atmosphere for almost two hundred million years. Could it be that the planet accomplished this through some kind of self-regulation, much as a healthy person regulates his or her blood sugar or blood pressure? Lovelock found solid evidence for one kind of planetary self-regulation.

When the temperature warms, marine algae proliferate, then die, emitting a compound known as dimethyl sulfide, or DMS. Tiny droplets of DMS then evaporate and float in the air over the seas, where they glom onto molecules of water vapor. This aggregate of water and DMS helps seed clouds, which reflect incoming sunlight and send rain down to the ocean, thereby cooling the atmosphere and the ocean below them (figure 32). A cooler ocean causes marine algae to grow more slowly, thereby keeping the algae population in check. In this way, living things work together with the inanimate environment—air, clouds, and seawater—to form a thermostat that regulates the temperature near the surface of the Earth. This is a classic example of negative feedback—self-correcting, dampening feedback that maintains homeostasis.

Lovelock also noted another such negative feedback that helps stabilize carbon dioxide levels. Carbon dioxide in the air reacts with water to form a weak acid called carbonic acid. Although not nearly as corrosive as acid rain, carbonic acid is still able to eat away at rocks, and, over geologic time, this reaction locks away carbon on the seafloor in the form of chalk. More CO_2 in the atmosphere heats the planet, which causes more evaporation, which in turn makes the atmosphere wetter. When it's warmer and wetter, CO_2 also reacts more quickly with rocks. This draws it down and locks it away faster, thereby curbing carbon dioxide's warming effects. This chemical weathering is part of a crucial feedback mechanism that helps stabilize CO_2 levels and the planet's temperature. Unfortunately for us, it takes millions of years to act—far too slowly to correct for all the greenhouse gases we're pumping into the atmosphere.

Lovelock realized that life on land could also play a role in maintaining Gaia. For example, plants' roots help break up rock, and roots make the areas around them moister, so they could speed up chemical weathering, drawing down CO_2 that much faster. The idea that life helps regulate these basic reactions of rocks, soils, and atmosphere, and so influences the planet's temperature, is now well accepted. Although Lovelock's hypothesis was ridiculed and attacked by many scientists for years, his basic insight about self-regulation has stood the test of time.

HOLDING GAIA STEADY

The planet as a whole doesn't seem to be alive, since it has parts that are living but also others—like rocks and air—that clearly are not. So instead of calling it a superorganism, it would be more appropriate to call it a supersystem—a large system with many smaller systems embedded in it, all interacting via positive and negative feedbacks. Still, the central idea of Lovelock's Gaia hypothesis, which has since been elevated to Gaia theory, is useful: The planet shares a key quality with living creatures, and that is the ability to regulate itself.

The Earth's feedback mechanisms that Lovelock took as evidence for the Gaia hypothesis—cloud-forming algal emissions and weathering of rocks—form negative feedbacks loops that help stabilize the climate. But the Earth also has positive feedback loops that amplify and reinforce small changes, potentially destabilizing the climate. These include the potentially catastrophic release of methane from frozen underwater stores, as previously mentioned.

A second important positive feedback involves sunlight glinting off ice. The Arctic Ocean's cap of sea ice and Antarctica's snow and ice cover reflect most of the incoming sunlight at the poles. But, like a black baseball cap that will warm your scalp on a sunny day, exposed areas of ocean are darker, so they absorb most of the incoming rays, warming the Earth. If the ice shrinks just a bit, the ocean absorbs more heat, which melts more sea ice, thus opening up more ocean, and so on in a self-reinforcing loop. The same goes for ice anywhere on the planet, from the glaciers of Mount Kenya and Asian and Andean mountain ranges to Greenland's ice sheet. When ice melts away, the darker land and ocean revealed underneath soak up more rays, heating the Earth. *Albedo* is the technical term for the reflectivity of the Earth, so this process is called the ice-albedo feedback.

Today Earth's albedo is about 30 percent, meaning that 30 percent of incoming sunlight is reflected back out into space, primarily by ice and clouds. If the Earth's albedo drops a percentage point or two, the climate system could reach yet another tipping point. We do not know the albedo threshold that would trigger runaway warming, but future examination of ice core records could yield some clues.

A third positive feedback involves water vapor, which acts as a powerful heat-trapping greenhouse gas, just like CO_2. Though water vapor is in the air only transiently before it rains back down to earth, it accounts for a remarkable two-thirds of the greenhouse effect. The atmosphere can hold more water vapor as it warms up—as mentioned previously, about 7 percent more per 1°C of warming. So, when the atmosphere warms a bit, evaporation increases, and the atmosphere can then hold onto this vapor longer before it falls as rain. So, more heat means more water vapor, which means more heat and still more water vapor, and so on. Our CO_2 emissions thus create much more of a greenhouse effect on Earth than they would on a perfectly dry planet.

A fourth positive feedback involves the world's ocean. Roughly a third of the CO_2 emitted from factory smokestacks and automobile tailpipes now dissolves into the oceans, leaving less of this greenhouse gas in the atmosphere than there otherwise would be. Warmer water can hold less CO_2 than cooler water, which is why a warm can of soda loses its fizz faster than a cold one. As CO_2 warms the planet, the oceans warm, then absorb less of that gas, and they release some of what they've already soaked up. Then the level of this greenhouse gas climbs, warming the planet even more.

Throughout our planet's history, these kinds of runaway positive feedbacks have been kept in check by stabilizing negative feedbacks. In this way, Earth's feedbacks maintain a stable climate for long periods of time. The ice ages that our planet have cycled through over the past two million years are great examples of this kind of regulation. They've been imperfect at keeping the planet's climate stable, yet great at keeping it livable—and, indeed, conducive to human evolution.

When all the science is considered, a picture emerges of a planet that has oscillated for probably two million years between two states: a cold state with large polar ice caps and about 180 parts per million CO_2, and a warmer state with medium-sized ice caps and about 280 ppm CO_2—where we stood before the industrial revolution. Today's atmospheric CO_2 levels and temperatures are outside the limits that they've maintained for more than two million years, and the negative feedbacks that stabilize the climate are being pushed beyond their capacity. We may even be headed for a third stable state with elevated CO_2 levels, much smaller ice caps, and significantly higher sea levels than we enjoy today. If we take some Gaian license, it's fair to say that the Earth has fallen ill.

WILD WEATHER AND THE ARCTIC DEATH SPIRAL

As levels of greenhouse gases climb, what are the odds they'll cause the climate system to flip? That question remains unanswered, although we do have some clues. To project changes to the climate from greenhouse gases, climate scientists use sophisticated computer programs called models. Models depict how we think parts of a system work together, and how much change in one part will influence the others. Today they're run on powerful supercomputers, and they have grown very accurate.

To create a climate model, scientists divvy up the world's air and oceans into a set of virtual three-dimensional boxes. Then they change some aspect of climate, such as CO_2 levels, and calculate what happens to the temperatures, air pressures, or water flow in each box. Next, they put these numbers into a mathematical formula that can quantitatively help predict what happens to the climate. For example, a model might calculate from physical principles how much heat from the atmosphere penetrates the ocean. Today's sophisticated climate models account for CO_2 as it moves through air, ocean currents, growing plants, volcanic eruptions, and more, using principles of physics, chemistry, and biology.

When temperature and precipitation measurements match model projections, including the impacts of the greenhouse gases humans create, we say we've found a "fingerprint" of human activity. And when many fingerprints appear, as they have, the evidence becomes overwhelming.

Before researchers attempt to predict the future, they verify each model by hindcasting—that is, by feeding the model past conditions and checking to see if it can accurately describe the present. Today's climate models can. They simulate the rising temperatures measured over the twentieth century, even including most of the bumps and dips over the decades due to rare events such as volcanic eruptions. If we omit the effects of greenhouse gases from the simulation, then the predictions are way off what we see occurring. That's how scientists can conclude confidently that the greenhouse gases we've added—in combination with natural factors such as orbital cycles and solar fluctuations—are causing the globe to warm.

In recent years, scientists have noticed that the planet is changing more, and more quickly, than even the best climate models had predicted. Scientists don't know why; it's possible that these models account for CO_2 but not all other greenhouse gases and feedbacks influencing the planet. It's also possible that current mathematical limitations keep the models from forecasting jumps and lurches that typify the behavior of real systems. Either way, it's worrying because it means that the models may do a poor job of predicting abrupt climate change—the kind of change we see so frequently in ice core records of the past.

Although climate models aren't yet up to that task, there are still characteristics of a system that may indicate how likely it is to change abruptly. The key is to watch for instabilities and variations. Just as stocks fluctuating on the Dow Jones may warn of looming big drops or leaps in the stock market, an increase in volatility may warn us that the planet's climate is becoming less stable.

This is already happening. Intense downpours are on the increase, as previously mentioned. The ice sheet covering Greenland is starting to slip away, and West Antarctica's is showing similar signs of instability. The permanent sea ice covering the Arctic has shrunk from 7 to 4.25 million square kilometers (2.7 to 1.6 million square miles), reaching record lows in the summers of 2007 and 2008. One expert says it's in a "death spiral," on pace to disappear as soon as 2030.

Also on the rise are novel events and outliers—extreme events that are so rare, they had only a miniscule chance of happening during our lifetimes. The European heat wave of 2003 was a major outlier, part of the hottest European summer since at least 1500. In 2004, an unprecedented five feet of rain fell in a day and a half on the Caribbean island of Hispaniola, home to Haiti and the Dominican Republic. The same year, the first-ever hurricane in the South Atlantic spun toward Brazil. Meteorologists hesitated to name the storm, since they thought its occurrence wasn't possible.

Such signs of instability may augur bigger climate surprises to come, and we need to keep close tabs on them. As scientists continue to examine climate history for clues to our planet's future, they wrestle with data to make accurate predictions. But as we push CO_2 levels higher than they've been for millions of years, the past can only be a rough guide to the future. We are truly in uncharted waters.

Gaining Green by Going Green

The view from the Rüschlikon Centre for Global Dialogue is breathtaking. Looking from the pavilion past manicured hedges and a spacious stone patio where visitors dine, one can see a panoramic view of the blue waters of Lake Zurich, with snowcapped Alps beyond. The effect on visitors to this gorgeous conference center, built on the grounds of a grand Swiss villa, was intentional, for visitors are invited to the center to explore global issues and think thoughts as sweeping as the view.

In June 2004, the global reinsurance company Swiss Re, which owns the retreat center, hosted one hundred scientists, business leaders, United Nations officials, and environment leaders for three days at the Rüschlikon Centre. The occasion of the conference was the midway point of our two-year Climate Change Futures project. We had expanded the scope of the project to allow deeper exploration of the projected economic effects of climate change and their links to health and ecological effects.

After a round of introductions by all the participants, I presented an overview of the project to the conference attendees, who sat in a ring of tables in a large, high-ceilinged pavilion with glass walls overlooking the lake. John Coomber, Swiss Re's chief executive officer, was next to step to the podium. For years afterward, Swiss Re employees at the conference that day would refer back to Coomber's speech, citing it as a moment that crystallized their worldview, describing how much it had motivated them.

In his speech, Coomber offered some big-picture perspectives on insurance to the guests, who included many from academia and non-governmental organizations who would not have been familiar with the workings of the insurance industry.

Insurers are risk professionals: they gauge the odds of a claim, they write and price policies to account for those odds, and they try to sell those policies at prices that give them financial rewards that outweigh their risks. In doing so, insurers provide several services to society. Insurance provides a way to spread risk over geographic areas and among industries, individuals, and communities, which provides safety, security, and social cohesiveness.

Insurance is also required for economic development. Banks and other funders won't often lend the serious sums of money for large ventures unless they have assurance that the borrower will be able to pay back the loan in case an unforeseen problem, human-induced or otherwise, prevents the builder from realizing the return on his investment. Without insurance, it would be difficult to build buildings, bridges, or much else.

Insurance companies keep large pools of capital available to pay claims when needed, Coomber said. But when extreme weather or other natural disasters strike, and thousands of people file insurance claims, the run of claims can bankrupt an ordinary insurance company. Insurance companies protect themselves by buying insurance against such a financial disaster from one of several large multinational companies, including Swiss Re, Munich Re, and others. That insurance is called reinsurance, Coomber said, and it serves as the insurance industry's financial shock absorber.

Coomber was a trained actuary, and he had worked in that capacity for years, climbing gradually to the top rung on Swiss Re's management ladder. As such, he understood numbers and he understood risk, giving him great credibility with the audience that day when he described which threats worried Swiss Re most.

He mentioned the demographics of an aging population, which would dramatically drive up the costs of health care and the claims that insurers would be called upon to pay. He mentioned terrorism, which could quickly ring up a hugely expensive run of claims, as it did in the aftermath of September 11, 2001. But, he told his audience, the number-one long-term threat, to the insurance industry and to society, was climate change.

THE COSTS OF CLIMATE CHANGE

When I began my career as a physician and public health practitioner, I'd focused on providing medical care and public health services to underserved communities, whether in Mozambique or Boston. I could hardly have imagined then that I'd one day be sitting in a posh Swiss villa, listening to the wealthy CEO of a global financial firm expound on his company's concerns. Yet, in retrospect, my central goal of contributing to the protection of public health had never changed.

As early as 1992, I'd received an important clue from the devastating 1991 Peruvian cholera epidemic that had opened the door to the manifold ways that climate change might harm human health. A year later, the same epidemic opened my eyes again. At the Earth Summit in Rio de Janeiro, only a few reporters had come to the press conference Eric Chivian and I had held on the health dimensions of climate change. One of them was from the *Wall Street Journal.*

Because of the epidemic, the reporter wrote, Peru had lost millions of dollars in shrimp exports and tourism revenues, both huge hits for a developing country. This demonstrated how climate change and its health impacts could profoundly affect economies.

To better understand the climate–economics connection, I began collaborating during the mid-1990s with forward-thinking leaders in the insurance sector. The first such person was Frank Nutter, president of the Reinsurance Association of America. In my discussions with Frank, I gradually realized that if you thought about the global economy as a living and dynamic system, then the financial industry, which includes insurers, may be thought of as its brain: it senses and integrates signals, including pain, from all sectors of the economy. And insurance loss data, it turns out, is one of the best indices to gauge how the extreme weather linked to climate change harms businesses and threatens economic stability.

The scenario of a regional heat wave illustrates that. As temperatures soar, people suffer from dehydration and heatstroke. Some die of heatstroke, prompting life insurance claims, but most recover after some time in the hospital, prompting health insurance claims. More smog forms in the heat, causing an upsurge of respiratory illness, and the combination of heat and air pollution prompts more heart attacks and strokes, prompting more health insurance claims.

Meanwhile, people turn up their air conditioners to escape the baking temperatures. The regional power grid is overwhelmed, causing a blackout lasting two days. Workplaces across the city shut down, prompting claims from business owners who are insured against business interruptions. Refrigeration fails in some grocery stores and restaurants, causing the loss of tens of thousands of dollars of perishable food—and more claims. The flood of claims from such business interruptions hits property and casualty insurers hard.

In nearby rural areas, crops wither in the heat, prompting farmers to file claims for crop insurance. Chickens roast and livestock are lost. Forests turn into tinder and ignite, and forest fires spread. The fires are gradually brought under control, but not before destroying fifty homes, whose owners file large claims with their homeowners insurance companies.

Such a scenario, based loosely on events during the 2003 European heat wave, illustrates that when extreme weather and natural disaster cause economic pain, societies and insurers feel it. But what happens when a disaster is so devastating that it sets the entire insurance industry reeling? That's what residents of South Florida discovered beginning one summer day in 1992.

A WINDY WAKE-UP CALL

In mid-August of that year, a tropical depression formed off the west coast of Africa. Over the next ten days, it moved ominously westward across the Atlantic Ocean, picking up strength from the warm seas below. By August 24, 1992, when it hit land near Miami, it was a ferocious storm with 275-kilometer-per-hour (170-mile-per-hour) winds, and it had a name: Hurricane Andrew.

Hurricane Andrew demolished entire towns, including Florida City and Homestead, and tore apart sections of Miami. It crossed over Florida, then made its way to the Louisiana coastline, where its rampage continued. More than seven hundred thousand people evacuated as the storm approached, undoubtedly saving hundreds of lives. Those who remained behind later struggled to describe the sound the storm had made: a freight train in the living room, a fighter jet on the roof. One man reported that the storm sounded like his house was being blasted by a giant shotgun.

Hurricane Andrew killed fifteen people directly, and twenty-five indirectly. It flattened the Homestead Air Force Base, forcing thousands of airmen to work out of a tent city for months afterward. A quarter of a million people were left homeless, and one hundred thousand residents of southern Dade County never returned.

Hurricane Andrew delivered a rude wake-up call to thousands of residents—both those occupying luxurious homes with gorgeous coastal views and those living in less-expensive housing on otherwise unwanted, swampy land. They couldn't just board up their windows and hope, as they had in years past. Their properties were terribly damaged at best, and at worse, completely destroyed.

Hurricane Andrew also delivered a rude wake-up call to insurance companies. The hurricane prompted a record seven hundred thousand claims from locales spanning thousands of miles, totaling $16 billion ($25 billion in 2010 dollars). Eleven insurers became insolvent, abandoning thousands of homeowners. Surviving companies struggled to pay these bills, turning to reinsurers for the money they needed to survive.

Over the next decade and a half, insurance companies scrambled to avoid another loss on the scale of Hurricane Andrew. Florida law mandated a ten-year moratorium that limited the number of policies that a company could drop in a given county at the same time. When the restrictions expired, companies reacted by trying to drop customers at high risk of hurricane damage. Companies that stayed in these markets hollowed out coverage through increased deductibles, reduced limits, and new coverage exclusions.

This wholesale retreat by mainstream insurers forced Florida's state government to step in, creating a joint underwriting association as a last resort for people unable to obtain commercial insurance. Despite premiums higher than the market rates, by 1996 nearly one million Floridians had enrolled in the initiative. The state also instituted stricter housing codes to dissuade people from building in the most vulnerable coastal areas, but development in these areas still continued at a rapid pace.

For more than a decade, Hurricane Andrew remained the most-expensive natural disaster on record. That changed in summer 2005, when Hurricane Katrina toppled and breeched the levees in New Orleans; its sister storms Rita and Wilma landed follow-up blows to the Gulf Coast, and insured property losses topped $61 billion. Insurers pondered how they could avoid suffering the same fate as the wrecked city of New Orleans.

. . .

The insurance industry relies on being able to calculate risks, then to spread those risks over numerous policyholders. Lately, they haven't been able to do this as well as in the past. In particular, they are ill-prepared to manage the skyrocketing worldwide losses from natural catastrophes.

In the 1960s through the 1980s, losses from natural disasters averaged approximately $4 billion a year, and only about 10 percent of that was insured. In the 1990s, losses jumped dramatically to about $40 billion a year, and a higher percentage of those losses was insured. In the 2000s, losses kept rising, to more than $225 billion in 2005, with more storms hitting the United States, Europe, and Japan, and $83 billion—a third of the total losses—was by then insured. In 2008, losses from extreme weather surpassed $200 billion, suggesting that the greater losses have become the new norm.

Since disasters cause runs of claims on property and casualty insurance, the insurance industry has invested a lot of brainpower to determine why weather-related losses have skyrocketed, as have academic researchers who study the industry.

The obvious suspect is climate change, and beginning in 2007, our center and the Insurance Information Institute co-organized the Catastrophe Modeling Forum, which brought together insurers, reinsurers, insurance brokers, and modeling firms that serve the industry. Our goal was to probe the link between climate change and weather-related disasters.

In our forum, we began with the understanding that, as with heart attacks or disease outbreaks, there are multiple factors that raise the costs natural disasters impose on insurance companies. More flood- and fire-prone areas had been developed in recent decades, and more people were living in such places. More were relying on vulnerable electric power grids, and real estate was getting costlier (at least until the housing market crashed in 2008), so it cost more to fix or replace what was damaged. Extreme weather events have also become more frequent and more severe.

Climate is now changing so rapidly that past temperature and weather-related statistics no longer serve as good indicators for future events. For example, based on past flood frequencies, so-called five-hundred-year floods are expected to occur only once every five centuries, on average. However, floods of that magnitude swamped Iowa and other parts of the Midwest in 1993 and again in 2008. Winter weather, too, is becoming more erratic in the United States, Europe, and Japan. Winter precipitation is more likely to fall as rain instead of snow, which can freeze and spawn destructive and dangerous ice storms. As my colleague Evan Mills of Lawrence Berkeley Laboratory has written, "History has shown that society in general, and insurers in particular, are often caught unprepared for ostensibly 'inconceivable' disasters. This reflects, in part, the recurring social miscalculation of using the past to predict the future."

While insurers profit from risk, the risks must be manageable, and some industry leaders are now well aware that climate change's unpredictable consequences could jeopardize their business. The chairman of Lloyd's of London joined John Coomber, the chief executive officer of Swiss Re, in calling climate change the biggest threat facing his company. Europe's largest insurer, Allianz, projected that "climate change will drive up insured losses 37 percent per year on average within a decade." The brain of the global economy, it appears, is feeling the pain of climate change.

A DOWNHILL PATH

How much economic pain will climate change deliver in the future? When my collaborators on the Climate Change Futures project—Swiss Re and the United Nations Development Programme—launched the project, most investigations of climate change's impacts had focused on the natural sciences, with less attention on human health effects and even less on potential economic impacts. However, forward-looking business leaders like those at Swiss Re had been considering this question for a while. "As a reinsurance company, our goal is to evaluate and plan for the long term," Jacques Dubois, former chair and chief executive officer of Swiss Re America Holding Corporation, has said.

We had organized the Rüschlikon conference in part to bring in more expertise to the project from the financial industry to help us forecast economic impacts. To that end, we'd invited thinkers and leaders from financial firms like Goldman Sachs and JPMorgan Chase, as well as leaders from other sectors, such as the oil company BP and the health care and pharmaceutical giant Johnson & Johnson. By mixing business leaders with a wide range of researchers from academia, UN agencies, and conservation groups, we hoped to gain a new integrated view of the health, ecological, and economic effects of climate change. After three days of breathtaking views and productive cross-fertilization, project participants returned home to finish their case studies and forecasts. Just over a year later, we had completed the study.

Each of the ten teams of experts had conducted a case study in its area of expertise, focusing on one health or ecological effect. In each case, the team also forecast the economic effects—of malaria, heat waves, forest degradation, and so on—and summarized how each of our two scenarios would play out, including for the insurance industry.

As mentioned previously, both of our scenarios assumed a business-as-usual approach, in which we continue to burn fossil fuels, drive gasoline-powered cars, heat our houses and supply our electricity with coal, and cut down and burn forests. If that occurred, carbon dioxide levels would reach 550 parts per million—an amount double preindustrial levels—by 2050. In both scenarios, climate continues to change as it has been: temperatures continue to warm and extreme weather becomes more common. But the responses differ.

In the first scenario, ecosystems respond gradually but remain intact. Even so, the economic damage is significant. With more frequent extreme weather, insurers' reserves are strained, and the more variable weather makes insurance harder to price, leading to more severe stress in the industry. A whole new class of health and life insurance loss arises from more extensive heat waves, reduced water quality, air pollution, food insecurity, and injuries and deaths from weather-related disasters. Warming-driven biological responses, such as the proliferation of bark beetles, increase the vulnerabilities to expanding fires. Extreme events cluster in time or place more than they used to, and combinations of events—floods and disease, for example, or heat waves and droughts—cause more damage than either would alone. When several extreme weather events hit in succession, insurers struggle to pay, prompting them to sell off huge blocks of assets, thereby disrupting financial markets.

In this scenario, struggling insurers react defensively by raising premiums and deductibles, and lowering limits. People demand that government step in to insure them, which it struggles to do, adding wildfire and windstorm insurance to the existing government flood and crop insurance. This scenario leads to greater economic instability than before climate change kicked in.

The second, harsher scenario involves the same degree of climate change as the first, but in this scenario, ecosystems buckle. The world is chaotic and uncertain. Killer heat waves occur two to four times as often as they do today, with death rates five times higher. Coral reefs weaken and teeter on the verge of collapse. Pests and pathogens run unchecked in ecosystems, and epidemics spread. Drought and stronger winds spread more forest fires. Whole regions are affected, and more events occur akin to what happened around Kuala Lumpur, Malaysia, in 2005. That summer, noxious smoke from forest fires descended into the valley in which the city lies, sending many people to the hospital with respiratory and eye ailments, temporarily shutting down the harbor and many businesses, and stopping flights to and from Subang Airport.

Even in developed countries, storms shut down businesses or the electrical grids on which they depend with increasing frequency. Insurers withdraw from markets, stranding development projects in the developing world. Travel, trade, and tourism all suffer big hits. Insurance covers much less than it did in the past and costs more; governments struggle to pick up the slack. Companies fold, and the insurance industry shrinks.

As we were finishing our assessment, the prominent British economist Sir Nicholas Stern, a former chief economist at the World Bank, was conducting a broad study of the economics of climate change, commissioned by then British Chancellor Gordon Brown. It reinforced the conclusions of Climate Change Futures and offered cause for both alarm and hope.

The Stern Review warned that under business-as-usual greenhouse gas emissions, there's a one-in-two chance the Earth could experience a 5°C (9°F) average temperature increase by the end of this century. The same temperature difference separates our current climate from the last ice age. This rise would trigger a host of extreme economic and social consequences, including the displacement of two hundred million people by rising sea levels, floods, and droughts; the extinction of up to half of all species; massive crop failures due to drought; and water shortages for one in six people worldwide. Even a more moderate rise of 2°C–3°C (4°F–6°F) could trigger some of these effects, as well as a collapse of the Amazon rain forest and a doubling of damages from hurricanes in the United States.

Such change would wreak havoc comparable to the great wars and economic depression of the twentieth century, Stern wrote. Stern and colleagues estimate that "business as usual" emissions could cause total losses from climate change of from 5 to 20 percent of global GDP, annually, meaning $1.75 to $7 trillion a year, with the world's total GDP roughly $35 trillion at the time the report was issued. This magnitude of loss would mean a dramatic reduction in well-being and quality of life, disproportionately hitting those who already are the world's most vulnerable and destitute.

But *The Stern Review* wasn't all gloom and doom. Though our current actions may have little effect on the climate in the coming three decades, Stern argued, what we do over the next decade or two will have profound effects for the latter half of the twenty-first century and the century that follows. Spending just 1 percent of the global GDP, or about $350 billion, a year could reduce emissions while creating economic development opportunities. "The world does not need to choose between averting climate change and promoting growth and development," the authors wrote.

Stern pegged realistic stabilization at between 450 and 550 parts per million CO_2. (Since 2007, NASA scientist James Hansen and others have revised this to call for a return to 350 ppm.) Stern concluded that the economic and social benefits of strong, early action on climate change greatly outweighed the costs. The most cost-effective actions identified by Stern are reducing power industry greenhouse gas emissions, increasing energy efficiency in buildings and other venues, and preventing deforestation.

"It is still possible to avoid the worst impacts of climate change, but it requires strong and urgent collective action," Stern wrote. "Delay would be costly and dangerous."

LOOKING UPSTREAM FOR SOLUTIONS

Back in 1988, after two years of meetings and serious discussion, an expert committee of the Institute of Medicine issued a 240-page report called *The Future of Public Health*. The committee that wrote the report considered various definitions of the field and settled on one. The mission of public health, they wrote, was "the fulfillment of society's interest in assuring the conditions in which people can be healthy."

As it dawned on me in the 1990s that climate change posed a fundamental threat to public health, I began to think broadly about the best ways to help advance conditions that fostered good health. I drew on my experiences as a physician and social activist battling epidemics in Mozambique and Boston, on my academic training in public health, and on my then-newfound appreciation of the link between ecosystem health and human health.

Although it was essential to pursue downstream methods to counter the threat and help treat those afflicted, I felt I would do more for public health if I could devise midstream methods to predict epidemics so health officials on the ground could intervene. By the late 1990s, I'd helped advance this use of climate information through efforts such as our project on epidemics in coastal oceans.

But if climate change posed a mounting public health hazard (and it seemed certain that it did), we would have to turn to upstream measures of prevention that got to the root of the problem. With the threat of such major disruptions to health and well-being, it was critical to focus on stabilizing the climate.

That meant drastically slashing greenhouse gas emissions, and doing it quickly. How could my efforts contribute to this larger project?

By then, I'd begun exploring the web of connection between health, ecosystems, and the economy. In addition to sensing and processing signals from the outside world, the brain also sends signals that determine our actions. And so it was with the economy's brain—the finance sector.

The financial industry—which includes insurers, commercial and investment banks, other institutional investors, and brokers—signals companies in other industries by choosing where to invest capital and by placing conditions on insurance, loans, and investments. If financial companies sent the right signals, other industries would be more likely to raise their energy efficiency and lower their greenhouse gas emissions. For that reason, I'd begun meeting with corporate executives like those from Swiss Re to learn more about their views of climate change and to help them sort through the new risks that their businesses now faced.

I believe that the best way to prod a person or organization to change behavior is to use a combination of carrots and sticks. For businesses, the sticks were regulations and product and performance standards. But corporations are motivated by profits and benefits to their shareholders. Was there any way that reducing emissions could also bring economic rewards?

GAINING GREEN BY GOING GREEN

In 1999, Swiss Re gave Chris Walker a groundbreaking assignment— to explore potentially profitable business ventures the company could undertake that would simultaneously fight climate change. Just thirty-six years old at the time, Walker had been working as a lawyer at the reinsurance giant for three years. His career path had made him the right man for the job. As a law student at St. Johns University in the late 1980s, Chris did a stint at the U.S. Environmental Protection Agency. There he helped sue multinational companies for violating federal Superfund laws, in an effort to compel them to pay the huge sums needed to clean up polluted sites. His Superfund litigation experience helped land him a job with the New York law firm Standard Weisberg, which represented insurance companies. The same multinational companies he'd successfully helped sue at the EPA were now suing their insurers to cover the massive costs of the cleanups, arguing that they were covered. The insurers denied the claims, and Walker helped defend them in court.

Such battle-of-the-titans lawsuits were standard operating practice in asbestos and Superfund cases, and sometimes insurers had to pay out, at which point they'd often sue their reinsurers, who would also resist paying. After leaving Standard Weisberg, Walker spent four years litigating such suits on behalf of the insurance giant AIG. But asbestos and Superfund claims were so huge, they could knock even reinsurers for a loop, so reinsurers demanded money from *their* reinsurers—a handful of European firms, including Lloyd's of London. Walker handled disputes between Swiss Re and Lloyd's for his first few years at Swiss Re.

Although Walker was a lawyer by trade, he was an environmentalist at heart, and he realized that this traditional command-and-control approach to environmental regulation, combined with litigation, wasn't working. About three-quarters of the vast cleanup sums for asbestos and Superfund sites was going to lawyers and accountants, major industries were being hammered, and, most importantly, toxic sites were not getting cleaned up. "If you did that in the atmosphere, it would not solve anything on climate change," he said. "Having had that experience, I was very receptive to the idea that there's got to be a better way."

In 2000, the Kyoto Protocol, which required nations to reduce emissions, had been passed, but it had not yet been ratified by enough nations to take effect. Both Al Gore and George W. Bush, who were running for president that year, promised to ratify the treaty and regulate carbon dioxide emissions. So, Walker and his colleagues believed that the United States would ratify the treaty in 2001 and that Europe would quickly follow, causing the emissions-reductions treaty to go into force. Walker pitched a feasibility study to his bosses to see how Swiss Re could make money by reducing emissions, and he received the assignment. He was given a budget of a million Swiss francs, six months, and authority unprecedented at Swiss Re to roam across the corporation's many divisions to pull in expertise and information. "The company had been talking about this from a risk point of view. My job was to create a business opportunity," Walker said.

Walker had been working in a unit of Swiss Re that developed innovative financial products. Such insurance products are similar in principle to the standard health, homeowners, or auto policies familiar to Americans, in that they cover the customer if something goes wrong. But they're often specialized policies, developed to cover a particular industry's equipment or infrastructure. For example, an oil rig is hugely expensive to install, and, since it's like a giant skyscraper in the water, high winds and high seas can damage it severely. No bank would lend the money to develop an oil rig unless that rig was insured. Swiss Re sold insurance to a rig's owners, which allowed them to borrow money from a bank to develop it.

Besides selling insurance, big insurance companies like Swiss Re invest huge pools of capital that they hold from premiums. They also serve as financial services companies, creating funds into which people invest their money.

Walker assembled a team of nine experts on financial topics such as insurance products, commodities trading, and investment vehicles. Gradually the team identified four or five areas that seemed ripe for development.

On the insurance side of the company, Swiss Re could insure renewable energy projects such as wind farms. At the time, Walker knew, developers of offshore wind farms in Europe were forced to buy expensive policies designed for oil rigs, but wind farms, which might consist of a hundred pilings spread over a couple of miles, were far less vulnerable to loss. By developing an affordable new policy for wind farm developers, Swiss Re could ensure that more wind farms would be built, more of their owners would buy insurance, and Swiss Re would make money.

Also on the insurance side, Swiss Re could modify an important type of policy known as directors and officers insurance. A corporation's officers and board of directors have a fiduciary responsibility to the company's shareholders; as such, they can legally be held liable for decisions they've made that damage the company and cost those shareholders money. To protect their directors and officers, companies buy insurance from Swiss Re or other large insurers to cover them if they are sued and found liable.

Walker knew from his Superfund litigation work that such policies covered almost everything related to the insured business but criminal actions. But, because of the expensive history of Superfund litigation, they now had exclusions for pollution. Say a worker on the shop floor was dumping barrels of toxic chemicals in a river by the factory. These actions damage the company by exposing it to fines from regulators, damages from civil suits, and lost business from a lousy corporate reputation. The company president, who told the worker to dump the chemicals, could be held personally liable. Insurers didn't want to be stuck with the bill in such cases, so they excluded pollution from their coverage.

As the public learned more about the threats of climate change, courts might regard excess carbon dioxide emissions as pollution, opening companies that failed to reduce their emissions to costly lawsuits for environmental negligence, Walker and his colleagues surmised. For that reason, Swiss Re became the first insurer to consider an exclusion for directors and officers of companies that did not act to reduce emissions. "The fastest way to get a company to move is when the directors and officers think they're personally liable," Walker explained. And by protecting themselves from climate change–related claims, such an exclusion in their directors and officers policies could save Swiss Re money.

(At Walker's urging, Swiss Re later surveyed the directors and officers it insured about how they were addressing climate change risks, a move that served as a shot across the bow to corporate executives, warning them that they needed to address climate change's risks and opportunities. But to date, neither Swiss Re nor any other reinsurer has followed through by requiring this of the directors and officers they insure—a policy that would send a strong signal throughout the corporate world.)

On the financial services side of the company, Swiss Re could create a fund that would take investors' money and invest it in what financial professionals call project finance. That is, they would invest money in the bricks and mortar a company needed to build a clean energy project like a wind farm or solar array. In exchange, Swiss Re would own a piece of the project, and when the wind farm was generating electricity, Swiss Re and the investors would make money.

If a policy known as cap and trade were to go into effect, Swiss Re could also get into the business of emissions trading. In a cap-and-trade system, governments would place national or worldwide caps on carbon emissions. Companies would get a certain number of credits for each ton of emissions they reduced below a certain cap. The more they reduced, the more credits they could sell to other companies that were not able to reduce their emissions enough. Walker proposed that Swiss Re act much like an investment bank, but instead of buying and selling stocks, they would buy and sell carbon emission credits. As emission credits were bought and sold, Swiss Re would make money by taking a small fee for each transaction and by collecting fees to help its clients manage their carbon portfolios.

Two months after Walker's team concluded their study in January 2001, President George W. Bush broke his campaign pledge and refused to submit the Kyoto Protocol to the Senate for ratification. "We were counting on ratification of Kyoto to create this huge market," Walker said. But as the treaty's prospects dimmed, Swiss Re backed away for a time from implementing any of the ideas that Walker's team had uncovered.

Reinsurers, such as Swiss Re and Munich Re, and insurance companies can push for beneficial policies that reduce risk to the public; in fact, insurers have a long history of doing so. The industry pushed for fire codes early in the twentieth century after suffering extensive fire-related losses. Similarly, the industry pushed for stricter building codes and stepped up vehicle safety testing, making the public safer while simultaneously reducing the claims they had to pay.

Although climate change is a much more complicated problem than urban fires or unsafe cars, the industry could still press to reduce risk. After losing a projected $7 billion in insured losses from flooding in 2007, the Association of British Insurers leaned on the UK government to increase investment in flood defenses as a condition for maintaining insurability. The Reinsurance Association of America and insurers have championed wetland and barrier island preservation, which reduces risks of injury and property damage. Swiss Re has lobbied for policies that combat climate change, including the 2003 McCain–Lieberman bill, which would have established a cap-and-trade mechanism in the United States.

Insurance companies in the United States decide for the most part who to insure and what to cover, and they can use this leverage to demand change. After Hurricane Andrew, insurers adjusted policies and the cost of premiums to discourage rebuilding in the most vulnerable areas and to encourage retrofitting or rebuilding with less-hurricane-prone designs. Those efforts paid off. After Hurricane Charley struck Florida in 2004, homes constructed to modern, post-Andrew building codes suffered 60 percent fewer claims, and those claims were 42 percent less severe.

Auto insurers could promote new pay-as-you-drive auto policies that reward people who drive less. In Canada, Germany, and Israel, insurers use in-car GPS tracking systems to verify mileage, and low-mileage drivers receive discounts of up to 50 percent. Such policies are scarce in the United States, but they could save the country $30 billion per year from reduced accidents and congestion, and two-thirds of drivers would save money in premiums, according to a 2008 Brookings Institution report. They'd also slash driving miles and carbon emissions by at least 7 percent, as much as an 81-cents-per-gallon increase in the gas tax.

It took Europe a few years to move on after Bush rejected the Kyoto

Protocol, but by September 2004, Russia signed on and the treaty went into effect. The agreement mandated emissions cuts of at least 5 percent compared with 1990 levels, and Europe implemented a cap-and-trade system to help achieve that goal. By then Swiss Re, too, had gotten over its initial hesitation and had begun implementing most of the business ideas Chris Walker had proposed.

BANKS STEP UP TO BAT

I was introduced to Charles O. "Chuck" Prince, former chief executive officer of Citi, at a meeting the giant financial corporation hosted in August 2007 at their midtown Manhattan headquarters. Three months earlier, the company had pledged to invest $50 billion to fight climate change—to this day, the largest investment of its type by any company. That summer day, they'd invited several academics and representatives of about a dozen green groups, including major players like the Climate Group and the Nature Conservancy, to help them decide how to do the most good with their environmental investments.

By then Citi had invested $10 billion in projects, including solar energy, biofuels, and a wind farm in Baja California, Mexico; it planned to spend $10 billion more reducing its own carbon footprint at 14,500 or so offices and branches worldwide. Another $30 billion was supposed to be invested over the next decade.

In announcing its $50 billion initiative three months earlier, Citi had been cheered by many environmental leaders, but it had also faced a fair amount of skepticism. Some bankers whispered that Citi had gone green to hide other business woes, and other critics accused the company of greenwashing, since the company was simultaneously investing hundreds of billions of dollars in projects that amounted to business as usual, financing oil companies, coal-fired power plants, and polluting industries.

I viewed Citi's investment as a hopeful sign, and I was glad to offer advice. In a well-appointed conference room, we listened all morning as people from half a dozen divisions of the huge company presented their plans. After lunch, we returned for more discussion. Prince seemed relaxed as he stood to speak, and he put the room at ease. "I hope you don't think anybody believes the editors of the *Wall Street Journal* or reads them anymore," he said. This got a laugh from the audience, but the message came through: even the conservative bankers at Citi no longer gave any credence to hard-core global-warming deniers like those that write the *Journal*'s editorials. Instead, the company was serious about fighting climate change.

Citi's move could help set the pace for Wall Street, spurring other banks to realize that such an investment would not only generate good public relations but would also help their bottom line. This process has begun. In March 2007, Bank of America committed $20 billion toward green investments over ten years. It has also allocated $1.4 billion to making its own buildings greener and has spent $100 million on energy conservation in its offices.

The economic crisis of 2008–2010 has obviously affected banks' abilities to lend and invest, but in late December 2008 both Bruce Schlein, Citi's vice president of environmental affairs, and a Bank of America spokesperson said they do not expect major cutbacks to their green investment plans over the next decade. Indeed, thanks to expected government guidelines and support, clean energy and other green initiatives may fare better than other investments. For example, divisions of companies like General Electric and 3M that help build solar panels, concentrated solar arrays, wind turbines, and materials for an improved electrical grid are proving to be those with the greatest growth.

Large financial companies can also push other companies to adopt green practices by requiring them as a condition for receiving a loan or other investment. That's the idea behind the Equator Principles, in which banks commit to scrutinizing recipients of loans and investments with regard to environmental risks. Many banks adopted those principles after years of intense pressure by the Rainforest Action Network.

And in 2007, Bank of America and four other major institutions— Credit Suisse, Wells Fargo Bank, JPMorgan Chase, and Morgan Stanley—adopted the Carbon Principles, which commit these banks and others that signed on later to taking a much harder look, called enhanced due diligence, at environmental and health consequences when considering whether to finance new coal-fired power plants.

Prince thanked us for coming and asked the attendees for input on what the company was doing. For forty-five minutes, attendees peppered him with questions and offered advice. Someone asked about lending guidelines, and we discussed how to invest in forestry efforts that sustain rather than destroy forests. Someone else asked about innovative financing methods to further Citi's green goals. I had the opportunity to say to Prince that when sustainable technologies are chosen, they must be examined with an eye toward their health and environmental effects.

Otherwise, funds could be lost if technologies or energy sources were abandoned, which would costs lenders dearly.

Overall, I believed that Citi's initiative was too big a change to write off as greenwashing. While they certainly intended to profit from their investments, Citi's corporate sustainability group seemed like well-intentioned people who were sincerely trying to make a difference.

A CLEAR AND PRESENT DANGER?

As the world's largest banks and insurance companies were acknowledging the threat of climate change and embracing new solutions, the Bush administration kept up its role, trying to belittle and quash reports of the likely effects of climate change.

In 2005, the news came out that the White House Council on Environmental Quality's chief of staff, Philip A. Cooney, a lawyer and former lobbyist for the American Petroleum Institute, had edited government scientific reports extensively to water down strong warnings of harm from climate change. In 2006, James Hansen of NASA and Columbia University went public with a scandal. Hansen described how a twenty-four-year-old political appointee in NASA's public relations office had blocked media requests to interview him and had warned him of "dire consequences" if he continued to make stark statements to the media concerning the warming climate. Later in 2006, NASA revised its annual mission statement to delete any mention of what astronauts working in the space agency have often fondly called our "home planet."

In 2007, atmospheric CO_2 levels had reached 384 parts per million, 12 ppm higher than when Bush took office. The IPCC reported for the first time that global warming was "unequivocal" and that human activity had "very likely" caused most of the rise in temperatures since 1950. Looking forward, they warned the world of centuries of warming temperatures, rising seas, and shifting weather patterns but said that many harmful consequences could be avoided or eased by acting promptly to avoid the most dire impacts.

By then, ten U.S. states, frustrated by the Bush administration's intransigence on climate change, had sued the U.S. Environmental Protection Agency to try to force it to regulate carbon dioxide emissions. In the vanguard was my home state of Massachusetts. The ten states, joined as plaintiffs by three large cities and more than a dozen health and environmental advocacy groups, argued in court that by failing to regulate carbon dioxide, the agency had violated the Clean Air Act, the most important U.S. law in combating air pollution. The law's unambiguous language requires the head of the EPA to set emission standards for any air pollutant from motor vehicles that "may reasonably be anticipated to endanger public health or welfare." The plaintiffs, drawing from numerous peer-reviewed journal articles on health and climate change, argued that good science supported the conclusion that carbon dioxide emissions from tailpipes and smokestacks do exactly that.

The Bush EPA, however, had argued that it did not have the authority to regulate CO_2 as an air pollutant under the Clean Air Act, because fuel-efficiency standards were the purview of the Department of Transportation. And they argued that carbon dioxide was not an air pollutant because it did not threaten public health. To bolster that argument, they cited the 2001 Institute of Medicine report, *Under the Weather,* discussed in chapter 9, which found that the impacts of climate change on disease were "highly uncertain."

On April 2, 2007, the U.S. Supreme Court issued a landmark decision on the case. The court found 5–4 in favor of Massachusetts and the other plaintiffs. Writing for the majority, Justice John Paul Stevens wrote that the agency had violated the Clean Air Act and that the overlap of responsibilities between the EPA and the Department of Transportation "in no way licenses EPA to shirk its duty to protect the public health and welfare." The court required the EPA to evaluate whether greenhouse gases "may reasonably be anticipated to endanger public health or welfare" and, if so, to implement regulations.

Six months later, on October 23, 2007, Julie Gerberding, director of the Centers for Disease Control and Prevention (CDC), the nation's top disease-fighting agency, was scheduled to testify about the health impacts of climate change to a Senate committee. She submitted six pages of written testimony, but reporters quickly obtained copies of the twelve pages of written testimony she had originally planned to deliver and uncovered evidence that the White House had censored it. In the testimony she presented, as edited by others for her, Gerberding focused on the CDC's efforts on a broad range of public health impacts associated with climate change but listed few details.

The six pages of censored testimony, in contrast, laid out in stark detail the harm that could await us. "Catastrophic weather events such as heat waves and hurricanes are expected to become more frequent, severe, and costly," read the deleted section. Some vector-borne diseases and diseases that jump from animals to humans could spread. More frequent heavy rains and flooding could overwhelm sanitation infrastructure, spread disease, worsen allergies, and lead to a possible scarcity of food for some Americans. The testimony "was eviscerated," an unnamed CDC official told the Associated Press.

Political appointees in the Bush administration claimed otherwise. "What needed to be said, as far as we're concerned, was said," stated a CDC spokesman. "It was not watered down," insisted White House spokeswoman Dana Perino. A White House science official explained that the testimony was simply modified to ensure that it reflected current science in the IPCC report. But in fact the science in the deleted testimony had not deviated from IPCC-sanctioned science.

Six months later, in April 2008, Gerberding's full testimony was finally presented to Congress. But the apparent censorship had prompted a Senate investigation, led by Senator Barbara Boxer of California. In a breakthrough in the case the following summer, Boxer obtained testimony from an EPA official turned whistleblower.

The whistle-blower described events in the office of EPA chief Stephen L. Johnson at the time of Gerberding's testimony. The Council on Environmental Quality (CEQ) and Vice President Cheney's office "were seeking deletions to the CDC testimony," the whistle-blower said. "CEQ requested that I work with CDC to remove from the testimony any discussion of the human health consequences of climate change." The former official had checked with EPA scientists, concluded that the draft testimony was accurate, and refused to push for changes. But changes were made anyway. A spokeswoman for Cheney said about the testimony, "We won't discuss internal deliberations."

MILES AND MILLIONS STILL TO GO

The 2008 Ceres Conference promised far more in the way of environmental wisdom than your typical gathering of business leaders and investors. The food was organic, mostly vegetarian, and locally grown. The conference brochure and program were printed on 100 percent postconsumer waste, chlorine-free recycled paper with vegetable-based inks by a union shop committed to green principles and social responsibility. Ceres, a national network of investors, business leaders, and environmental activists, would also purchase enough carbon offsets with fees from attendees to make the conference carbon neutral.

If you took a stroll around the airy second-floor lobby of the Renaissance Boston Waterfront Hotel on the cool April morning in 2008 when the conference kicked off, you might have been in for a surprise. True, exhibits there sported images that would not be unexpected at such an event, such as a watery Earth seen from space, or snowcapped mountains under a bright blue sky. And some of the companies exhibiting did indeed seem to come from central casting, such as BigBelly Solar, whose solar-powered trash compactor—a large, deep-green metal contraption—sat at stadiums, in downtown shopping areas, and in the lobby of the hotel, ready for a demonstration. There were educators from Marlboro College, touting their MBA program ("Managing for Sustainability"); there were investment fund managers such as Progressive Asset Management ("Balance Your Investment with Your Values") and Pax World Mutual Funds ("For Tomorrow").

But others on the roster of participants suggested that some very mainstream businesses were going green. Swiss Re's exhibit reminded the more hard-nosed types in the crowd that "climate change is both risk and opportunity," while a few feet away, General Motors had a display that highlighted its corporate social responsibility. Across the way sat a representative from Pacific Gas and Electric, the giant California utility that was defendant in the toxic-pollution lawsuit made famous by the movie *Erin Brockovich*.

In her welcoming remarks the first morning to a crowd in a large hotel ballroom, Ceres president Mindy Lubber spoke with pride about her organization's accomplishments, and she reminded the crowd of what they'd come through together. From its start in 1989, Ceres had challenged business and capital markets to promote the well-being of human society and protect the Earth's resources. Lubber spoke proudly of the Global Reporting Initiative, which Ceres helped launch in 1997 and which was now used by 1,500 multinational corporations to augment traditional financial reports with an annual "corporate social responsibility report" on the company's environmental and social performance.

Lubber extolled Ceres's Investors Network on Climate Risk, which included pension funds and other large institutional investors managing $22 trillion in assets and used its financial clout to push companies to take positive steps on climate change. She also shared good news about the Carbon Principles: how they had stopped more than fifty coal-fired power plants from being built in the previous year in places like Florida, Kansas, and Texas. "Every five hundred megawatt coal plant that's not built is the equivalent of taking six hundred thousand vehicles off the road," she said.

Lubber reminded the audience of the huge opportunities for businesses to profit while tackling climate change, water shortages, and other environmental woes, and she spoke with a controlled passion about the urgent need for effective action. "This is the first time in history that we are about to leave a planet that is weaker, more damaged, and sicker than the one we inherited," she told a rapt audience. "We have an extraordinarily long way to go."

. . .

For the rest of that day and part of the next, conference attendees enthusiastically discussed such topics as alternative energy and corporate governance, green-collar jobs, and ecosystem services. The next morning, Brendan May, a whiz kid with a British accent from the giant public relations firm Weber Shandwick, took to the podium to emcee Ceres's annual sustainability awards, which were given to companies who had produced what the judges regarded as exemplary corporate social responsibility reports. These documents covered environmental metrics such as energy saved or water conserved, emissions reduced, or waste averted. They also covered social performance—labor–management relations, how well they maintain a healthy workplace, and more.

May was blunt about the long-term prospects for business under climate change. The world's supply of freshwater, its forests, and its fisheries are quickly becoming depleted. By 2050, nine billion people will walk the earth, he said. "Unless we make some very fundamental decisions, by 2050 our planet will be a very unpleasant place to live and a hard place to do business."

He was equally blunt about how well businesses were doing. "The first step is for companies to report their progress becoming sustainable, and most companies don't even do that," he said. However, even if every corporation issued thorough corporate social responsibility reports, May said, "reporting is just the beginning."

After the corporate reporting awards were given, two pioneering CEOs took the stage for a panel discussion: Jeffrey Swartz of Timberland and Gary Hirshberg of Stonyfield Farm. The new emcee was Andy Savitz, an author and consultant on corporate sustainability.

He introduced Swartz, CEO of the Timberland boot company, who talked about his family business; how he learned from his grandfather, who founded the company; how to treat workers fairly; and what he hoped for his kids. "When I die, I can't just leave behind for my kids the legacy of the best boots on earth. I have to do better than that."

Then Savitz introduced Hirshberg, the self-proclaimed "C-E-Yo" of Stonyfield Farm. The fact that he uses such an irreverent term, Savitz says, "means he's so successful he doesn't care what anyone else thinks." Hirshberg, who seemed at ease, smiled and gave a thumbs-up.

"Stonyfield is my definition of a sustainable business," Savitz said. In the late 1970s, Hirshberg, who trained in ecology, had worked at the New Alchemy Institute, a pioneering ecological research and education center on Cape Cod. Later he worked at the Rural Education Center, a struggling organic farming school in Wilton, New Hampshire. He and his partners began selling yogurt as Stonyfield Farm, starting in 1983 with seven cows.

The company's mission has remained mostly unchanged since the beginning—to serve the highest quality organic yogurt and other products, to educate people about protecting the environment, and to serve as a model of a socially and environmentally responsible business that also makes a healthy profit.

Hirshberg described how Stonyfield became the first U.S. corporation to track greenhouse gas emissions through their entire supply chain, and how this had led to new savings. He described how, instead of shipping truckloads of waste manure from their cows, they used an old Chinese technology that let the sludge decay without air, thereby producing methane, which they then sold. Finding such green but profitable savings "almost became a sport," Hirshberg told the attendees.

For Stonyfield, Hirshberg said, doing the right thing ecologically had paid off beyond all expectations. The company had grown an average of 17 percent a year since 1983, becoming an international operation that had surpassed its largest American rival, Kraft, and had enjoyed $320 million in annual sales.

Savitz, the moderator, turned to Hirshberg. "What does sustainability mean to you?"

"There are three myths that drive our economic system," Hirshberg replied. "One, the myth of externalities," which occur when a deal between a buyer and seller affects the well-being of other people in ways that, according to society's laws and customs, don't affect the price. Pollution is the classic externality. An externality, Hirshberg said, "is a fabrication of economists.

"Two, the myth of away, as in, 'send waste away.'

"Three, that the solution to pollution is dilution."

Stonyfield's very success, which came without compromising either principles or profits, puts the lie to all these myths. It proves, as Hirshberg said that day to his audience, that "it is possible for business to be part of the solution."

11

Healthy Solutions

Spend enough time pondering climate change, and the magnitude of the challenge can begin to overwhelm anyone. There's a daunting array of numbers and trends, choices and consequences. Much of what we do as modern humans contributes to the problem, little by little by little. Our appetite for high-carbon energy has unquestionably put the world and its inhabitants at risk, and we appear to be hurtling toward a very unsettling conclusion. But we must not lose sight of a very simple and reassuring fact: we have already invented virtually everything we need to get us out of this crisis. The job won't be easy, and we could certainly use a few more clever tools. But we can build a low-carbon society. Indeed, it's already happening.

In 1999, David Riecks and Anna Barnes faced a dilemma familiar to many homeowners. Barnes was working at home as an editor and Web designer, but their 1,200-square-foot house in Champaign, Illinois, lacked adequate air conditioning to ward off the stifling summer heat. Their furnace had seen better days. Any way they looked at it, they were about to spend a lot of money. They could have easily invested in a high-efficiency conventional furnace and air conditioner and felt comfortable with their choices. Instead, they went underground, installing a ground source heat pump.

Just a few feet below the ground, the earth's temperature remains a nearly constant 13°C (55°F). The heat pump taps this natural energy source, reaching as far as 150 feet down with plastic tubing with a circumference about that of a garden hose. A nontoxic liquid that works like antifreeze cycles through the tubing, slowly attaining the temperature of the surrounding earth. When the weather is cool, the heat pump uses electrical compressors to deliver that underground heat into the house, helping to provide heat and hot water. In the summer, the system runs in reverse, removing heat from the house and helping to provide free air-conditioning.

Adding a traditional central cooling system might have doubled their utility bills; instead, Barnes and Riecks estimate that they've saved hundreds of dollars in heating expenses per year. While ground source systems cost more to install than a typical furnace or central air conditioner, they typically save buyers enough on their power bills to pay for themselves within three to seven years. Such savings have helped sell more than a million ground source heat pumps in the United States and a similar number in China, reports James Bose, executive director of the International Ground Source Heat Pump Association.

Choosing a more efficient heating and cooling system simply made sense to Barnes and Riecks, just like riding their bikes or buying local produce. "I think it's a personal responsibility," Barnes said. "Just like milk does not magically get into a milk carton, electricity just doesn't jump into a light switch. It's not lost on us that every Btu that we're buying is coming from coal." And she knows that coal carries a heavy carbon price. "Everything has a cost. It's a matter of accepting your responsibility for that cost."

. . .

For years, Ron Later scoured the marketplace looking for an alternative to his monthly $300 electric bill. That's what he paid to power the 4,500-square-foot home he shares with his wife, Edna. The Laters live in Hinkley, California, an unincorporated community in the Mojave Desert where summer temperatures often soar above 38°C (100°F), and staying cool isn't cheap. But the monster on their utility bill is the well pump that waters their home and a nearby stand of two hundred pistachio trees.

To save money on his electricity bill, Later, a 58-year-old freight train mechanic, decided twenty years ago to try alternative energy sources. "I'm tired of giving people my money," said Later. "The wind is out there, the sun is out there, so why should I have to pay for it?"

When he first looked into alternative energy sources, he suffered sticker shock. But Later, a do-it-yourself type who grows his own vegetables and fruit and built his own house, kept his eye out, finally spotting an advertisement for a small wind generator manufactured by Bergey Windpower in Norman, Oklahoma. Later finally managed to install his ten-kilowatt wind generator, complete with a 120-foot tower, in 2003. He estimates that he now pays just $600 a year for electricity.

The generator cost $47,000 to purchase and install, but state and federal incentives cut the costs by more than half. No one has ever complained about the site of the tower or any noise. In fact, the Commonwealth Edison meter reader didn't even notice it. She did notice that the meter wasn't racking up the watts, and the Laters received a call from the utility wanting to replace their "broken" meter. "Well, ma'am," Later told her, "did you see the wind generator out there?"

The turbine has become something of a novelty, with friends stopping by when it's windy simply to watch the meter run backward. And two neighbors have since installed their own systems. "It's quite trippy," Later said. "It's very exciting. You did something and it's actually working for you."

. . .

Jerry Brous and his wife, Pat, were accustomed to keeping an eye on their appetite for electricity. They used to take long trips in an oceangoing trawler, and all appliances—including vital systems like their radio—ran off a bank of batteries. "We became our own utility," said Brous, sixty-eight, who was retired from a job in management at U.S. Steel. "We learned what each thing costs you in terms of amp-hours." So one day in Sequim, Washington, when Jerry heard about GridWise, he jumped at the chance to participate.

GridWise is a nationwide consortium of research labs, high-tech companies, and utilities working to increase the efficiency of the entire electricity grid, from the power plant to the plug. Its goal is to make a smart grid, to automate the kind of keen awareness of power use that the Brouses developed at sea.

The problem is that the two sides of the grid—supply and demand—run independently of each other. Demand is constantly waxing and waning: up for morning showers and after-work air conditioners; down at night. But on the supply side, the grid has virtually no capacity to store electricity. Balancing supply and demand is a tightrope act, and right now it's accomplished with special power plants called peakers that get cranked on and off, up and down, according to demand. Compared with the average watt of electricity, peak power costs more and emits more carbon dioxide. "That's the most expensive power you can buy," said Rob Pratt of the Department of Energy's Pacific Northwest National Laboratory.

In some areas, customers can allow their utility to power down major appliances such as air conditioners during peak demand. The utilities can avoid firing up the peakers, and cooperating customers get a discount on their utility bills. This is a start, but it supplies only crude, on–off intelligence. The GridWise team plans to refine this.

During a one-year trial in Washington and Oregon, Pratt's team put computer chips into appliances and linked them to the Internet. Home owners could set their heating and cooling to choose either comfort or energy savings, or some balance in between. For example, if three homes wanted to heat some water during peak demand but there was only enough for one, the home set for the highest comfort setting would get the electricity but also pay a bit more for it. "We literally auctioned off the available electricity that we could deliver to that neighborhood to the highest bidders," said Pratt.

Every five minutes the grid talked to the appliances, giving them feedback about how taxed it was. When demand was low, current flowed, but when it was high, the GridWise system might shut off the dishwasher or power down the air conditioner for twenty minutes. The appliances talked back to the Brouses' computer. Jerry could sit down at any computer and monitor their energy use and control their appliances. "It was exceptionally easy," said Brous.

The Brouses went for maximum energy savings, though they occasionally overrode the settings for comfort. They barely noticed the difference in their lives, yet during the one-year study, they cut their overall energy use by 15 percent—a typical result, said Pratt—and they could reduce it up to 50 percent on short time scales. The GridWise system opened their eyes to how much electricity they were using, which inspired them to save energy in other ways, such as opening the curtains on cold mornings to let the sun's rays warm up the house. "That made a huge difference," said Brous.

THINKING BIG, THINKING POSITIVE

David Riecks and Anna Barnes installed a ground source heat pump. Ron and Edna Later erected a ten-kilowatt wind turbine. Jerry and Pat Brous helped test the smart grid. Not one of these people qualifies as a professional environmental crusader. Yet each cared enough to invest a little extra time and some resources into making a good choice. They all saved money and reduced their carbon footprints at the same time.

Scale these efforts up, compound them by the millions of opportunities out there, and the world begins to look like a very different place. The ground source heat pump industry has set its sights on increasing its share of the heating, ventilating, and air-conditioning market from about 2 percent to 30 percent by 2030. This will significantly reduce the largest source of residential greenhouse gas emissions. An ambitious deployment of 1.5 million two-megawatt wind turbines in the United States could be accomplished by 2020, meeting 40 percent of global electricity needs and repurposing dozens of shuttered automobile assembly plants to manufacture the turbines in the bargain. A national smart grid like GridWise could save customers 10 percent on their bills and $70 billion over twenty years and eliminate the need to build thirty large coal-fired plants. We can change our dangerous trajectory. But with a problem as big as climate change, how can we frame the challenge so that it can be best understood, discussed, and ultimately solved?

In 2004, ecologist Stephen Pacala and engineer Robert Socolow from Princeton University's Carbon Mitigation Initiative unveiled a framework for dealing with climate change called the stabilization wedge. Though we'll be debating the details for decades, it's generally accepted in the climate change community that preventing a doubling of the preindustrial concentration of CO_2 is feasible and that it has a chance to stave off the most devastating changes. When you draw this target line on a graph alongside today's trend of a steep and steady increase in greenhouse gases, the area between them is essentially a big triangle. Pacala and Socolow cut the triangle down to size, dividing it into seven equal wedges, each worth roughly a gigaton (one billion tons) of carbon emissions avoided annually by midcentury (figure 33).

Then they assessed the current array of potential fixes, calculating what was needed to equal a wedge. The initial list compiled by Pacala and Socolow included simple but ambitious conservation wedges such as doubling fuel economy or halving vehicle travel; utility-scale wedges such as replacing 1,400 gigawatts of coal-fired power with natural gas generation; alternative energy wedges such as increasing wind generation fiftyfold; and ecological wedges such as stopping deforestation and ramping up reforestation.

In all, Pacala and Socolow tagged fifteen different technologies that were already "beyond the laboratory bench and demonstration project." Their list was not exhaustive, and more potential wedges have been

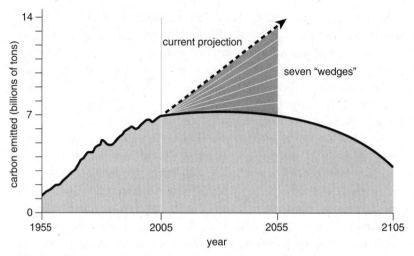

FIGURE 33. Here a wedge, there a wedge. At current rates, carbon emissions will double by midcentury, significantly raising the odds of climate catastrophe (current projection). Each stabilization wedge represents large-scale implementation of a technology or policy that could cut 1 billion tons in annual emissions by 2055. Seven wedges would keep emissions steady; twice as many, or wedges twice as large, would keep greenhouse gas levels steady. (Recent calculations show that wedges need to be somewhat larger.) (From S. Pacala and R. Socolow, "Stabilization Wedges: Solving the Climate Problem for the Next 50 Years with Current Technologies," *Science* 305 [2004]: 968–72)

identified using this framework. For example, increasing recycling and routinely harvesting methane from decomposing material in municipal landfills could save more than one billion tons of carbon emissions a year, enough for a wedge. In their report on the wedge concept in the prestigious journal *Science,* they concluded that "humanity already possesses the fundamental scientific, technical, and industrial know-how to solve the carbon and climate problem for the next half-century."

TO BUILD A WEDGE

From a public health standpoint, the concept of stabilization wedges offers a marvelous template to help reduce greenhouse gas levels in the atmosphere, thereby tackling the upstream cause of climate change. And since their introduction in 2004, stabilization wedges have inspired and challenged those of us who are working toward such solutions.

The concept is flexible as well as powerful, which is important because the stabilization wedge itself is something of a moving target.

Using 2004 calculations, seven wedges would offset projected increases in greenhouse gas emissions to 2054. But to stabilize or actually reduce greenhouse gas concentrations in the atmosphere would take more wedges, larger wedges, or both.

In choosing which technologies and policies to use, we must put first things first. No matter how well a technology or policy reduces greenhouse gas emissions, for both ethical and practical reasons we must make sure that it doesn't inadvertently harm humans or damage the health of ecosystems. For lessons in putting human health and the environment first, we can draw from the many occasions we have failed to do that.

Asbestos provides a good example. It was first mined in 1879 and was used for most of the twentieth century in building insulation, roofing, fireproofing, automobile brake pads, and many other products. Yet its fibers lodge in the lungs forever. Asbestosis, lung-tissue scarring that impedes breathing, was first reported in 1898; asbestos-linked lung cancer in 1932; and asbestos-linked mesothelioma, a lethal cancer of the tissue that encases internal organs, in 1955. Although its use was scaled way back by the 1980s, asbestos fibers take decades to cause disease, and as many as ten thousand Americans, many of whom handled the material on the job, die each year from asbestos-related diseases.

Still, industry fought asbestos bans for decades, denying that it was dangerous. But science and the law caught up with them. In 2004 the Rand Institute reported that companies paid out about $70 billion on 730,000 asbestos claims between the 1970s and 2002. For decades, as former workers aged, got sick, and died from asbestos-related disease, the insurance claims kept coming. The insurance industry calls this delayed wave of insurance claims a long tail.

The case of asbestos illustrates how far astray we can go when we choose technologies strictly for economic or utilitarian reasons. Similar tales of woe could be written about leaded house paint and a variety of industrial toxins. In each case, industrial leaders scaled up a technology or material that seemed effective and economically viable without adequate testing, but that turned out to be dangerous to health or the environment, which in turn led to enormous economic costs. In today's era of declining natural resources and faltering environmental resilience, we can no longer afford long tails, if we ever could.

Consider, too, compact fluorescent lightbulbs (CFLs). These coiled, white bulbs have become symbolic of the fight against global warming because they use 25 percent as much energy and last ten times longer than a typical incandescent bulb. The federal Energy Star program has been running a "Change a Light, Change the World" campaign, an initiative that has been picked up by Walmart, countless hardware stores, and hundreds of businesses and organizations and Web sites, all enthusiastically trying to green themselves. As a nation we like the elegant simplicity of the solution, the utter attainability.

And yet compact fluorescent bulbs contain mercury, a potent toxin that can damage the central nervous system, the endocrine system, the gastrointestinal organs, and the kidneys. Fetuses are five to ten times more sensitive than adults. While exposure may not be a significant issue for consumers, workers making or disposing of the bulbs may be exposed to mercury. And in most communities, proper recycling of the new lightbulbs is far from convenient, which means the bulbs may end up in landfills, break, and contaminate groundwater.

Although the risks of compact fluorescent bulbs to users are probably minimal—if all 290 million compact fluorescent lightbulbs sold in 2007 broke, they would release less than 0.1 percent of the mercury Americans release annually—the development of compact fluorescent bulbs nevertheless illustrates a crucial flaw in the way new green technologies are implemented. They were developed and promoted with all best intentions to offer consumers an affordable way to cut energy use and to fight climate change. But the mercury hazard was not factored into the decisions to commercialize. What if it had turned out to be more hazardous?

As these cases indicate, technological and economic feasibility, while critical, are not the only issues to weigh. Some technologies may prove unsustainable because of serious health and environmental damage. Some may inadvertently enhance global warming. Safer and more energy-saving alternatives may not have been considered ahead of time. (In the case of CFLs, a longer-lasting and safer alternative that provides excellent light—light-emitting diodes, or LEDs—is now available.) And the market is hardly free from powerful and deep-pocketed special interests that will try to install policies that boost their profits.

Indeed, these forces are already at work. Despite the exciting commitment of President Barack Obama to alternative energy, we are in danger of fast-tracking some potentially perilous choices. Nanotechnology may hold great promise for solar power and batteries, but research suggest that nanoparticles made from petroleum may mimic the effect of asbestos particles on the body. (Nanoparticles made from biodegradable plant-based material, in contrast, can deliver healing medicine directly to the body's organs.) Schemes to geoengineer carbon sequestration by fertilizing the ocean or shooting cooling sulfates into the atmosphere require far greater scrutiny to prevent unintended consequences. For example, repeated injection of sulfates into the atmosphere may add to ocean acidification. Wave and tidal energy generation on the scale necessary to effect climate change could possibly damage the coastal ecosystems that invigorate the ocean and feed millions.

In such a fraught landscape, we need some powerful guiding principles. Fortunately, a single one will do. The precautionary principle, which is central to public health work worldwide, simply means that we should avoid or minimize risky practices, particularly when the consequences could be great. Before we move forward, we must consider the outcomes that we can imagine and weigh their impact on human health and the health of the ecosystems that support us. It's only by viewing them through this health and environment lens that we can best ensure we're doing the right thing.

If we don't, we might be in for some unintended consequences.

AFTER THE FLOOD

Just after midnight, on December 22, 2008, an earthen retaining wall collapsed in Harriman, Tennessee. The dike had held back an eighty-acre lake of fly ash collected since 1958 from the coal-fired generators at the Kingston Fossil Plant, one of the largest facilities run by the Tennessee Valley Authority. In one wet night, more than 4.1 million cubic meters (5.4 million cubic yards) of waste laced with heavy metals surged forward, damaging houses, felling utility poles, and ultimately covering three hundred acres. As the spill oozed into the Emory River, so did huge amounts of five heavy metals: arsenic, which can damage the skin, lungs, blood vessels, and nerves; lead, which impairs brain function; cadmium; barium; and selenium. In testing done more than two weeks after the spill, these toxins were present at more than twice safe drinking water standards, and exposed fish were shedding scales and had damaged gills. The Emory flows directly into the Clinch River, which supplies Chattanooga's drinking water.

Coal helped lay the foundation for modern industrial society, and the United States, with the largest coal reserves of any nation, has been called the "Saudi Arabia of coal." It generates just over half of our electricity—more than twice that provided by nuclear or natural gas, the next largest sources.

At current burn rates our stores could last for more than 130 years,

coal life cycle

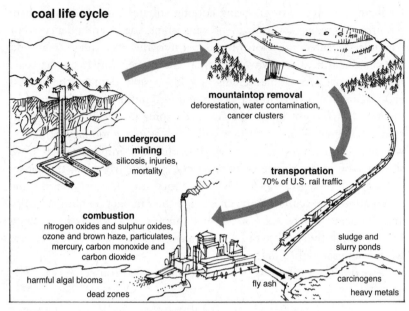

FIGURE 34. Coal's toll. Coal sickens us and poisons our environment in multiple ways, starting when it's mined and ending when it's burned.

according to federal government estimates. In an era of steeply rising energy prices and increasing concerns over energy security, coal seems cheap and readily available (although new analyses indicate there may be far less of it than we think). Even if we decided to kick the coal habit tomorrow, our existing and planned coal infrastructure is significant enough to guarantee it would have a prominent role in our energy economy for several more decades.

But as the Tennessee accident, the worst of its kind in the United States, illustrates, coal has always had a dark, lethal side. Scattered across the country are more than 1,300 similar dumps, many unregulated, that are filled with coal combustion waste. But fly ash is just a part of the problem.

At each stage of production, coal extracts a blood price (figure 34). The toll begins in the ground, where mining coal gouges both the workforce and the landscape. In the United States, more than 20,000 former miners have died since 1990 from black lung disease. In China 4,700 miners died in mining accidents in 2006 alone.

Black lung and silicosis are of less concern today in Appalachia, one of the coal-producing regions of the United States, but only because so-called mountaintop removal is now the mining method preferred by the coal industry. *Mountain range removal* is a more apt term for the method; in Appalachia, one of the most biodiverse regions in the United States, more than 450 mountains have been decapitated for their coal deposits.

First, forests and soils are bulldozed into adjacent valleys and streams. Then holes are bored hundreds of feet into the bedrock and packed with explosives. Once the explosives are blown, the debris from the shattered mountain is again shoved into valleys and over streams. Well over seven hundred miles of Appalachian streams were buried under rubble between 1985 and 2001, according to a study by the U.S. Environmental Protection Agency. The water in surrounding communities often runs toxic with cancer-causing chemicals and heavy metals and reeks with contamination that offends the senses.

Coal thus obtained is then washed and treated, and the slurry—a nasty brew of water, coal dust, clay, and toxic chemicals such as arsenic and lead—is stored in retention ponds. Shipping then adds another toll, as coal transport trains make up as much as 70 percent of U.S. rail traffic.

Finally, burning coal releases a wide range of bad actors, including particulates, nitrates, sulfates, and mercury, which pollute soils and water, including the aquatic food chain, where they concentrate in fish. As discussed in chapter 4, the resulting air pollution contributes to asthma and allergy attacks, heart attacks, and lung cancer. Particle pollution alone from power plants leads to thirty thousand premature deaths each year (nearly twice the seventeen thousand homicides that occur annually in the United States). Finally, coal emits 50 percent more carbon dioxide per unit of energy produced than oil and twice that of natural gas, making it the worst of all fuels for the climate.

In the past, we've often chosen energy sources like coal that seem cheap or easy to mine and burn, disregarding the true costs in terms of pollution, habitat loss, and people and other creatures sickened or killed. Coal may seem cheap, but the overall price we pay is huge.

THE TRUE COSTS OF COAL

Many argue that coal must be rapidly phased out, and even coal's advocates believe that coal technology must evolve to remain viable in a carbon-challenged world. The American Coal Council, which represents the coal industry, has responded by promoting the idea of clean coal. As this wealthy and powerful industry goes whole hog to promote the idea, it would be worthwhile for us to pause to consider exactly how clean "clean coal" really is.

The notion of clean coal first emerged in the late 1980s. By then it was clear that sulfur from coal-fired smokestacks reacted in the air to form sulfuric acid, which traveled downwind to cause the acid rain that was poisoning lakes and streams throughout New England. Technology had been developed to scrub sulfur from coal-fired smokestack emissions, which was an important breakthrough. Although the industry cried foul, claiming that it would be too expensive, the 1990 overhaul of the Clean Air Act, the country's premier air pollution law, forced the industry to install scrubbers.

A decade later, George W. Bush promised clean coal during his first campaign for president. In 2001, Dick Cheney's controversial National Energy Policy included a recommendation for a ten-year, $2 billion federal research program in clean coal technologies. The short-term goal was to meet existing and emerging environmental regulations; long term the aim was low-cost power plants that emitted no carbon dioxide and generated nearly twice as much energy per lump of coal as today's plants.

There are two prongs to the clean coal strategy. The first prong is to improve the efficiency of the burn. This is commonly done by coal gasification, in which coal is treated with heat and pressure in the presence of steam, thereby pulling gaseous elements from the coal to create a mix of hydrogen and carbon monoxide called syngas, which is then burned to generate power. Proponents of gasification and other novel combustion technologies say they burn cleaner than coal and harvest more energy from the coal as well.

The second prong, known as carbon capture and storage or carbon capture and sequestration, is a completely different kind of bet. The idea is to scrub CO_2 from the smokestack and store it—underground, or perhaps under the seafloor. That's not simple, and capturing the CO_2 takes a lot of energy; using current technology, it uses between 25 and 40 percent of the energy produced from the burn. That either cancels out projected gains in combustion efficiency or burns a lot more coal to generate the same amount of electricity.

What are the potential consequences of such underground sequestration? Proponents of carbon capture and storage cite pilot sequestration projects in Norway and Canada as evidence that sequestration is safe, despite several reported leaks from the Norwegian site. But if the process

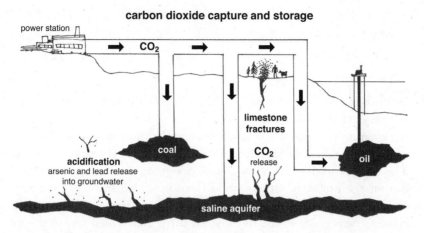

FIGURE 35. Burying the problem. In theory, capturing and burying dangerous carbon dioxide emissions from power plants seems like a great solution. In practice, it could potentially acidify groundwater, leaching toxic heavy metals and dissolving enough limestone to escape and suffocate people, livestock, and vegetation.

were scaled up to the volumes needed to generate a stabilization wedge, it could have unforeseen consequences.

The infrastructure needed would be enormous: to sequester just 10 percent of the CO_2 from coal-fired plants, we'd need new pipes and infrastructure on a scale matching what we now use for natural gas. We'd also be injecting billions of tons of CO_2 each year into underground saline aquifers, turning certain aquifers acidic. This could make them dissolve toxic heavy metals like arsenic and lead from rocks, possibly polluting groundwater, according to a special IPCC report on the subject. The acidified water could also drive fractures in limestone (chalk), leading to CO_2 leaks and releases. Large releases of very concentrated CO_2 are toxic to trees and animals, including us (figure 35). Ocean storage has also been proposed, but since oceans are already becoming more acidic, thus threatening marine life, it doesn't seem wise to add billions of tons more CO_2 to the ocean each year. Insurers and financiers are already taking a hard look at such risks.

Even if we could bury all the CO_2, coal combustion releases other pollutants—mercury, fly ash and other particles, oxides of sulfur (SO_x) that cause acid rain, and nitrogen oxides (NO_x) that react in the air to form ground-level ozone (smog). NO_x from combustion also floats through the air to pollute bays and estuaries, contributing to the dead zones that are destroying coastal fisheries worldwide.

With all these hazards, it's clear that—even if one could bury all the CO_2—clean coal isn't clean at all.

DEAD-END FUELS

From start to finish, coal takes a huge toll on the environment and our health. And oil is no better, a point that was driven home to the world the evening of April 20, 2010, when the Deepwater Horizon oil rig exploded in a massive fireball fifty miles south of the coast of Louisiana. The explosion killed 11 crewmen. Workers on the rig screamed and ran for the lifeboats. Others stood there in shock, staring at the fire, or jumped into the Gulf of Mexico to escape the flames. "I was pretty certain I was going to die so I just sat there and waited for what was going to happen," roustabout Stephen Stone told the *Christian Science Monitor.* A supply boat eventually plucked 150 surviving crew members out of the Gulf or pulled them from lifeboats and returned them to land.

But the catastrophe was just beginning for Gulf Coast residents—and for the Gulf of Mexico. Because BP's so-called fail-safe mechanisms had not sealed the well, for months the well spewed more than nine million liters (sixty thousand barrels) of oil a day into the Gulf, creating plumes up to a kilometer deep and many kilometers long, causing the worst environmental disaster in United States history. Cleanup workers and residents complained of reactions to the toxic fumes, and in June over one hundred workers were hospitalized with heatstroke. The oil fouled marshes and oyster beds, poisoned fish and seabirds, from Louisiana to Florida. It idled many in the tourism and fishing industries, causing millions of dollars of economic damage (figure 36).

Not coincidentally, the Deepwater Horizon accident occurred at a rig that was drilling in the seafloor a full one and a half kilometers (one mile) below the surface of the Gulf. In recent years, easy-to-tap oil fields have become scarce on land and in shallow seas, and oil companies are scouring the planet for new reserves. Their difficulties are a symptom of a slow-moving but extensive fuel crisis commonly known as peak oil. As a nation and as a civilization, we're almost entirely dependent on oil to keep parts of our economy, particularly our transportation sector, humming. But the global oil supply is finite. Sooner or later, supplies will peak, then decline and become scarce.

When they do, it won't be pretty, according to peak oil theorists. Oil-intensive industries will be hammered. The era of cheap airfares will end. Trucking and international shipping will become more expensive.

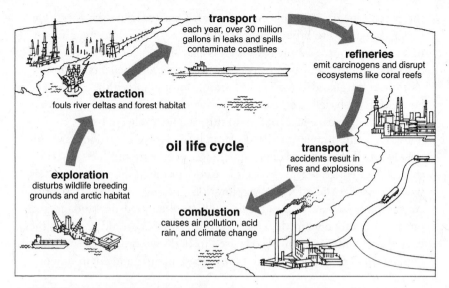

FIGURE 36. The real price of oil. From exploration to combustion, oil harms human health and fouls the environment.

Conventional agriculture, which relies on oil to make fertilizer and to fuel tractors, will suffer. As farmers' costs and the costs of shipping rise, so will food prices. Some authors and activists predict even more severe consequences, including blackouts, food shortages, and suburbs morphing into slums.

Some geologists say we have already passed the peak. Geologist M. King Hubbert's prediction was dead on in 1956 when he famously said that U.S. oil production would peak in the early 1970s; he may prove to be right about the world's oil production, which he predicted would peak in 2004. A quick look at the behavior of oil companies today supports this view. They're still looking for and finding oil, but often in places that they once would have scorned as too difficult, too expensive, or too dangerous to drill. These include the deep sea, as well as the so-called tar sands of Alberta, Canada, where they clear trees, topsoil, wildlife, and water off large swaths of land, then scrape up the sandy soil to extract a very heavy form of oil. Referring to the tar sands and similar projects, journalist Mark Hertsgaard has warned, "If peak oil arrives before the addiction is treated, the junkie will seek even more dangerous ways to get his fix."

But what about the alternative energy sources we hear so much about? When evaluating them to power our society, how can we apply the precautionary principle to make safer and healthier choices? The key once again is to broaden our vision. Rather than choosing a fuel or other energy source simply because it is easy to mine or burn, we should consider broadly, and in advance, its full costs and benefits, including its impact on global warming pollution, on human health, and on the land- and sea-based ecosystems that sustain us. Fortunately, scientists have devised an excellent tool to do this: the life cycle analysis.

As its name implies, a life cycle analysis is a tool to look systematically at the health and environmental aspects of a product or service through its entire life cycle. In the case of coal, a life cycle analysis would consider the effects of exploration, mining, refining, transport, consumption, waste disposal, and pollutants emitted on humans and on the environment. In the case of wind power, it would consider the effects of obtaining the materials used to build wind turbines and their manufacture, transport, installation, infrastructure, and environmental impacts.

True life cycle analyses are extensive, formalized studies. Our center produced a life cycle analysis of oil's impacts that included the contributions of eleven graduate student researchers and four peer reviewers; it ran seventy-three pages with more than two hundred references. In 2008, Stanford University engineering professor Mark Jacobson reported the results of an extensive life cycle analysis of potential sources of energy touted for the future. He considered an energy source's effects on climate change, air pollution mortality, and energy security, along with water supply, land use, wildlife, resource availability, thermal pollution, water chemical pollution, nuclear proliferation, and undernutrition.

The alternatives lined up like this, from first to worst: wind, concentrated solar, geothermal, tidal, solar photovoltaics, wave, hydroelectric, and then a last place tie between nuclear and coal with carbon capture and storage. "Coal with carbon sequestration emits 60 to 110 times more carbon and air pollution than wind energy," Jacobson concluded.

Nuclear power, which is enjoying a revival in some policy circles, still produces twenty-five times more carbon than does wind. What's more, the safety, storage, security, and cost concerns that have brought the industry to a standstill are by no means solved. The cost can be prohibitive: the price tag of a nuclear plant recently doubled from $6 billion to $12 billion, while renewable technologies continue to drop in price. In addition, a nuclear plant needs prodigious amounts of cooling water, and one-quarter of the 104 nuclear power plants in the United States are in areas where aquifers are already overdrawn and underfed. Cooling

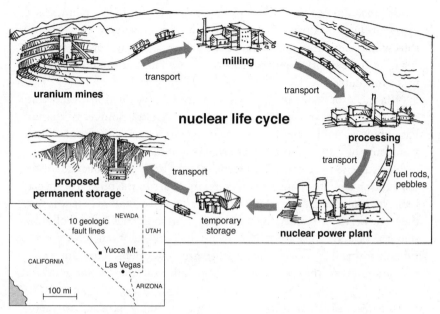

FIGURE 37. Nuclear energy's heavy footprint. Often touted as a clean energy source, nuclear energy actually poses hazards to humans and the environment throughout its life cycle. Everyone who handles the radioactive fuel must be protected, huge amounts of carbon-heavy energy are used to build nuclear power plants, and no one has found a truly safe way to store radioactive spent fuel.

water for these plants is thus already scarce and will get even scarcer as droughts and heat waves increase (figure 37).

The nation's designated permanent nuclear waste repository in Yucca Mountain, Nevada, has been the subject of endless legal and political wrangling, which has prevented the site, located 101 miles northwest of populous Las Vegas, from storing waste. As a result, dangerous radioactive spent fuel from nuclear power plants is instead stored on-site at dozens of plants around the nation, raising the odds that nuclear material will get into the wrong hands, notes Jacobson. Scaling up nuclear power to create a climate wedge would require opening a Yucca Mountain–size facility every five to ten years until midcentury.

Biofuels, a current policy darling, did not fare much better than nuclear in Jacobson's analysis. Vehicles powered by E85 (85 percent ethanol, 15 percent gasoline) would cause more air pollution–related deaths than any other technology, with the exception of a catastrophic accident involving nuclear fallout, Jacobson found. Both ethanol and biodiesel combustion produce chemicals that cause dangerous air pollution, including ground-level ozone (smog), corroding our lungs and aggravating respiratory illnesses. Ozone is also a heat-trapping gas, and thus wider use of E85 will enhance the urban heat island effect, making it harder to keep city dwellers healthy as we adapt to the climate change that's inevitable.

Whether from corn or prairie grass, ethanol production would consume more land and water than any other energy source. "Ethanol-based biofuels will actually cause more harm to human health, wildlife, water supply, and land use than current fossil fuels," said Jacobson.

Nuclear energy, clean coal, biofuels: these have been policy favorites for fighting climate change, thanks in large part to the powerful special interests promoting them. But Jacobson's life cycle analyses suggest that they are between twenty-five and one thousand times more polluting than the best available options. Despite the backing of our most powerful political and economic players, they are policy duds. They don't pass the simplest test: they're not good for people, and they're not good for the planet.

To forestall further climate change and handle what's already coming, new energy and transportation infrastructure must be strong and flexible. Completely rethinking these core systems might seem like a setback, but in fact it's an opportunity. All we need to do is circle back to the wedges. We do have other, healthy, choices.

THE HOME FRONT

Bill Moomaw was a lead author on the IPCC's 2005 *Special Report on Carbon Dioxide Capture and Storage.* But it's just possible that the home he and his wife conceived with their architect could ultimately have as much impact on climate change. Bill is a professor of international environmental policy at Tufts University and has worked with me to refine my thinking on healthy solutions to climate change. His wife, Margot, is an independent environmental consultant with a background in public health. After decades of cutting home energy costs for themselves and others, in 2007 they began constructing their retirement home, putting their experience to the test.

That home in the Berkshires of Massachusetts became something of a science experiment, and, like all good experiments, it started with a question: Can we build a house that uses half the energy of a regular house? Then, how about a house that uses less than one-fourth the energy of a regular house? And finally, how about a net-zero energy home, one that produces as much energy as it consumes?

Cutting energy by two-thirds is fairly easy, they say. But to reach carbon neutrality, they applied a variety of special measures. To make full use of the sun's energy, they needed eight hours of sun, four hours on either side of noon, even in the wintertime. They used a topographical map of their property and a computer simulation of the sun to align the house properly, ultimately placing it on an east–west axis, with large windows facing south-southwest for maximum passive solar benefit. Just the placement of the home saves 20 to 30 percent in energy annually.

To reduce the energy used for shipping, when possible they used materials sourced within five hundred miles of their Williamstown, Massachusetts, home. They chose nontoxic and sustainable materials where they could. The walls are twelve inches thick, twice the norm, and filled with cellulose insulation made from recycled newspapers. Triple-paned windows help keep out the cold.

The Moomaws placed sixty-six solar panels on the sloped south roof of the house and a storage shed to convert sunlight to electricity, and they installed a ground source heat pump that taps the earth to heat and cool their house. The result is a sleek and stylish two-thousand-square-foot, two-story house with a five-hundred-square-foot guest house. It burns no fossil fuels whatsoever.

The Moomaws' overall costs were 10 percent above a conventional home of the same size, and they did make some aesthetic sacrifices to keep costs down, such as omitting expensive moldings and other such frills. But they managed to install some very desirable features in a new house that fit in with the turn-of-the-century New England–style houses nearby. "We wanted it to be respectful of the historical neighborhood," said Bill. Added Margot, "We were in the price range of a lot of custom-built houses in the Berkshires. We just made different choices."

Different choices, but healthy choices. They chose materials and design principles that would promote their own health and well-being. And by considering the energy costs of both construction and long-term upkeep, they are promoting a healthy environment.

It's only one home, but it is part of a small study on zero-emission homes by the U.S. Department of Energy's National Renewable Energy Laboratory. What happens when you take these ideas to the next level? That's a question worth a few billion dollars. Then there's this one: What happens when an entire city begins to focus on efficiency?

COMMUNITY COUNTS

When Rob Pratt (no relation to Rob Pratt of GridWise) talks to audiences about energy and climate change in the Boston area, he asks them to first imagine themselves on top of the city's famous John Hancock Tower, surveying the urban expanse. About 80 percent of the region's greenhouse gas emissions come from buildings, a typical percentage for an urban area. If you cross that with our need to reduce overall greenhouse gas emissions by 50–80 percent, the task becomes clear. "We've got to affect every building that you see," he concludes.

Sitting in the audience one day was a councilor for the city of Cambridge. Across the Charles River from Boston, that city of one hundred thousand is home to both Harvard University and the Massachusetts Institute of Technology. While it had one of the most ambitious climate action plans in the state, it was also in the midst of a building boom. Its carbon footprint was going up, substantially. Cambridge officials asked for a meeting with Pratt, a veteran of three decades of working with efficiency and renewable energy. Pratt recommended an aggressive plan for energy efficiency as the most cost-effective solution. And he aimed high: $100 million worth of efficiency investments to dramatically retrofit most of the city's housing and building stock.

"It's a little mind-boggling, but you have to start someplace," he said. "Let's start in Cambridge, and start showing it can be done." So after some planning and coalition building, the Cambridge Energy Alliance was born in April 2007. The five-year plan is to cut overall electricity use by 10 percent and reduce peak power demand by 15 percent—aggressive goals given Cambridge's rapid rate of growth. Audits of energy and water use in buildings will set the priorities; then comes weatherization, replacing energy-guzzling appliances, even replacing toilets. Reducing water use is key, as pumping water to and from buildings takes a phenomenal amount of energy; in Cambridge, as in many municipalities, the wastewater facility is the largest energy consumer in the city. The project will stabilize energy costs for residents and businesses, reduce pressure on the regional grid, and create new jobs and economic development.

Why such a big budget? Why not start modestly? To begin with, Pratt was then senior vice president at the Kendall Foundation and head of its climate change initiative. (He has since become president of Cambridge Energy Alliance.) The elite team Pratt assembled has achieved more than $1.5 billion in renewable and efficiency installations, so *modest* does not seem to be in its vocabulary. And by pooling the demand for efficiency work, you get more interest from potential suppliers. When the Cambridge Energy Alliance requested proposals, a solid group of nationally known energy-service companies bid for the projects.

Financing strategies for the retrofitting projects differ depending on the type of building (residential or commercial) and its size (small or large). But they're innovative. For example, an owner of a home or condominium would borrow money for the project, but an energy-service company would make all the payments. In exchange for paying up front for the retrofit, the company would then reap all the energy savings until the customer paid off the loan and provided them some profit. By simply agreeing to participate, the homeowner would have the satisfaction of doing something substantial to fight climate change. He'd pay no money down and make no payments. And once the retrofit paid itself off, he'd cut his energy bills in perpetuity.

To boost participation, the campaign has a political flavor, with yard signs, door decals, and strategic outreach to local business and political leaders. "Efficiency is fundamentally boring, but a $100 million campaign is not," explains Pratt. "We want to show people that it can be done."

"Climate is going to be a tough thing to solve, but you can't just wring your hands and say, 'Johnny and Susie, we just ruined your planet.' You've got to do everything you can do to try to make a difference," said Pratt. "This is going to be a national model. We've got to be doing Cambridges all over the country."

THINKING BIG

Halfway across the country, atop the Willis Tower (formerly called the Sears Tower) in Chicago, is another critical vantage point on climate change. "The Best Place to View Tomorrow" is the marketing slogan of the nation's tallest building, and for once the marketers might be right. On a clear day you can see nearly fifty miles from the observation deck. To the east lies the inviting expanse of Lake Michigan. The city looms over the lake, a huge standing wave of glass-and-stone skyscrapers cresting over a beachhead of parks and breakwater. To the west, buildings step quickly downward and spill into a tabletop landscape that's gridded nearly to the horizon. Trees fill in between buildings.

Chicago, like any large city, is complex and full of problems. For decades municipalities in the wider region have competed against one another for investment and aid, jobs and residents. New infrastructure at the fringe has trumped repair and maintenance in the hollowing urban core. These are hothouse conditions for urban sprawl, and the result is painful to the planet: Chicago has the third most clogged road network in the country, where drivers waste 253 million hours and 151 million gallons of fuel each year stuck in traffic. That's $4 billion lost and more than 1.4 million metric tons of carbon dioxide equivalent annually, about 4 percent of Chicago's total greenhouse gas impact.

In defiance of this gridlock, Mayor Richard M. Daley has committed Chicago to fighting climate change by reducing its impact. It's the next logical step in his evolving green agenda. During the nearly two decades of Daley's administration, more than four hundred thousand trees have been planted. The city already has more than two hundred miles of bike lanes and plans to have five hundred miles by 2015. It aims for a bike lane within a half mile of every Chicago resident and hopes eventually that one in twenty trips under five miles will be taken by bike.

But it is the green roof that has become the city's signature of sustainability. In 1998 Daley was traveling in Europe and decided that the rooftop gardens he saw in Germany could help cool Chicago and cut energy costs. City Hall came first, with 1,800 square meters (20,000 square feet) of prairie, including beehives. The city now has more than two hundred thousand square feet of green rooftops, with ten times that in the pipeline.

If your goal is strictly cost-effective energy conservation, a green roof is actually down the list. But by focusing on green roofs, Daley, whose term expires in 2011, has set the bar high. If you're going to go so far as to put a green roof on your building, you're probably going to do everything else, too. The roof becomes emblematic of the paradigm shift we're striving to make.

Rooftop gardens include a diversity of plants and specially constructed bases to trap and use rainwater. Like all the best solutions, they have multiple benefits: they cool buildings; reduce carbon dioxide, toxic chemicals, smog, and heavy metals; absorb noise; shield rooftops from damaging ultraviolet rays; attract birds that control insects; decrease the urban heat island effect; and create enterprises and jobs. They make life more pleasant.

Building stock contributes 71 percent of Chicago's greenhouse gas emissions, so greening those buildings is a cornerstone of the city's climate strategy. But green roofs are just the beginning. All new municipal buildings must be LEED (Leadership in Energy and Environmental Design) certified by the U.S. Green Building Council. Since time is money for developers, the city awards permits faster for green buildings. As a result, Chicago has more LEED projects under construction than any other city. As with the retrofit buildings in Cambridge, LEED buildings quickly return the initial investment. In a Minnesota study, fourteen of sixteen high-performance buildings examined—which included schools, libraries, offices, and stores—paid themselves off with energy and other savings within three years. Once they're paid off, the buildings keep giving.

But is green building enough? Chicago would like to cut carbon dioxide emissions 80 percent by 2050, but if you project current growth patterns to that year, emissions soar to about fifty million metric tons from commuting alone. "Sprawl is a huge, huge problem for us," said city environment commissioner Sadhu Johnston. "We could do an incredible job, and regional emissions could still go up."

The answer, Johnston says, is smart growth—one of the most important and overlooked tools we have to slow climate change. Smart growth means favoring compact building design and choosing a mix of buildings and land uses to make neighborhoods walkable and diverse. It means directing development in existing communities and creating a range of housing and transportation choices. And it means preserving open space and other features of a healthy environment. Surprisingly few of these elements are explicitly green.

"Environmentalists can't just look at the environmental piece, or we're not going to be persuasive enough," said Kaid Benfield, director of the smart growth initiative at the Natural Resources Defense Council. "We need cities to work. We need them to be energy efficient, resource conserving, and nonpolluting as much as possible. But we also need them to be places where people want to be."

Smart growth is more challenging than constructing LEED-certified buildings or retrofitting existing buildings. It must be guided by an active community with good aesthetic and design standards, and it involves rewriting the zoning codes, business assumptions, and financial ground rules that encourage sprawl. Still, "it really won't make life more difficult for people to tighten up the development pattern, but it will have huge results in the next fifty years," said John Norquist, former mayor of Milwaukee and now head of the Congress for the New Urbanism in Chicago. "The convenient remedy to the inconvenient truth is better land use planning."

Chicagoland could be an ideal incubator for smart growth. The region expects two million new residents and 1.5 million new jobs by 2030.

New homes, schools, shops, and factories must be built, and business as usual will worsen the tangle of mobility and environmental problems. Growing smart could help fix them. The region already has unparalleled (if neglected) rail infrastructure and an active design and architectural community. Like-minded projects are succeeding in the marketplace.

One key question remains: Can Chicago muster the political will? And it's not just Chicago. It's being asked of us all.

PUTTING THE PIECES TOGETHER

In 1893, Chicago captivated the global imagination with an astonishing World's Fair, a fabulous and forward-thinking mock-up of prosperity and innovation. At a time of economic upheaval, the fair showcased better days to come. Engineering marvels such as the Ferris wheel and the future of consumer electric power were on display. Consumer products such as Pabst Blue Ribbon beer and Shredded Wheat cereal were both introduced to the public.

Chicago has already begun the hard work of turning itself into a green city. With a little help from the imagination, it's not too difficult to complete the job by adding a few extra ingredients.

Let's start with Chicago's classic urban DNA, and let's project a few decades into the future. Its passenger rail system has been buffed and rebuilt, both inner city and regional routes. High-speed light rail connects to Milwaukee, Minneapolis, Detroit, Cleveland, St. Louis, and Indianapolis. Because coal is being rapidly phased out, more room is left for passenger trains, and new lines have been established. There is also more room for other rail freight, which has decreased truck traffic on roads and highways. Not a single lane has been added, but cars, trucks, and buses now move more freely. In the event of a backup, they do not sit idling, because most are plug-in hybrids equipped with batteries that can carry them five hundred miles between charges.

Marching up the eastern margin of Lake Michigan, out of sight of the shore, legions of windmills churn ceaselessly. To the west, in Iowa, still more tumble in the air. Every south-facing roof has a solar collector. At night, when demand is low and wind is high, hundreds of thousands of vehicles silently and cheaply recharge. During the day those same vehicles, plugged in at their parking stalls, discharge back into the system, smoothing out spikes in the smart grid.

The regional grid is now designed according to the principles of a healthy ecosystem, with a web of diverse and distributed power sources

smart, cleanly powered grid

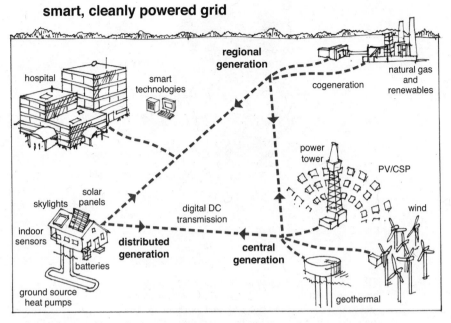

FIGURE 38. Smart grid, clean design. To power society cleanly, our grid should mimic a healthy ecosystem: It should rely on a diversity of renewable energy sources, including wind, photovoltaic (PV) solar panels, and concentrated solar power (CSP). It should produce electricity regionally and near the point of use, which will lend resilience in case one source temporarily fails. And it should include smart technologies and an improved capacity to store electricity.

lending the system resilience and using computerized feedbacks and control mechanisms to maintain homeostasis. This all makes the grid, and the society that draws power from it, not only smarter but more resilient (figure 38). We have at long last learned, when designing our technologies and infrastructure, to take lessons from nature rather than trying to bend it to our will.

In this future version of Chicagoland, every new building in the region is a zero carbon emitter, and most others have been retrofit. Few people struggle to pay their utility bills. Almost everyone can walk or ride a bike to the nearest store, getting daily exercise and enjoying life on the street with neighbors; for longer trips, people take public transport or drive plug-in hybrids and electric vehicles of all sorts (figure 39). Citizens are healthier and more comfortable than the people of Chicago were in the earlier era. They suffer from far fewer heart attacks, and asthma is much reduced. They are fit and less prone to diabetes. The air

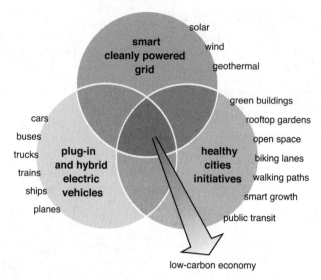

FIGURE 39. Road map to a low-carbon economy. Using a battery of existing technologies and policies, we can quickly make huge strides toward a low-carbon economy. The ingredients include greener transportation options, renewable energy sources, smart grids, and healthy city initiatives that simultaneously cut greenhouse gas emissions and make cities more livable.

they breathe is cleaner and cooler in summer, and their city is built to dampen the impacts of heat waves. Lake Michigan is cleaner, and the trees planted in the Daley era are now mature, majestic, and beautiful.

One of the most pleasing aspects of this march into the future is how transforming the primary systems that power our world will open up new realms of opportunity for creative businesspeople and innovators of all kinds. Where people of an earlier age argued that protecting the environment would cost jobs, in the United States of the future that notion will seem quaint and outdated, a relic of a dirtier era. We'll have spent decades by then building a robust, resilient energy infrastructure. The nation will be safer than ever, as wars for oil will have faded to a historical footnote—a blemish on our storied past, but one we will have overcome. We'll have phased out coal when we saw the mortal danger to our planet that it posed. We'll recall our short-lived mistake growing biofuels to power polluting vehicles, and our croplands will be once again used to grow food. We'll burn less of everything to move ourselves around the globe, conserve our resources far better than today, and reduce and recycle our wastes far more. We'll have a more sustainable approach to the land, water, and air that support all life.

As our shift to sustainable energy technology accelerates, its benefits will spread globally. In the 2010s, scores of developing nations, which did not cause the climate crisis, nevertheless will have become central to its solution, for global funds and redirected investments will have made them major consumers, and ultimately producers, of green technologies.

In rural energy-poor areas all over the developing world, people will have stopped razing forests and burning them for fuel. Instead, they'll have resilient, cleanly powered smart grids, augmented by stand-alone generators powered by the sun, wind, and even human and bicycle-assisted energy, to pump, decontaminate, and desalinate water; irrigate land; power clinics; light homes; and run small businesses. This in turn will give developing regions a greater capacity to cope with the climate-related challenges that will by then be inevitable, such as water shortages and volatile weather, while granting them self-sufficiency and far better health.

Achieving this future will be neither easy nor automatic. But it will create new enterprises and jobs that promote economic security and equity that is not handcuffed to the price of a barrel of oil. In solving the climate crisis intelligently, we can create the engine for a healthy new global economy.

Of Rice and Tractors

Throughout this book I've described how climate change harms human health and how it menaces humanity. To maintain the health of populations, we should prevent health problems before they arise, and to do that, we urgently need to address the root causes of climate change—deforestation and the burning of fossil fuels.

Today, we know far more about the devastating impacts of burning gas, oil, and coal than we once did. We have workable alternative energy technologies available, and more efficient ones in the research pipeline. All of which raises an essential question: Knowing what we now know, why do we continue to power our civilization with fossil fuels?

Although I didn't know it at the time, I received an important clue as a medical *cooperante* in Mozambique in the late 1970s. I saw a country then that was desperately poor but that had an effective and inspiring leader, Samora Machel, who worked hard to provide his people with a decent life. Despite the efforts of this government, I worked in the only major hospital for eight hundred miles, an aging facility with few respirators and rudimentary surgical, monitoring, and laboratory services. Few medically trained Mozambican colleagues worked with us because the country's educational system had provided almost all Mozambicans with far too little education to pursue advanced training in modern medicine.

Mozambique in the late 1970s was a nation whose wealth lay in its land, which provided bountiful harvests of cashews, and in its access to the sea, which supplied fishers with boatloads of shrimp. The country exported and sold these products and several others, yet it was mired in foreign debt, spending what profits it made from exports simply to service its debt. Indeed, the nation was barely able to mount a budget large enough to provide the services needed to maintain a healthy population.

In Mozambique I treated dozens of patients who had been poisoned, literally, by neighboring Rhodesia's brutal attempts to maintain an unjust apartheid system, and I witnessed South Africa begin violently destabilizing Mozambique's government to help maintain its own apartheid society. The latter effort left a legacy of damaged infrastructure and uncleared land mines that would harm ordinary Mozambicans for years to come.

My experiences in Mozambique gave me a deep appreciation that even sovereign nations could not entirely direct their own fate. Instead, what went on in nations was shaped by geopolitical forces, and it was shaped even more so by the nation's role in the dominant economic order. The health of entire nations, I realized, came down to money.

HONDURAS

On a hot, sun-baked early afternoon in Choluteca, Honduras, Dr. Juan Almendares pulled his pickup to the side of the road. He looked across the highway where an old woman sat in a meager pool of shade near two small shacks. Several thin children played nearby.

"Señora!" Almendares called out to the woman, which also roused two younger women inside, one holding a newborn infant. All three women crossed the highway, approaching warily. They were trailed by five children (figure 40).

"I am a medical doctor," Almendares said in Spanish. "I am here to help."

Eva Ambrosia, seventy-five, was grandmother to all nine children who dwelled in these roadside hovels cobbled together from sticks, cardboard, and plastic sheeting. Lucy Aguilera, twenty-four, and Natividad Aguilera, thirty, were her daughters.

A decade earlier when Hurricane Mitch slammed the country, Ambrosia's two farmworker sons had drowned, depriving Ambrosia, her daughters, and their children of both income and food. Lucy Aguilera, then just fourteen, lost her first two children to the flood, as well. Hundreds of their neighbors drowned. The river running through Choluteca was transformed into a raging torrent thirty feet deep. It

FIGURE 40. Lucy Aguilera stops some of her children and her sister's children from venturing onto a busy highway. The family—the two women, their mother, and nine children—were living in a makeshift roadside shack built from sticks, cardboard, and plastic sheeting in Choluteca, Honduras. Extreme poverty makes families like these vulnerable to the impacts of climate change. (Photo by Jason Lindsey/Perceptive Visions)

sluiced over bridges and flooded the land for a mile beyond its banks, scouring twenty feet of soil from the floodplain. Over three days, an estimated 1.8 meters (six feet) of rain fell.

Like many people in the developing world today, this family faces a more subtle threat than drowning: persistent poverty. They cook and heat their home with wood, which can cause respiratory disease. They pull their drinking water from the shallow Rio Choluteca, which runs with garbage and untreated sewage from upstream Tegucigalpa. If there's a wood shortage, they cannot boil the water. They are vulnerable to pneumonia when it rains and suffer from chronic stress and depression.

As the doctor left, one child after another wanted to touch him, hold his hands.

"Adios, adios," the children called. "When are you coming back?"

. . .

Hardships such as the Aguilera family endure are common to people throughout Honduras. Here, thousands of poor families dwell in make-shift houses constructed from the detritus of civilization. Their plight cannot be blamed on a single disaster like Hurricane Mitch but on many disasters, some natural, some the result of climate change, and some the result of global forces beyond their immediate control.

Shifting El Niño conditions have made the Pacific coast of Honduras warmer. In 1972, Choluteca's climate was 6.6°C (12°F) cooler on average than it is now. The region around Choluteca is also much drier as a result of the wholesale removal of mangrove forests in the nearby Gulf of Fonseca. The mangrove forests, like forests everywhere, kept the region moist. They formed a bountiful wetland system about half the size of Rhode Island, harboring exotic creatures like the mangrove warbler and the mantled howler monkey. The mangrove swamps also served as spawning grounds for fish, crabs, and shrimp that helped feed local fishermen and their families.

For the last several decades, however, large aquaculture companies, encouraged by the Honduran government and reinforced by foreign investment, have been clear-cutting mangroves to carve out ponds for commercial shrimp farms. The shrimp farms, in turn, have polluted the gulf's water with levels of nutrients too high for the depleted mangroves to soak up, thereby causing red tides that contaminate fish and sicken human beings. Overgrazing by cattle and monoculture growth of cotton, melons, and sugarcane, all crops that are heavily treated with pesticides, have contributed to the water pollution as well.

In December 1998, a mere four weeks after the torrent of Mitch devastated Honduras, the country's legislature passed the General Mining Law, which marked the beginning of a disaster of much greater proportions. The law gave North American mining companies carte blanche to mine the country's rich veins of gold.

Written into the law were bonuses in the form of low taxes to encourage the companies to begin their open-pit mining practices. Perhaps predictably, the law also relieved the foreign-owned mining companies from the burden of environmental concerns. As further enticement, Honduran law afforded these companies the right to simply relocate entire communities if these communities were situated on land below which the precious metal lay. Where and how these relocations might occur were not spelled out.

A major gold-mining company quickly moved in to take advantage of these generous terms. Eventually, this profitable company was subsumed by the Canadian mining giant Goldcorp. In order to build their enormous open pit mine in El Porvenir, a municipality in Honduras's central Siria Valley, the Canadian firm tore out a 150-year-old village and relocated its residents to shabby prefab houses adjacent to the mine.

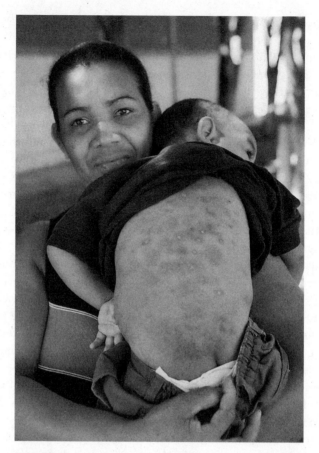

FIGURE 41. In El Pedernal, a mining village in Siria Valley in central Honduras, residents have spent years drinking water from wells contaminated by heavy metals, including arsenic, from a nearby gold mine that uses cyanide and is subject to few environmental rules. Residents young and old have developed similar skin problems, and investigations point to the mine's contaminated water as the cause. (Photo by Jason Lindsey/Perceptive Visions)

To extract gold from the rocks, the mining operation uses cyanide, a process that's known to leach other toxic chemicals into the groundwater. Residents of the relocated community now drink, cook with, and clean with water from a well that draws from groundwater that runs under the mine.

Although Goldcorp denies that the water is contaminated, the Honduran government has fined the company for pollution, and lead, mercury, and arsenic have been found in tissue samples from neighbors of the mine. And Juan Almendares makes regular trips to the communities near the mine to provide medical treatment to local residents, both adults and children, who suffer the effects of arsenic and heavy metal poisoning (figure 41).

Since the passage of the General Mining Law, an incredible 30 percent of the land area of Honduras has been claimed by North American mining companies. Hondurans who must work for these companies to survive receive low wages. Their labor is conducted in an environment that is entirely unregulated by workplace health or environmental laws.

All of which raises the question: What drives the government of this Central American country to condone the violent exploitation of its natural resources and its own citizens?

In the pages that follow, these questions will be explored in depth. For now, it is enough to point out that mining, shrimp farming, and monocrop agriculture in Honduras have several things in common. Each of these industries extracts resources from Honduras. Each is owned by wealthy foreign companies, most often located in the United States. Each generates products—meat, melons, shrimp, gold—for export. And in each case, the profits from these endeavors are shared among a handful of politically connected Hondurans and the North American companies that own the operations.

There is nothing coincidental about any of this.

THE THIRD WAY

Proponents of globalization often cite the theories of Adam Smith, a Scotsman who wrote his famous 1776 text, *The Wealth of Nations*, in an era that social historians call the Age of Reason. Smith argued that free trade was essential for maximum development of a nation's wealth.

The brilliance of Smith's sophisticated arguments can never be diminished. His protégé, David Ricardo, expanded on Smith's work, showing that free trade benefited both trading partners under a wider variety of circumstances than even Smith envisioned. Economists have used Smith's and Ricardo's arguments to persuade national leaders who feared losing trade wars to lower their shields (tariffs) and trade freely. Nations practiced what Smith and Ricardo preached, and for centuries free trade has been the way of the world.

However, Smith's and Ricardo's conclusions rely on a crucial assumption about the nature of trade and capital flow that is ignored by today's boosters of globalization. They lived in an era marked by strong allegiances to nation-states, when transportation among these states was a difficult and costly undertaking. So the two economists made a reasonable assumption for their time: goods would cross national borders freely, but investors' money (capital) would not. Neither man could have envisioned a future in which transnational banks' and corporations' only true allegiance would be to shareholders instead of nations, or one in which they would have the ability to move capital almost instantaneously to anywhere in the world their owners believed they could profit. Under these conditions, free trade between two countries would no longer necessarily be a win-win proposition. Instead, one country could win and the other could lose.

As the nineteenth and early twentieth centuries unfolded, international trade accelerated. During the first four decades of the twentieth century, the world economy became volatile and was marked by dramatic booms and busts. Nations' coffers were drained by World War I and the Great Depression, and economic fear prompted protectionism. To gain advantage over other countries, in the 1930s one country after another restricted trade and devalued its currency to make its exports cheaper so they could sell more of them. But such protectionist efforts were a futile form of economic warfare that only deepened the Depression.

In this milieu of economic warfare and desperation, the British economist John Maynard Keynes published a book that altered economic thinking forever. Keynes's dense but innovative 1936 work, *The General Theory of Employment, Interest and Money,* sold widely and influenced policy makers as no social science treatise has since.

Keynes proposed a method of economic development he called the Third Way, which rejected pure socialism and pure laissez-faire capitalism. Instead, he called for a market economy in which government should intervene in the business cycle whenever necessary to ensure economic well-being. Specifically, he proposed that governments employ both fiscal policy (the use of government spending and taxation) and monetary policy (control of money supply and interest rates) to fine-tune their economies. That approach is standard today.

Keynes recognized that the main driver of the business cycle was investment. He argued that government needed to step up investment in the public and private sectors to sustain full employment during a recession or depression, even if doing so created budget deficits. Significantly, he was careful to distinguish between productive investment—money that paid for a factory or its equipment, for example—and speculation, in which, for example, investors buy an asset in the hopes of selling it and turning a quick profit when its price rises. President Franklin Roosevelt followed Keynesian principles during the New Deal by spending government money to stimulate the economy. President Barack Obama's economic stimulus has drawn heavily on Keynesian principles as well, highlighting their continuing importance.

In 1944, eight years after his pivotal book was published, Keynes used the influence and respect he had garnered to mastermind the most radical, conscious restructuring of the global economy ever undertaken. That summer, delegates from forty-four Western nations convened at Bretton Woods, a resort region in the White Mountains of New Hampshire. World War II was a year away from its shuddering end, and the old world order had degenerated during two world wars and a prolonged economic depression. The countries present recognized the time had come to frame a new world order.

In the two years leading up to this conference, Keynes and Harry Dexter White, the U.S. assistant secretary of the treasury for international affairs and a Harvard-trained expert in international finance, had worked out a tentative plan to reshape the international financial system. Their plan became the starting point for discussion in the clear mountain air of New Hampshire. Delegates designated Keynes their chairman; he was assisted by White. Keynes was a lively man who in the 1920s was part of a literary circle that included Virginia Woolf and Bertrand Russell. In 1944, he was a sixty-one-year-old member of the British House of Lords and editor of a venerable economics journal. For three weeks, hundreds of delegates met in committees, negotiating each aspect of the plan.

It could be said that Keynes's genius was that he figured out he didn't have to figure everything out. Instead, he realized, to change the functioning of the system, one had to change the behavior of the actors, and to do that, one had to change a few key operating rules.

Keynes had intuitively grasped one of the fundamental properties of systems. In both natural and social systems, if we understand the underlying forces that drive a system, and a few simple rules by which it operates, we can understand the way it functions.

Despite the incredible complexity of the Earth's climate system, we can understand its workings to a first approximation by comprehending just a few simple principles: how it's driven by sunlight that shines in and is reflected out to space as heat, how greenhouse gases trap some of that energy, how humans produce greenhouse gases and plants and algae soak them up, and how key feedback loops regulate the system. By changing just one key element—the amount of greenhouse gases in the atmosphere—we've altered the behavior of the entire system.

Similarly, changing just one or two key rules can change the behavior of social systems—our families, communities, organizations, governments, even our global economy—just as changing the rules of baseball can alter the flow of the game.

In their three weeks at Bretton Woods, Keynes and White persuaded the leaders to run the global economy according to three rules. First, goods were to be traded freely, which meant that nations could not erect tariffs or provide subsidies to block trade or protect their industries. Second, exchange rates were to be fixed using the dollar as the primary currency. The dollar itself would be pegged to the price of gold, which was fixed at $35 an ounce. Third, capital transfers from one country to another were to be reined in by regulations. Nations could stipulate that money invested in the country remain to do the work for which it was intended, such as building housing, bridges, schools, or clinics, and they could penalize speculators with taxes when it didn't.

The delegates also called for the establishment of two permanent international bodies. The International Bank for Reconstruction and Development, known today as the World Bank, would help speed postwar reconstruction by providing financial assistance to nations for development. The International Monetary Fund (IMF) would maintain an international monetary system that would promote foreign trade by stabilizing exchange rates and would reconstruct the world's international system of payments.

"As an experiment in international cooperation, the conference has been an outstanding success," Keynes wrote in a letter home to England.

THE PEACE GENERATION

An era of optimism, peace, and prosperity followed the Bretton Woods agreement and the end of World War II. In 1945, the United Nations was born out of the ashes of Woodrow Wilson's League of Nations. Three years later, in 1948, Eleanor Roosevelt shepherded the Universal Declaration of Human Rights through the nascent UN General Assembly, proclaiming the rights of every human being to life, liberty, security, fair legal treatment, education, and "a standard of living adequate for the health and well-being . . . including food, clothing, housing and medical care and necessary social services."

A new global economy soon arose that would drive postwar development. This economic system was shaped by three key elements: new rules, new institutions, and new funds. The rules approved at Bretton Woods guided the system, providing the sticks that kept participants' behavior in line. The new institutions designed there provided the scaffolding for the system. These included the World Bank and the International Monetary Fund, formally launched in 1945. A third international financial institution, the General Agreement on Tariffs and Trade (GATT), was created in 1947 as an international forum to encourage free trade and resolve trade disputes. Today GATT's successor, the World Trade Organization, founded in 1995, continues as the arbiter of international trade.

The funds came mostly from the United States. To further peacetime goals, the Serviceman's Readjustment Act, commonly known as the GI Bill of Rights, was enacted in 1944, providing an enormous boost to the national economy by subsidizing housing, education, and jobs for returning war veterans. The European Recovery Program (the Marshall Plan) provided $13 billion over four years—$159 billion in 2009 dollars—to help fund postwar European reconstruction, thereby building healthy European economies that would also serve as robust trading partners for the United States. These funds provided "carrots" that boosted Western economies and helped drive the behavior of entire nations.

Testament to the success of Bretton Woods, the three decades following World War II were among the most prosperous in history. In the United States, annual economic growth rates exceeded 4 percent, and the total output of the economy tripled between 1945 and 1973. Returning GIs went to newly formed colleges, moved into solid middle-class jobs, and bought houses in burgeoning suburbs. (In the 1950s, the U.S. government provided huge subsidies for airports, highways, and coal and oil exploration, with the result that air and ground transportation industries became, along with housing, the primary drivers of the postwar economy.)

Wall Street participated, too, investing private capital in industries that produced tangible products and services, in contrast to the speculative financial options that would become so popular in contemporary United States. Unions ensured that working Americans earned a living wage.

TRICKY TIMES

In 1971, the good times began to unravel. Responding to mounting U.S. debt incurred by the expense of a continuing Vietnam War, President Richard Nixon opted for a quick fix. In a move that would lead to seismic shifts throughout the global economy, the wartime president broke one of the paramount rules established at Bretton Woods by uncoupling the value of the dollar from gold. No longer could dollars be traded for gold at a fixed sum of $35 per ounce. This allowed the United States to print more dollars, which helped pay for the war. But the world's currencies, which had been pegged to a gold-standard dollar, became unmoored. Exchange rates among countries began floating like ships in a turbulent sea. The world financial system, intact since Bretton Woods, lost its stability.

With currencies of nations fluctuating widely, investors began speculating on currency itself. Currency speculators would buy a nation's currency, then sell it when the price moved—in essence placing bets on currencies' relative values, such as yen versus rubles or pesos versus dollars. Currency speculation became a highly profitable practice in the financial world.

In addition, corporations and investors began moving money to whatever location promised the greatest return on their investment, effectively ending constraints on the international movement of capital. Free trade in capital reigned, in antithesis to the recommendations of Smith and Ricardo as well as Keynes.

Nixon's decision led to more dollars in circulation, but the amount of gold in Fort Knox was finite, and as a result the price of gold skyrocketed from $35 an ounce to over $600 an ounce. More importantly, the price of oil—the substance on which every nation in the world had come to depend for economic growth—rose tenfold between 1971 and 1976, from $3 per barrel to $30 per barrel. As energy costs went through the roof worldwide, the costs of producing goods and services rose as well, causing their prices to skyrocket. This more than anything else caused the inflation that characterized the 1970s economy.

As Smith, Ricardo, and Keynes all recognized, free trade in goods stimulates the real economy: it allows people to obtain tangible goods and services that they need. But Nixon's changes had the opposite effect. Inflation made everything more expensive, and the flow of capital across national borders worked against long-term investments. Instead it favored short-term investment, even speculation.

Inflation caused discomfort in industrialized nations in the 1970s, but it hammered poorer nations, forcing them to go hat in hand to the World Bank to borrow money to run their countries. At the same time, oil-producing nations were realizing enormous profits from the inflated oil prices, and they were suffusing the coffers of the World Bank with this money, which we can call petrodollars. "The Bank," as it's known to Beltway insiders, thus had plentiful funds to lend to countries suffering from inflation and devalued currencies.

In a few short years, the Bretton Woods system had been swept aside. This unraveling not only undid an era of optimism, peace, and prosperity. It also intensified an ongoing battle for control of developing nations' vast wealth of natural resources, including gold, ivory, crops, uranium, fisheries, and forests.

This battle for control took place in handsome corporate offices, in quiet corridors of the Washington, DC, headquarters of the World Bank and International Monetary Fund, and in the capitals of dozens of developing countries worldwide. On one side were the citizens of developing nations, who were fighting for their fair share of the emerging global economy. On the other side were multinational firms like Halliburton, Brown and Root (known as KBR today), Bechtel, and their patrons in the U.S. government.

It was a battle fought not by armies of soldiers but by armies of financiers, and it was never a fair fight. In part that's because the multinational firms had a secret weapon: highly trained economists whom author John Perkins has called "economic hit men."

ECONOMIC HIT MEN

Beginning in the early 1970s, Perkins worked for one of those large firms, a then-powerful international consulting firm called Chas. T. Main. In his eye-opening book *Confessions of an Economic Hit Man,* published in 2004, Perkins laid out in graphic detail how he and other economists did their dirty work. Perkins's job was to create economic models that estimated the costs and benefits of huge development projects such as large dams, highways, seaports, airports, electricity grids, and industrial parks. As an expert in quantitative economic modeling, he knew how to tweak his models to produce the results he wanted, and they inevitably offered forecasts of strong economic growth for years to come. Soon Perkins was plying leaders from Indonesia to Ecuador with his sunny forecasts.

"I was to justify huge international loans that would funnel back to ... U.S. companies (such as Bechtel, Halliburton, Stone & Webster, and Brown & Root) through massive engineering and construction projects," Perkins wrote.

Perkins's pitches usually succeeded, and the government leaders he met borrowed billions to pay for such undertakings. These loans typically required the country doing the borrowing to hire U.S. companies to build those projects. As a result, huge sums of money were simply transferred from U.S. banks to U.S.-based construction and engineering firms. The developing countries never touched the money, but they were required to pay the loans back with interest.

Throughout the 1970s and into the 1980s, one developing nation after another borrowed heavily from the World Bank, regional development banks, and private banks to fund these sorts of undertakings. Yet, as one nation after another fell deeper into debt, each paid more and could borrow less.

But the damage didn't stop there. According to Perkins, economic hit men (and women) served the common interests of what he calls the "corporatocracy." This may not have been a conspiracy, but it was at minimum a loose alliance of multinational corporations, development banks, and local elites in underdeveloped nations. Given the large sums of money being passed among a select few hands, many of the local elites soon fell prey to corruption, and the citizens of the country suffered. Debt was more than a by-product; it was the way the largest post–World War II superpowers, particularly the United States, exerted economic leverage and control, accruing riches from underdeveloped nations' natural resources at bargain-basement prices. Poor nations like Honduras remained on the other side of the divide, sold out by their elites and almost hopelessly ensnared in an unending spiral of debt and interest payments owed to the very nations that were exploiting their natural bounty.

And when nations default, Perkins explained, "we demand our pound of flesh. This often includes one or more of the following: control over United Nations votes, the installation of military bases, or access to precious resources such as oil or the Panama Canal. Of course, the debtor still owes us money—and another country is added to our global empire."

MISSION CREEP

What had become of the International Monetary Fund, founded to preserve the stability of the international economy to prevent another Great Depression? By the early 1980s, both Bretton Woods institutions—the World Bank and the International Monetary Fund—were nearing forty.

The IMF in particular had suffered profound mission creep, its original purpose having receded into some utopian past.

As originally defined, the International Monetary Fund's mandate was to inject money into countries with slumping economies to stabilize them and, by extension, the global economy. But loans and grants from the IMF to developing countries were now being offered with myriad strings attached, often in the form of "structural adjustment programs," also known as conditionalities. The structural adjustment programs required nations to continue paying principal and interest on their loans, even if they had to impose fiscal austerity to ensure those repayments were made.

The fiscal austerity measures enforced by the IMF led to tax hikes and layoffs of government workers in health, education, housing, and sanitation, often with devastating effects on human health. Equally ruinous, these agreements demanded that export crops like the melons and sugarcane planted near Choluteca, Honduras, which generate cash flow to pay back the national debt, take precedence over subsistence agriculture that once kept poor families fed. Higher unemployment and hunger followed.

Under the structural adjustment programs, which continue to be enforced today, the IMF also pushed for privatization, forcing struggling nations to sell state-run companies and utilities, including railroads, airlines, steel companies, and water systems, to private investors. Investors who could afford these entities were almost invariably foreign corporations with few ties to the communities they now served. Whereas governments of developing countries sometimes employ people in order to fight poverty, as Franklin Roosevelt did in the United States during the New Deal, these corporations were instead obligated to maximize profits for the benefits of shareholders in the United States and other developed countries. The result of privatization was often higher unemployment and more poverty.

Not surprisingly, given the corporate agenda that it served, the IMF also pushed countries hard to deregulate private sector companies. Health, worker safety, and environmental rules were among the first to go. Even when the rules remained intact, many countries had debt payments so massive that they had no money left in their budgets to enforce them. The results can be seen in Honduras's Siria Valley, where Dr. Juan Almendares treats mine workers and nearby residents for arsenic and other heavy-metal poisoning and where mining has contaminated soils and drinking water of entire regions, and in the area bordering the Gulf of Fonseca in the south of the country, where corporate shrimp farms have decimated essential mangrove forests with scarcely a peep from the Honduran government.

Finally, the true shape of the post–Bretton Woods system had become clear. True to the view held then by leaders in Congress, the White House, and the three international financial institutions—the World Bank, International Monetary Fund, and World Trade Organization—the top priorities when setting international economic policy were three: trade liberalization, privatization of state enterprises, and deregulation. By the late 1980s, this combination of policies was known as the Washington Consensus. All three policies were and still are remarkably effective at maximizing the short-term profits of multinational corporations.

These policies, combined with the austerity measures imposed on debtor nations, imposed a staggering burden on the citizens of poor nations. The abundant natural resources of most developing nations were and still are extracted and exploited by the corporatocracy. Poverty rises, increasing the income gaps between rich and poor. Land and forests are degraded. As health care and public health systems are eviscerated, the incidence of disease mounts. These stressors, in combination, have left many nations weakened and vulnerable. When stressed further by climate change and a global financial crisis, such nations can become bankrupt, failed nation-states whose main exports are migrants, drugs, and violence. Today, the list of failed and teetering nation-states is growing.

Tragically, the very institutions created at Bretton Woods to prevent disastrous imbalances of wealth and power have transmuted into vehicles that promote these same ills, undermining development instead of catalyzing it. By 1983, the amount of money, much of it interest, collectively being paid by developing nations to service their debts exceeded all the funds flowing into them in the form of aid, loans, and investments. The debt crisis has grown ever since.

The financial order of the post–Bretton Woods era has enshrined rigid free-market ideals and has hewn to corporate profit, cheap goods, and cheap labor as guiding principles. It has done a much poorer job at serving the health of populations and the health of nations.

RICE AND TRACTORS

Even when developing nations create the successful export markets the International Monetary Fund and World Bank require, economic realities stemming from their relative powerlessness can drag them into debt all over again.

Chief among these forces is the vast disparity between the prices of a country's imports and exports. If developed countries pay Vietnam little more for its rice exports than they did in the 1960s but Vietnam must pay inflated 2009 prices to import tractors, then the terms of trade are tilted against Vietnam. Such unfair terms of trade are the root cause of recurring national debt in developing countries. Even if a country manages to pay off its debt—and even if its debt is forgiven—it will be forced to borrow again to afford imports whose prices have multiplied in recent decades, and it will rapidly reaccumulate debt.

Unfair terms of trade are maintained in two ways. First, large corporations in rich, industrialized countries use their size and market reach to hold down the prices paid for raw goods such as rice or coffee from around the world. Second, the World Trade Organization, which sets the rules of trade, tilts those rules to favor the same corporations. It does this by permitting wealthy governments to provide large subsidies for their crops, thereby lowering their prices and underselling goods produced by farmers in developing nations.

To protect themselves, these nations sometimes try to shield themselves with protective tariffs that raise the prices of imports from developed nations, thereby allowing their farmers and other domestic industries to compete fairly. But the World Trade Organization prohibits poor nations from erecting such tariffs, and the International Monetary Fund and the World Trade Organization often respond by choking off investment and development aid. With swords allowed to rich nations and shields denied to developing nations, many poor farmers in the latter have been forced out of business.

This ignoble combination of an unpayable debt cycle and unfair terms of trade drives poverty in the developing world. It also drives environmental destruction by forcing underdeveloped nations to log, mine, and develop monoculture export crops to garner funds to repay their debts.

No amount of financial aid can boost a nation's development and stop the environmental degradation until these perverse incentives, enforced by international financial institutions like the World Trade Organization, are dismantled.

HOT MONEY

As developing nations accumulated crushing debt loads through the 1970s and 1980s, yet another disastrous consequence of the collapse of the Bretton Woods rules was playing out: a rapid rise in currency speculation.

As long as different national currencies exist, some mechanism of currency exchange is necessary for international trade. If a U.S. importer wants to sell Swiss-made watches, he pays his bill in dollars. But the Swiss watchmaker pays his workers in Swiss francs, so he needs someone to pay him francs to buy his dollars. Likewise, if Toyota builds an assembly plant in Indiana and pays its workers with revenues from car sales in Japan, it must find a currency trader who will pay dollars to buy its yen, so it can use these dollars to pay its Indiana workers. Such transactions benefit manufacturing companies, shareholders, employees, and consumers. Neither of these exchanges is currency speculation.

In contrast, if you invest $50 million in a South Korean bank in the form of the country's currency, the won, then sell that currency a week later simply to turn a profit, you're not investing—you're speculating.

After the Bretton Woods system collapsed and exchange rates became more volatile, there were huge profits to be made this way, and currency speculation proceeded to grow to almost mind-numbing proportions. In the early 1970s, currency speculators exchanged a mere $18 billion a day. By the 1990s, however, networks of modern computers made it possible to electronically transfer capital in and out of countries with dizzying speed, and currency speculation grew as well. By 1997, currency speculation had reached $2 trillion a day, the rough equivalent of the U.S. gross national product being turned over via currency speculation every week. Today, the total has reached $3 trillion a day. Economists call speculative transfers of currency among countries "hot money," a term that suggests just how close to the limit of moral behavior the practice comes.

How do these hot-money transactions affect the fortunes of developing nations? When billions of dollars move in and out of a country in periods of days, governments cannot possibly know for certain whether money will be available for long-term initiatives like building schools or health clinics or protecting national forests. They are unable to plan. Instead, they are forced to seek short-term profits from the only material they have that can be rapidly transformed into income: their country's natural resources. Tragically, such economic pressures have led the government of Indonesia to raze vast swaths of tropical forests in Java and Sumatra to create palm tree plantations, which yield palm oil, a lucrative export crop used in food additives and biodiesel. Lost are irreplaceable tropical forests that, not incidentally, soak up carbon dioxide and fight climate change.

To understand how currency speculation wreaks havoc in developing countries, one need only look to the Asian financial crisis of the late 1990s, which devastated the economies of several Asian nations and left the world holding its breath in fear of a global financial meltdown. At the time, international investors viewed the Asian tigers—Thailand, South Korea, Hong Kong, and Singapore—as emerging markets, or engines of economic growth. They invested accordingly in these four countries, which made possible the factories that produce the electronics, shirts, and hats with tags that read "made in Singapore" or "made in Hong Kong."

At the time it was industrializing, Thailand was also in debt, having borrowed from the World Bank in the 1980s to pay for fuel and to fund development projects. As long as international investors were sinking money into the country, it was able to make payments on its debt. Nevertheless, Thailand's debt mounted.

For two years, beginning in 1996, currency speculators at hedge funds and other financial institutions moved billions of dollars worth of hot money into and out of the Asian tigers. There was so much currency speculation at the time that the prices of these countries' currencies often depended less on the underlying strength of their real economies than on the fickle opinions of herds of currency speculators. Concerns about an economy could be amplified into wholesale panic, sending these herds stampeding out of a country, taking their money with them.

Unfortunately, that's what happened in Thailand. In May 1997, speculators sold off Thailand's currency—the baht—in huge quantities. The country then spent billions of dollars worth of foreign currency to buy back its own currency in an attempt to defend it from these speculative attacks. This didn't protect the baht, so two months later, Thailand tried another defensive tactic, choosing to float its currency. But the baht nevertheless went into freefall, settling at half its previous value.

The fall of the baht set off a chain reaction that led to economic chaos in countries throughout East and Southeast Asia. The value of South Korea's currency, the won, plunged, forcing the country to borrow $57 billion from the International Monetary Fund. But the IMF demanded in exchange that businesses and the Korean government lay off thousands. For several months in 1998, South Korean companies were laying off ten thousand people a day, prompting a two-day nationwide strike. Fathers stole rice to feed their children, workers went on hunger strikes for back wages, and a labor activist burned himself to death to protest the layoffs.

Of all the Asian countries whose economies were at risk, only Malaysia defied the onslaught. Its leaders had restricted the flow of capital across its borders by exacting strict penalties for money suddenly exiting the country, a move that World Bank and IMF officials roundly criticized at the time. A decade later, however, World Bank officials would admit that Malaysia's economy had fared far better than those of other Asian countries precisely because it had imposed these controls.

As the Asian financial crisis was playing out, another speculative frenzy was under way in the stock markets of developed countries: the dot-com boom. For six years beginning in 1995, exuberant investors drove up the stock prices of Internet-based businesses and hardware and software companies, bringing the stock market along for the ride. The same era saw a wave of deceit in industry. Enron, for example, manipulated the energy market and used fraudulent accounting to gouge consumers before it collapsed.

Deceit and the gaming of markets ran rampant on Wall Street as well. Extensive investigations by Eliot Spitzer, when he was attorney general of New York state, revealed how "ostensibly independent 'stock analysts' were really pumping up recommendations so their investment-banker bosses could win underwriting business," as Robert Kuttner, coeditor of the *American Prospect,* pithily put it. Then, in 2001, the market caught on all too suddenly, and the era culminated with the dot-com crash—which slashed the value of U.S. stock and bond markets.

EXOTIC DERIVATIVES

Unfortunately, little of substance changed in the world of finance after the dot-com bubble burst in 2001. Hoping for a quick profit, speculators kept betting on currencies and other high-risk, high-return vehicles, including the market in U.S. home mortgages. Before long the finance "brain" of the economy was seduced into yet another speculative bubble, one of the most disastrous ever.

At the time, money was cheap, thanks to former Federal Reserve chairman Alan Greenspan's decisions to keep lowering interest rates, and to lots of capital from abroad. This made it easy to borrow. As housing prices rose, bankers and mortgage brokers who had once issued mortgages only to those with sufficient assets and good credit (prime mortgages) now peddled so-called subprime mortgages to customers who lacked enough money or solid credit.

While the Washington Consensus was playing out in foreign policy, the same laissez-faire ideology held sway in domestic politics, setting the stage for the subsequent crash. Democrats and Republicans alike had dismantled regulations and weakened watchdog agencies, on the theory that an unfettered market would allow capital to flow to its most efficient use.

The dismantled regulations included the Glass-Steagall Act, a Depression-era law that had separated commercial banks (the kind where you can open a savings account) from investment banks and securities firms, preventing conflicts of interest and risky investments of consumers' savings. They also included rules restricting trade in derivatives—a class of security whose value is derived from the value of another asset. This allowed derivatives trading to take off: in 2001, trade in derivatives equaled $900 billion; by 2007, it totaled $45.5 trillion, a sum greater than the entire global gross domestic product.

In 2007 the SEC removed another key regulation, which had controlled short sales—bets by investors that the price of a stock, bond, or derivative would fall. If too many people bet short on a company's stock, or if a hedge fund puts a lot of its money into shorting a stock, it suggests to other investors that something's wrong with the company and that it may fail. Investors then sell off the company's stock, driving the price of the stock down, which can cause a company to lose the capital it needs to invest in equipment and make payroll, leading it to fail. In this way excessive short selling, or shorting, adds greatly to economic volatility and waste and can cost workers their jobs.

In 2007, the Securities and Exchange Commission repealed a Depression-era regulation called the uptick rule that had prevented excessive shorting by allowing short sales of a stock only after its price moved up (an uptick). Now short bets could follow drops in stock price, contributing to downward spirals. Derivatives markets were now subject to no rules at all.

Key to the subsequent crash were two particular kinds of derivatives: mortgage bonds and credit default swaps. From a bank or mortgage broker's perspective, a mortgage loan is an asset, in that it produces revenue from the borrower's payments. Investment banks such as Bear Stearns bought up big piles of loans from lenders, bundled them together, then sliced them up like a hunk of bologna. Then they sold these derivatives—mortgage bonds—to investors, in the process collecting on the underlying loans now rather than later, which drove up profits. The value of a mortgage bond depended on the likelihood that the underlying mortgages would default—the safer the loan, the higher the value of the mortgage bond. Investors who bought these and other derivatives were permitted to borrow ever more money to finance their purchases, which in Wall Street parlance is called increasing leverage.

The success of these financial strategies relied on homeowners continuing to pay their mortgages, and on housing prices rising forever. Financial analysts, realtors, and a parade of financial authors and journalists all assured us that they would.

The investors who bought these mortgage bonds, however, realized uneasily that they were now on the line if the underlying mortgages began to fail. They realized even more uneasily that many of the underlying mortgages were issued to homeowners with shaky finances, buying more house than they could afford, in markets where houses were overpriced. To reduce their risk, owners of mortgage bonds, including hedge funds and investment banks, sought insurance. But instead of a conventional insurance policy, they bought a derivative called a credit default swap that served the same purpose. Buyers of credit default swaps were protected if their mortgage bonds became worthless because homeowners stopped paying. Sellers of credit default swaps, such as the insurance company AIG, were now on the line if mortgage holders defaulted across the board.

Then the house of cards collapsed.

By 2007, subprime mortgage companies had moved from bait to switch, jacking up interest rates on adjustable-rate loans, and subprime borrowers started defaulting en masse. Mortgage bonds based on those subprime mortgages became worthless, which caused the holders of these bonds to call in their credit default swaps. Issuers of the credit default swaps, like AIG, then the world's largest insurance company, had set aside far too little capital to back the derivatives, and they teetered.

Then housing prices plummeted, hedge funds went under, and the stock market plummeted. Bear Stearns, the nation's fifth largest investment bank, toppled in March 2008, when the eighty-five-year-old firm was bought out at fire-sale prices by JPMorgan Chase. In September, the venerable investment bank Lehman Brothers went bankrupt, sending a seismic shock into the global financial system. By the spring of 2009, the list of U.S. financial firms that had either failed or turned to the federal government for bailouts included the giant mortgage lender IndyMac Bancorp; six of the nation's largest insurance companies, including AIG,

Prudential, and Allstate; the investment banks Lehman Brothers and Morgan Stanley; the retail banks IndyMac and Washington Mutual; the auto lender GMAC; and the credit card company American Express.

Meanwhile banks had no money to lend to ordinary citizens or to businesses, which needed it to grow or make payroll. This slowed economic activity. About 3.6 million jobs disappeared in the country between the start of the economic downturn in December 2007 and the spring of 2009. Tens of thousands lost their homes, particularly in black and Hispanic communities that had been targeted by predatory subprime lenders, and neighborhoods reeled. American taxpayers were on the hook for hundreds of billions of dollars.

By early 2009, the economic contagion had spread via the world's interconnected web of finance, infecting countries around the world. Iceland's three largest banks all failed and were nationalized, and Latvia's economy shrank a whopping 18 percent in the first three months of 2009, forcing both countries to look abroad for bailouts. As developed countries struggled, developing countries saw private investments from abroad shrink by more than two-thirds, threatening fledgling economies. Worldwide, tens of millions of people in developing countries lost jobs, plunging them into poverty and threatening them and their families with malnutrition. Civil unrest boiled from the Baltic to the Balkans.

By unhinging exchange rates and constraints on capital flow, and by maintaining unfair terms of trade, world economic leaders had removed the last meaningful checks national governments had exerted over the behavior of global corporations through rules and institutions designed to foster societal well-being. These moves had unleashed decades of petrodollar-driven loans to poor countries, resulting in insurmountable debts and two prolonged rounds of frantic speculation. The era had finally ended when the financial head of the economy found itself severed from the body. It seemed the engine of productivity in the world had become money itself rather than the labor and creations of human beings.

The old order, the Bretton Woods rules of global finance, had clearly fallen into disarray. Just as clearly, the system that had stealthily replaced it, the government-enforced laissez-faire system known as the Washington Consensus, had steered the global economy off a cliff, damaged the environment, worsened climate change, and harmed human health. What was not clear yet was the shape of the new global economic order that would emerge, and whether the new system would do a better job at fighting climate change and creating a healthier world.

Rewriting the Rules

By the time the economy crashed in 2008, the Earth's climate system was also in critical condition. Climate change had begun to deliver more frequent droughts, heavier downpours, heat waves, and wildfires. The world's glaciers and ice sheets were dwindling, and permafrost was melting. Sea level was rising at more than one inch per decade, threatening low-lying island nations and coastal areas worldwide. Temperatures were well on the way to the 2.0°C (3.6°F) change that scientists generally agreed would be catastrophic. Carbon dioxide had risen to levels 40 percent greater than before the industrial revolution, and they were heading up fast.

In February of that year, I sat in the Trusteeship Council Chamber of the United Nations in New York City, listening as one of the world's preeminent climate scientists, John Holdren, then my colleague at Harvard University, addressed more than 450 financial and corporate leaders from around the world at a special summit meeting on climate risk. *Global warming,* he told them, was not an adequate term to describe what was happening to the climate. *Global climatic disruption* was more like it because the climate was changing rapidly, and in harmful ways.

Climate change had already begun to put at risk everything climate governs, said Holdren, who would go on to be President Obama's chief science adviser. That included the availability of freshwater; the productivity of farms, forests, and fisheries; the distribution of species; and the geography of disease. In less than three decades, climate change had gone from a worry for a handful of climate scientists to a clear and present danger to humanity. We already have "dangerous anthropogenic interference with the climate system," Holdren told the investors and business leaders. "The key question now is, can we avoid catastrophic interference?"

In December 2007, climate scientist James Hansen of NASA, who first alerted Congress and the world about the dangers of climate change during the hot, dry summer of 1988, described at a geophysics conference an analysis he and his team had conducted. As he had two decades before, Hansen forcefully stated a conclusion that other climate scientists had shied away from making. He and his colleagues had analyzed past climates and past temperatures using cores of ocean sediments. They show that 35 million years ago, when the planet was ice free, carbon dioxide concentrations stood at 450 parts per million.

Right now we're on track for 550 ppm carbon dioxide levels—double preindustrial levels—in four decades. A succession of expert panels, including the IPCC, have said we must reduce greenhouse gas emissions by 60 to 80 percent below 1990 levels by 2050 to stabilize our climate and that the climate system can tolerate no more than 450 ppm, and policy makers have used that level to craft emissions reduction policies.

Hansen challenged their views. "If you leave us at 450 ppm for long enough it will probably melt all the ice—that's a sea level rise of 75 meters," Hansen told the British newspaper *The Guardian* after his research was published in a peer-reviewed journal that spring. "What we have found is that the target we have all been aiming for is a disaster—a guaranteed disaster." Indeed, under business-as-usual scenarios, enough ice would probably melt by the end of the twenty-first century to raise sea levels two meters (more than six feet), which would swamp coastal regions worldwide.

To avoid such a fate, Hansen proposed, instead, a target level of 350 parts per million—1988 levels—as the highest level the planet can tolerate without risking a profound and potentially catastrophic climate shift. Scientists may quibble over the details of Hansen's analysis and his prescription. Even sympathetic public officials will grouse about the political difficulties, and special interests will howl.

But other scientific analyses support Hansen's conclusions, and the overall direction of the data is clear. By burning billions of tons of fossil fuels annually for so many decades, we've been poking the climate system, that angry beast, with a stick, as Wally Broecker of Columbia

University's Earth Institute has so graphically depicted. Today the beast is talking back to us, and it's saying this: Back. Off. Now.

. . .

Can we do it? To reduce our net greenhouse gas emissions enough to stabilize the climate, to preserve an environment conducive to good human health, we need to accelerate the inevitable switch from fossil fuels to low- or no-carbon energy sources. Societal changes of this magnitude have happened before. Hansen has likened the measures required to the effort nations spent to fight World War II. My colleague Bill Moomaw, director of Tufts University's Center for International Environment and Resource Policy, has pointed out that only 3 percent of homes had electricity in 1905; now just 3 percent have renewable energy. "Here in the early years of the twenty-first century," he said, "we're looking for an energy revolution that's as comprehensive as the one that occurred at the beginning of the twentieth century, when we went from gaslight and horse-drawn carriages to light bulbs and automobiles."

We have described the beginnings of this revolution—how individuals, businesses, and communities are acting on their own to slash their carbon emissions. But there's only so far that most people can go to tackle a long-term global threat, even an existential threat, if they think change is beyond their reach or that it may cost them too much up front. To go further, they need help.

After John Holdren spoke at the climate summit at the United Nations, Peter Darbee, chief executive officer of Pacific Gas and Electric, the giant California utility commonly known as PG&E, told the crowd of business leaders that businesses needed to step up and take responsibility for reducing their carbon emissions right now. But they too needed help. The federal government, he said, should provide targets for reducing greenhouse gases, specific timetables to reduce them, and national rules on carbon emissions.

At first glance, it might seem odd that the CEO of a major utility was pushing the government to regulate carbon dioxide emissions. But Darbee had plenty of company. Somewhat surprisingly, over the past few years many U.S. and European business leaders outside the fossil fuel sector have pushed for tighter climate and energy policies.

Many manufacturers, utilities, and other companies have acknowledged the urgency of the climate threat and realized that they won't be able to power their enterprises with cheap and abundant fossil fuels forever. They've known that carbon emissions rules were coming, but they haven't known when. And that uncertainty has made it hard for them to plan investments in their own firms, and it has made it hard for financial firms to direct investments and loans to others.

In arguing that federal policy changes were essential, Darbee cited California's decision in the late 1970s to create incentives for energy efficiency by decoupling utility profits from electricity sales. That policy change allowed PG&E and other utilities to push for energy conservation without worrying about cutting into their profits. Another policy, called net metering, allows individuals and small businesses in California to generate their own electricity via solar panels or windmills and sell it back to the grid, as Ron Later, the pistachio farmer with the wind turbine, now does. Because of these policies and several others, California has kept its per capita energy use flat for three decades, while the per capita energy use in the rest of the country, where decoupling and net metering are not common, rose by 50 percent.

By 2008, two major corporate coalitions were pushing for new federal rules that would help their members plan for the coming carbon-constrained world. A coalition of energy-intensive companies and national environmental groups known as the U.S. Climate Action Partnership (USCAP) wanted a law mandating greenhouse gas emission cuts of 80 percent below 2005 levels by 2050. A second coalition of progressive, less energy-intensive companies—Business for Innovative Climate and Energy Policy (BICEP)—wanted even more muscular regulations: mandatory emission cuts to 25 percent below 1990 levels by 2020, and 80 percent below 1990 levels by 2050. BICEP's proposal was in line with the recommendations of the Intergovernmental Panel on Climate Change.

States, too, had demonstrated their impatience with the logjam in Washington. By the time of the 2008 climate summit, they had begun moving forward on their own to fight climate change. California started the trend in 2002 when it passed a law mandating a 30 percent increase in fuel efficiency from the state's fleet of cars, sport utility vehicles, and light trucks by 2016.

Car dealers and automakers sued to block the law, delaying it from going into effect. That lawsuit was finally shot down in court in 2007, but California's pollution control efforts were soon stymied again when the U.S. Environmental Protection Agency denied it and sixteen other states the right to set their own greenhouse gas emissions rules, claiming it would create what Stephen L. Johnson, then the agency's administrator, called a "confusing patchwork of state rules."

On April 22, 2008, President George W. Bush planted a Shumard oak tree in a public square in New Orleans, gave an Earth Day speech, and proposed a new national goal. Instead of reducing its greenhouse gas emissions immediately, the United States would allow them to increase, stopping their growth by 2025—seventeen years later. The plan was widely panned by environmentalists and officials who were seriously tackling the issue. One of them, Representative Ed Markey of Massachusetts, said, "By the time President Bush's plan finally starts to cut global warming emissions, the planet will already be cooked."

POLLUTE HERE, GROW A TREE THERE

By the time Bush planted his tree, seventeen states had banded together to create three regional agreements to curb greenhouse gas emissions. The three regional groups—in the Northeast, Midwest, and West—had established regional climate-preserving programs known in policy circles as cap and trade.

In a cap-and-trade system, strict limits are placed on total emissions across the region or nations covered, which constitutes the cap. Then a market is set up in which polluters buy and sell permits that allow them to emit a certain amount of pollutant, in this case carbon dioxide. If they reduce their emissions, they can sell extra permits to others and make money, but if they pollute more, they have to buy more permits. Each year the cap is reduced, and permits become scarcer and more expensive.

The cap-and-trade system was pioneered under the 1990 overhaul of the Clean Air Act. The reforms motivated operators of coal-fired power plants to install smokestack scrubbers, which helped curtail the plants' sulfur emissions and tame acid rain, and at a much lower cost than utilities had predicted.

Cap and trade is an example of a market-based solution, in that it uses market forces to convince companies to pollute less, and in the United States it had emerged as the dominant policy solution. Some environmental groups like cap and trade because they believe an economy-wide cap will help rein in greenhouse gas pollution. Some social justice activists like it, too, because auctioning pollution allowances to companies, as some have proposed, would generate revenue that could be used to reduce heat and electricity rate increases for the poor and middle class (although owners of coal-fired plants might continue polluting poor neighborhoods, to the detriment of residents, while purchasing carbon credits elsewhere to offset their pollution). Many economists like cap and trade because it's a market-based solution. And many politicians like it because it offers political cover. Both cap and trade and a carbon tax that directly taxes fossil fuel consumers, producers, or electrical utilities would raise the price of fossil fuel–based energy. But for several years cap and trade benefited in public deliberations from not including the word *tax*.

Despite their advantages, carbon cap-and-trade systems are potentially full of loopholes, with carbon dioxide bubbling out of every one. Our domestic cap-and-trade program to reduce sulfur emissions was successful because most of the emissions came from two thousand smokestacks in the Midwest—a manageable number to monitor and enforce. Carbon dioxide emissions, however, arise from a multitude of sources, including power plants, offices, homes, and cars, making it tough to truly cap the emissions.

A global cap-and-trade system would be particularly difficult to implement. One of the biggest potential loopholes is carbon offsets, in which companies in wealthy countries, like utilities in poor neighborhoods, might be able to buy their way out of emissions reductions by paying to preserve forests elsewhere. There's also little to stop a factory in the developing world from starting to pollute so that someone else will pay them to stop.

Market mechanisms like cap and trade excel at spurring the least expensive climate solutions, particularly those with low up-front costs but high energy savings, which pay for themselves relatively quickly. Examples include replacing incandescent lightbulbs with LED bulbs or insulating your water heater.

But we also need solutions that market forces alone won't provide—those with high up-front costs but low operating expenses once established. The work of the Cambridge Energy Alliance provides a good example. By paying up front to retrofit houses and commercial buildings and allowing the owners to pay off the costs with their energy savings, they make possible energy-saving projects that would not otherwise have happened.

By investing money up front, we also stimulate fledgling markets and create new ones. After three decades of free-market fundamentalism, it may seem almost un-American to some to intervene this way in the market. But such investments are no different from how we built the interstate highway system or how we fund schools or maintain national parks.

Today we need equally substantial investments to build mass transit systems, to pay new workers to protect forests and watersheds, to retrofit public buildings. We also need to invest extensively in research and development to make renewable energy technologies that are even more effective and efficient than today's. Such investments will slow climate change, stimulate existing markets, and create new markets and new green jobs.

We need these and other bold, all-encompassing practices that reduce carbon emissions from burning fossil fuels and increase carbon sinks all over the world. This is our main task, and we must not lose sight of our goal.

PRODDING THE POLITICIANS

By 2008, the forces of global-warming denial seemed at long last in retreat, and leaders of many stripes called for meaningful action on climate change. These included U.S. religious leaders.

The Jewish Council for Public Affairs had urged "the rapid adoption of clean and renewable energy sources and technologies" and "the phasing out of reliance on fossil fuel technologies." The U.S. Conference of Catholic Bishops called for climate action to protect the world's poor, who will suffer the brunt of climate change's harmful effects. "The United States bears a special responsibility to lead and help shape responses that serve not only its own interests but those of the entire human family," the bishops wrote to the U.S. Senate.

More surprising to some, "creation care" movements had sprung up among evangelical Christians, a largely conservative, probusiness bloc that had been previously unengaged on environmental issues. "We dare to imagine a world in which science and religion work together to reverse the degradation of Creation," said Richard Cizik, then vice president for governmental affairs of the National Association of Evangelicals. "We will not allow it to be destroyed by human folly."

Others advanced the policy discussion. In 2008 Al Gore, having received the Nobel Peace Prize for his climate advocacy work, gave a major speech in Washington challenging the United States to produce all its electricity efficiently and from renewable sources by 2018. Best-selling environmental author Bill McKibben launched a global advocacy effort called 350.org to impel governments to put policies in place to lower atmospheric carbon dioxide to 350 parts per million, which James Hansen had commended as the highest safe level for human civilization. Activists young and old pushed to reverse our destructive course.

As climate anxiety grew, so did the pressure to act, and politicians of both parties fell in line with the public sentiment. Politicians mostly said the right things, and sometimes did them. Senator John McCain, the Republican nominee for president in 2008, had twice cosponsored meaningful federal cap-and-trade legislation, once in 2003 and again in 2007, although special interests blocked each bill's passage. During the 2008 campaign, he called for a cap-and-trade program and higher fuel economy for cars and trucks. "The facts of global warming demand our urgent attention," he said.

McCain's good works and nuance on the issue were lost on the public, unfortunately, after he chose Sarah Palin, then governor of Alaska and a climate change naysayer, as a running mate and after the crowd at the Republican convention famously chanted "Drill, baby, drill" when the subject turned to energy and oil.

The Democratic candidate, Barack Obama, also seemed serious about tackling climate change. He called for cap and trade, higher fuel efficiency standards, and research on alternative fuels, and he supported reduction in carbon emissions to 80 percent below 1990 levels by 2050, in line with scientists' recommendations. He called climate change "one of the most urgent challenges of our generation."

A WORLD IN CRISIS

As the climate and financial crises intensified in 2008, a global fuel crisis also festered, creating shortfalls of oil and other petroleum products that caused U.S. gasoline prices to spike to more than $4 per gallon. U.S. oil and gas production had declined in 2004 and 2005, after hurricanes Ivan, Katrina, Rita, and Wilma disrupted seafloor pipelines in the states bordering the Gulf of Mexico, as well as oil rigs and natural gas processing facilities. As these examples indicate, stronger and larger hurricanes fed by warmed oceans pose serious threats to offshore oil production, as do safety and environmental concerns.

A global food crisis also worsened in 2008. It had started when westerly winds that whip around Antarctica tightened their vortex, drawing moist air and rain away from southern Australia and creating a multiyear drought in this major wheat exporter. Many climatologists believe that these wind patterns were altered by both climate change and the southern hemisphere's hole in the ozone layer, so the climate crisis worsened the food crisis.

Fuel shortages also worsened the food crisis. Inflation-adjusted food prices had been falling for a century until 2004. But as oil prices spiked worldwide in 2008, they drove up farmers' costs and the costs of getting food to market. Maize prices tripled; wheat and rice prices quadrupled. As prices of staple foods soared, riots broke out across the globe. In March 2008, violent mayhem at government-subsidized bakeries in Egypt, a major wheat importer, killed at least ten. The next month, a week of food riots in Port-au-Prince helped topple Haiti's government.

Poorly thought-out fixes for the fuel and climate crisis worsened the food crisis. As corn-based ethanol took off in the market, propped up by $5 billion a year in federal subsidies, U.S. and Mexican farmers had switched millions of acres from corn for food or animal feed to corn for biofuels. The price of corn flour shot up, and the price of corn tortillas, a dietary staple for Mexicans, quadrupled. On January 31, 2007, tens of thousands of angry farmers, workers, and poor Mexicans took to the streets of Mexico City to protest, some clutching cobs of corn to drive home their point.

As the food crisis worsened, it had a dramatic impact on the health of millions. By June 2009, the number going hungry daily had surpassed 1 billion. In some countries, the food crisis could raise the mortality rates of children under five by up to 25 percent, the United Nations Environment Programme found.

It's no coincidence that these four crises—climate, fuel, food, and finance—were happening simultaneously. Burning oil and other fossil fuels worsens the climate crisis. Fossil fuel shortages raise energy prices, worsening the financial crisis, and drive up the price of food, worsening the food crisis. The climate crisis can threaten oil supplies, worsening the fuel crisis, and it breeds drought and extreme weather, reducing harvests and worsening the food crisis. High energy and food prices combine to force more people into poverty, which worsens environmental degradation, which harms health.

This web of relationships means that unless we choose carefully, our would-be fixes for one crisis can inadvertently worsen another. For example, biofuels, particularly corn-based ethanol, are being promoted as an effective way to reduce greenhouse gas emissions, but they're hugely inefficient at best, and they worsen the food crisis, as the Mexican food riots showed. Similarly, ramping up oil and gas exploration on the continental shelf or in the Arctic, or scraping away forests to find tar sands, may give us a temporary fix for the fuel crisis, but it worsens the climate crisis. Poorly designed economic policies, as we've seen, can also worsen other crises. The World Bank and IMF policies of the last three decades—privatization, liberalization, and globalization—have at times increased cash flow to nations from crop exports. But, as the situations in Choluteca and the Siria Valley, Honduras, illustrate, they deepen poverty, degrade the environment, and harm the health of millions. In the long run, such policies backfire.

To maintain and improve human health, we must choose solutions that address all four of the great crises of our time—the global food crisis, the energy crisis, the economic crisis, and climate change. What's more, we need to do it in a way that is fair to everyone, including the world's poor. This seems like a daunting task. The good news is that solutions that address these multiple crises are available.

SEARCHING FOR SYNERGY

On November 19, 2008, just two weeks after he was elected president, Barack Obama signaled the world that U.S. energy and climate policy had turned a corner. He'd prepared a four-minute videotaped message for a climate change conference that California governor Arnold Schwarzenegger had organized in Los Angeles. In the audience were governors of twelve states; regional leaders from Mexico, Canada, Brazil, and Indonesia; and more than seven hundred delegates from nineteen countries.

By then the United States had gained a reputation abroad as a profligate polluter that was unwilling to do its part to solve climate change. For the entire span of the Bush administration, U.S. negotiators had been unhelpful at best in establishing global measures to reduce greenhouse gas pollution. The United States was also the only industrialized nation to reject the Kyoto agreement.

In Los Angeles, Obama's taped message signaled a sharp break from the Bush administration's climate policies. "The science is beyond dispute and the facts are clear," he stated. "Delay is no longer an option. Denial is no longer an acceptable response." Obama's administration would chart a course toward slashing U.S. greenhouse gas emissions 80 percent by 2050, he said. The president-elect rejected the view, already circulating in some corners of industry and in Congress, that an economic crisis was no time to tackle the climate crisis. Any company or state that was willing to fight climate change would have an ally in the White House, he said. "And any nation that's willing to join the cause of fighting climate change will have an ally in the United States of America."

At his inaugural address in January 2009, Obama vowed to double alternative energy—solar, wind, biofuels, and "clean coal"—in three years, to expand the nation's energy infrastructure, and to save billions in energy costs through renewed efficiency. Then he followed through. He chose science, energy, and environment advisers so accomplished and respected that the League of Conservation Voters called them the "Green Dream Team."

In May 2009 Obama announced tough new rules on car and truck fuel efficiency that would make the car and light truck fleet 40 percent cleaner and more fuel efficient by 2016. These were even stronger than the standards that California had passed in 2002 and the automobile industry and Bush administration had blocked. But this time the crippled U.S. auto industry, which was looking desperately to Washington for a bailout, meekly applauded. The director of the Sierra Club's Safe Climate Campaign called it "the single biggest step the American government has ever taken to cut greenhouse gas emissions."

Obama made several other moves that addressed economic stagnation, the fuel shortage, and the climate crisis simultaneously. His $787 billion fiscal stimulus bill was designed to create jobs during the economic downturn, a move reminiscent of the New Deal. It included the largest federal renewable energy investments ever: $30 billion for energy-related projects that would create so-called green jobs, including $11 billion to modernize the electrical grid; $5 billion to weatherize low-income housing; and $2 billion to research batteries for electric cars. Obama's stagecraft communicated his priorities. He signed the bill in Denver, a hub of the nascent U.S. renewable energy industry, following a photo op at the Denver Museum of Nature and Science, where he and Vice President Joe Biden strolled amid rooftop solar panels.

Other world leaders joined Obama in promoting green jobs and investments: South Korea had invested $36 billion, or 3 percent of its gross domestic product, in green jobs, including in mass transit, energy conservation, and river and forest restoration. China was spending $140 billion of its $586 billion stimulus, or just under 2 percent of its gross domestic product, to give its renewable energy sector a push. To create jobs, reduce poverty, and lower climate and ecological risks, the United Nations Environment Programme had begun a serious push for what it termed a Global Green New Deal—a worldwide Keynesian spending program that focused on clean energy.

Obama recognized that reducing fossil fuel dependency is the single most effective way to combat climate change. As a savvy politician, however, he painted it most of all as an economic choice—sending the nation's wealth away to pay for oil, or exporting clean energy. "The nation that leads the world in creating new sources of clean energy will be the nation that leads the twenty-first century global economy," he told a crowd at a Nevada Air Force base. "That's the nation I want America to be, and I know that's the nation you want America to be."

POLLUTER PUSHBACK

By 2009, the world's top climate science body, the IPCC, had reported that humanity would have to start reducing greenhouse gas emissions by 2012 to significantly lower the risk of climate catastrophe. But climate-warming emissions were rising faster than anyone had anticipated, and nearly three times faster than the IPCC's previous worst-case scenario.

To head off climate disaster, more than 180 nations would meet in December 2009 in Copenhagen, Denmark, to craft a new treaty to replace the Kyoto agreement. Obama sought to influence the world's nations on climate policy, but he faced a huge hurdle. He first needed to overcome the nation's international reputation as a climate slacker and an obstacle to progress. That meant putting the centerpiece of his green energy push, the first federal cap-and-trade law for carbon, on the books by the Copenhagen talks. Carol Browner, Obama's top adviser on energy and climate, said that such a law was "absolutely essential to our position and what we can ultimately hope to achieve in Copenhagen."

In May 2009, Henry Waxman, a tough, experienced progressive congressman from Los Angeles, and his colleague Ed Markey of Massachusetts, long a champion of green policies, introduced a cap-and-trade bill that was designed to cut U.S. greenhouse gas emissions 20 percent below 1990 levels by 2020, even more than Obama had asked for. Permits were to be auctioned to pay for renewable energy research and to offset possible electricity rate increases for the poor and middle class. So far, so good.

Then the horse-trading started. Some of it seemed reasonable. The wind and solar industries wanted transmission superhighways; farm groups wanted payments for no-till farming, planting trees, and capturing methane from feedlots. Utilities, manufacturers, and energy companies wanted permits worth hundreds of billions of dollars, but they claimed they would use the money they saved to offset electricity rate hikes. But some of it was just nosing at the trough. The biofuel industry pushed to lift the cap on how much ethanol could be blended into gasoline. The nuclear industry pushed for billions to insure new nuclear plants. Big coal wanted billions for carbon capture and storage, its only hope to stay alive in a carbon-constrained world.

The bill that emerged from the House of Representatives that spring did cap emissions and would begin to reduce them, although not fast enough. But it had many weaknesses. It gave away most of the pollution permits to industries that produce dirty energy or use a lot of it, rather than auctioning them and using the money to reduce electricity rates. And, most injuriously, it stripped the Environmental Protection Agency of its authority to regulate carbon dioxide emissions—the authority that provided the "stick" that was prodding Congress to enact legislation.

Despite the bill's failings, many progressives saw it as a first step that could be amended later to ratchet it up, and the need to enact something before the pivotal Copenhagen climate conference led most to go along. But several major environmental groups considered it so compromised that they withdrew their support.

Theatrics followed anyway, as conservatives, following ancient instincts, sought to portray the bill as a killer of jobs, particularly manufacturing jobs in the heartland, which would ostensibly migrate overseas where environmental rules are lax. "This bill would impose enormous taxes and restrictions on free commerce by wealthy but faltering powers—California, Massachusetts, and New York—seeking to exploit politically weaker colonies in order to prop up their own decaying economies," Indiana governor Mitch Daniels fumed in an op-ed for the *Wall Street Journal*. "Quite simply, it looks like imperialism." This obstructionism and overheated rhetoric was too much for even Daniels's hometown newspaper, the conservative *Indianapolis Star*. Citing economic and environmental benefits, the newspaper recommended investing more in renewable energy and burning less coal. "The transition will be difficult, but obstinacy is not an option," the editors wrote.

As the fateful Copenhagen conference approached, the Senate, preoccupied with health care legislation, failed to take action on climate change and energy. Then, on November 17, 2009, just three weeks before the conference, a purloined trove of a thousand emails between climate scientists appeared mysteriously online. Stories spread like wildfire through the conservative blogosphere and right-leaning media outlets, accusing scientists of manipulating data to prove global warming, conspiring to suppress dissenting scientific opinions, and hiding their raw data and documents from the public. It was a scandal—Climategate!—one that discredited decades of climate science, critics insisted. The Climategate frenzy spread to Congress, where a conservative Wisconsin congressman tarred the climate scientists as "scientific fascists," and U.S. Senator James Inhofe (R-Oklahoma) demanded criminal investigations of seventeen scientists who worked on the IPCC report.

Climategate dominated the news cycle in the weeks leading up to the Copenhagen conference; it put climate scientists on the defensive, and it confused the public. A survey published on December 3, two weeks after the story broke, revealed that 52 percent of Americans believed there was significant disagreement over global warming among scientists (there wasn't), and 84 percent believed that it was somewhat likely that some scientists had falsified data to support their theories on climate change. Opponents of climate action played this potent political card to great effect. Nations meeting at Copenhagen failed to agree to a treaty with binding emissions targets to replace the expiring Kyoto treaty, and the U.S. Senate lost its appetite for a cap-and-trade bill.

But Climategate failed entirely to weaken the scientific case for human-induced climate change, according to investigations by five blue-ribbon panels, including the British House of Commons, the Royal Society, and a specially appointed commission, that were released in the spring of 2010. Investigators did chide individual scientists for being rude, boorish, and too willing to exclude climate naysayers. Years of coordinated, well-funded attacks on leading scientists may have contributed to the defensiveness. And perhaps we wanted to believe that scientists conspired to cook the books, that the unfolding nightmare of climate change was not real. Nevertheless, the panels that probed Climategate agreed with the twenty-five top climate scientists who wrote to Congress in December 2009, "The body of evidence underlying our understanding of human-caused global warming remains robust."

REWRITING THE RULES

The tussles over Climategate and cap and trade illustrate the huge political challenge we face in dislodging the power players of the old economy to create the healthy green economy we so urgently need. Oil companies like ExxonMobil and BP, oil-field services companies like Halliburton, coal companies like Peabody, and energy-intensive industries such as cement, steel, and petrochemicals have billions of dollars at stake in maintaining the status quo. They commandeered White House energy and climate policy for the first eight years of the decade and have done all they could to stall climate progress. Today these same corporations use their wealth and clout to cling to power. It will take a tremendous amount of will and political skill to move these power players to the sidelines, where they now belong.

Today we stand at the crossroads. In one direction lies business as usual, the road we've traveled for decades. Down that path, we'd forgo serious measures to rein in our oil consumption, we'd continue buying oil from the dwindling number of oil-rich nations to drive our SUVs a mile to the grocery store, and we'd scrape Alberta clean to find a few years of tar to put in our tanks. We'd decapitate more mountains in Appalachia to heat our homes and power our flat-screen televisions, pumping out climate-warming carbon with every watt.

Another path exists, however, that leads to a far brighter future. As we traveled that path, we'd finally learn the lessons of the late twentieth and early twenty-first centuries. In the realm of economics, we'd learn that no market system is truly free; instead each is steered to some degree by powerful special interests that may not have the public welfare at heart. We'd learn that if we don't regulate the market for the public good, then these special interests will covertly assert control, causing long-term harm to the planet and everyone on it.

In the realm of international politics and trade, we'd learn from the impasse in climate politics that developed as the crucial Copenhagen climate talks approached. Dozens of poor developing nations had united to demand that developed nations make far steeper cuts in their greenhouse gas emissions than the wealthy countries had proposed and also to demand that developing nations be exempted so they could fuel development and pull their people out of poverty. We'd realize that poor nations need the financial resources to buy and make advanced technologies to fight climate change and that these international markets are central to the clean energy transformation. More broadly, we'd realize that while all nations did not contribute equally to the climate problem, all can be part of the solution.

Today the hour is far too late, and humanity's footprint is far too large, for business as usual. Instead, we have an opportunity. As our current economic order crumbles, we can consciously shape a new and enlightened global economic order for the twenty-first century, much as the world's developed nations, guided by John Maynard Keynes, shaped an enlightened global order at Bretton Woods.

The goal would be a form of capitalism akin to what worked so well in the mid-twentieth century, with a highly regulated financial sector that would steer industry to protect the global commons—our air, water, fisheries, and forests. We would shape the economy not for the profit of the few but for the benefit of all humanity, rich and poor alike. We'd create a truly sustainable low-carbon economy, one that preserved and restored the environment in a manner that's fair to the world's poor. We'd shape it to create a healthier world.

. . .

What would such an effort look like?

In planning a new order, both developed and developing nations must be at the table. In Bretton Woods, an old boys' club from developed nations cloistered itself at a cozy retreat in New Hampshire's White Mountains, with no representatives at all from large swaths of the globe. When planning the new economy, delegates from rich and poor countries and representatives of corporations, labor, nongovernmental organizations, UN agencies, international financial institutions, scientists, and citizens groups will be seated at the table. The process must be democratic.

In building a new international financial architecture, we don't have to return to the old Bretton Woods rules. But we do need to consider the same three aspects that the Bretton Woods system included: rules, institutions, and funds.

We must start by recrafting the world's monetary rules to regulate world trade and commerce. Within nations, we must seek optimal interest rates. They must be high enough to keep the finance sector healthy and to keep banks from making a lot of careless loans, as they did when handing out subprime mortgages. But they must also be low enough for individuals and businesses to be able to afford loans, and low enough to ensure that money is not locked up in banks, which occurs when interest rates are too high. By optimizing interest rates, we can help optimize the productivity of the economy as a whole.

The new rules must tamp down the wild and destabilizing trade in national currencies and restrict capital movement that nullifies healthy competition and cooperation among nations. In this way, the new rules will encourage long-term investment in stable enterprises that produce tangible goods and services.

Other new rules will promote equity among nations. First, we must ensure that wages in the developing world are sufficient to meet basic human needs. Second, we must forgive the national debts of developing countries as reparation for past trade inequities and the wholesale extraction of natural resource wealth by industrialized nations. Third, we must replace unfair terms of trade with new rules that ensure fair prices for basic goods from developing countries, to make sure debt does not reaccumulate. These tasks comprise the greatest challenge because they mean distributing global wealth more equitably. By promoting equity, we can counter poverty and environmental destruction, which adds to climate change and encourages the emergence of new infectious diseases that can race around the globe.

SHOW ME THE MONEY

Besides rewriting the rules of finance and trade, we must also establish financial incentives such as tax breaks or subsidies that provide "carrots" to persuade the private sector to do the right thing. Such "carrots" will promote reductions in carbon emissions, in part by stimulating producers of clean energy and energy-efficient technologies.

Conversely, we must dismantle perverse subsidies that drive deforestation and subsidize the production and consumption of fossil fuels. About $300 billion, or 0.7 percent of the global gross domestic product, is spent on energy subsidies, and the vast majority of it goes to fossil fuels. We must switch these subsidies to renewables to reduce global greenhouse gas emissions and free up funds for renewable energy and clean technology initiatives that would simultaneously address poverty and help stabilize the climate.

We'll also need new funds and grants to support what the private sector for the most part cannot: building new infrastructure for the low-carbon economy and preserving our atmosphere, watersheds, and wetlands, our forests, fisheries, and oceans. They should come from a large global environment and development fund.

There is a precedent for a successful global environment and development fund, and its story is instructive. In 1987, international delegates met in Montreal to address the serious issue of chemicals depleting stratospheric ozone. The desire to achieve a consensus ran high. Delegates reached their mutual agreement to phase out these chemicals, however, only when they were assured of an equitable way to pay for the phaseout. The solution was an international fund that supported the transfer and manufacture of new methods of refrigeration and air conditioning, the main culprits of ozone depletion. The resulting agreement, the Montreal Protocol, helped ensure that the damaging chemicals were phased out of production.

In contrast, absent a global fund to support climate-preserving technologies, the Kyoto Protocol has consistently failed to achieve its goals. At the 1997 Kyoto climate conference, Brazil proposed such a fund, which it called the Clean Development Fund. Despite widespread international support for the idea, the U.S. delegation pushed successfully to reject it at the eleventh hour, replacing it with the relatively impotent Clean Development Mechanism, the treaty's carbon offsets market, which has barely made a dent in fighting climate change.

It's crucial that the new global environment and development fund provide enough money to save the Earth's remaining carbon sinks—the ecosystems that capture and sequester the world's carbon emissions. Roughly one-fifth of humanity's global warming emissions stem from deforestation. The fund would also underwrite a proposed United Nations initiative known as Reducing Emissions from Deforestation and forest Degradation (REDD) in developing countries. Funds for this essential measure would enable several rain forest–rich nations in Central Africa, Asia, and Latin America to pay farmers to end the culling of forests, to restore forests in areas that used to have them, and to plant trees on what is now barren land.

The global fund could also be used to preserve natural habitats that are currently being converted to crops, and to compensate countries whose native plants and animals are developed into crops and medicines, giving them powerful incentive to preserve the tropical forests that house such wondrous and potentially healing biodiversity. Neither is the case under existing trade rules, as enforced by the World Trade Organization.

We'd also need funds to help all nations—particularly the most vulnerable—prepare for the coming climate. These preparations, known in climate policy circles as adaptation, have several facets. We'd need money to assess vulnerabilities to the impacts of climate change, to develop and adopt early warning systems, and to create clean, robust, resilient energy systems. The last would include new grids and power plants that are resilient to severe weather, which can do serious harm to power supplies, generation, transport, and distribution.

By creating one global environment and development fund that's coordinated by a single agency, we'd bring all of these goals—mitigation and even prevention of further climate change, protection and restoration of forests, and climate adaptation—under one roof to optimize the impacts.

At the Copenhagen climate talks, negotiators made tangible progress toward these funding goals. They created a framework to fund efforts to reduce emissions from deforestation and forest degradation (REDD).

(In May 2010, Norway singlehandedly allotted $4 billion to prime that initiative.) They also made progress behind the scenes to protect the wood-gathering rights of indigenous peoples while reining in the main threats to forests—large timber, ranching, agriculture, and mining interests—and they agreed on the need for significant global funding to fight climate change. U.S. Secretary of State Hillary Clinton proposed, and delegates from other nations accepted, the idea of creating an international fund that would dispense $10 billion annually for the following three years to lower carbon emissions in developing countries, and $100 billion per year by 2020.

Although this was a good start, the plan provided far too little money and waited too long to invest it. Such a fund needs to be substantially larger than the fund created in Montreal because priming and sustaining the clean energy transformation is a far bigger undertaking than phasing out several chemicals. The International Energy Agency has called for $500 billion per year for twenty years to move the world toward clean, low-carbon energy. This figure is in line with what Sir Nicholas Stern proposed in his famous 2006 report on the economics of climate change: a global fund that dispenses 1 percent of global GDP, or $350 billion per year.

To come up with such huge sums, we'll need creative, equitable methods of financing. This will be challenging. Revenue from cap-and-trade programs would probably provide several billion dollars a year at most. Taxes on airline, ocean, or Internet traffic arouse strenuous opposition from industry and consumers. Nations could contribute—Norway alone contributed $1 billion to this fund in 2009—but persuading them to part with tens of billions of dollars could be tough, especially in a down economy.

Taxes on financial transactions are probably the fairest and most lucrative way to go. A so-called Tobin Tax, originally proposed by Nobel Prize-winning Yale economist James Tobin, offers several benefits. By levying just a quarter of a penny on every dollar used in currency transactions, which total more than $2 trillion per day, we could cool the hot-money market. Simultaneously, we could generate as much as $500 billion per year for global environmental initiatives—a reasonable investment in our common future. What's more, such a tax would be denationalized, drawing money from wealthy financial speculators rather than economically struggling nations.

Taxes on financial transactions must not be so onerous as to halt them, of course, since loans and investments are engines for economic activity. But taxing currency transactions can help prevent the speculative bubbles that have taken the economy on roller-coaster rides, and they offer a way to once again connect the brain of the global economy—finance—with the needs of industry and agriculture.

RETOOLING GLOBAL FINANCE

In addition to new rules and rewards, we must restructure and reform our current international financial institutions—the World Bank, the International Monetary Fund, and the World Trade Organization. As described in chapter 12, these institutions have elevated the rights of transnational corporations over the rights of sovereign nations and their citizens through their destructive lending and trade policies, at great social and environmental cost. Now they must be revamped.

For starters, we should revisit one of Keynes's ideas that never came to pass. In 1944, the economist envisioned a central bank called the International Clearing Union that would address global inequities. Such an institution would issue its own currency, called the bancor, which would be exchangeable at fixed rates with national currencies—a policy that would dramatically reduce currency speculation. It would also serve as a standard measure of every country's trade deficit or surplus.

The latter function is key. To balance the economies of rich and poor nations, Keynes proposed that any year a nation had excessive trade surpluses (far more exports than imports), the International Clearing Union would transfer the year's surplus to poorer nations with trade deficits (more imports than exports), whose goods are usually exported at low prices. Nations with a surplus would therefore have a powerful incentive to get rid of it, and these imposed transfers would automatically clear other nations' deficits. There was a bit more to it, but the result would be to adequately compensate resource-rich but economically impoverished nations for the extraction of cheap goods made with cheap labor.

Unfortunately for all of us, the United States rejected this proposal and instead proposed an International Stabilization Fund, which became the International Monetary Fund. It is now time to resurrect Keynes's grander vision for international financial governance, retooled to meet today's needs. Establishing an institution for enhanced governance over global trade and finance would control currency speculation and ensure that finance is directed toward investments that serve society and preserve the global commons (figure 42).

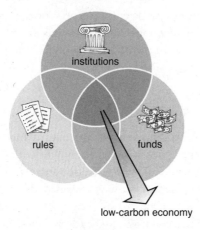

institutions

rules   funds

low-carbon economy

**policy
framework**

FIGURE 42. Policies for sustainability. To create a truly sustainable economic order, we need new rules, institutions, and funds.

Although the World Bank will undoubtedly propose putting itself in charge of funding the clean energy transformation, it is not the right institution for the job. As a bank, it makes loans and charges interest. It has presided over decades of deepening debt in the developing world. It can't be expected to oversee policies aimed at reordering the distribution of wealth between rich and poor nations.

We need instead a new institution that will dispense grants rather than loans. The Global Environment Facility, jointly supported by the UN Development Programme, the UN Environment Programme, and the World Bank, is a good model, and perhaps a candidate for this purpose. It already provides grants for biodiversity efforts, climate change initiatives, and global water projects. But it would have to be greatly expanded. Since its creation in 1990, the Global Environment Facility has been able to distribute just a few billion dollars. Hundreds of billions are needed.

TOWARD A SUSTAINABLE WORLD

If there were ever a problem that drove home the need for humanity to cooperate, climate change is it. We all share a single planet, a blue, spinning orb covered by just the thinnest skin of an atmosphere. For thousands of years, that atmosphere, along with the Earth's ocean, land, ice, and life-forms, has maintained a balance that made possible the world's grasslands, wetlands, and coral reefs, its forests, farms, and fisheries. It has maintained a balance that made possible advances in human health. It has made possible human civilization.

As our numbers grew and our civilization advanced, we tapped ever more of the planet's resources, until we took more than the Earth could withstand. And as we've burned through the Earth's supply of oil, coal, and gas to power our civilization, we've pumped so much carbon dioxide into our planet's thin skin that we have overwhelmed its capacity to absorb and recycle our wastes. And so today we face an unstable climate; tapped-out forests, soils, and oceans; an extinction crisis; deepening poverty; and a wave of climate-related ills. The signals could not be clearer that our current way of life cannot be sustained.

In the mid-nineteenth century, when London treated the River Thames like an open sewer, a seemingly never-ending epidemic of epidemics plagued the city. Public health reformers sought citywide authority for massive public works projects to clean up the water supply and dispose safely of sewage. Many Londoners resisted, protesting that an invasive government would threaten their rights as individuals and communities to make their own decisions about waste removal. The reformers won the battle; drinking water was piped in and treated, and modern sanitation systems were installed in cities throughout the developed world. The generations that followed lived longer and healthier lives.

Since then, we've been pumping the waste of the industrial revolution, our greenhouse gases, into the atmosphere. In doing so, we're changing the climate, threatening ecosystems worldwide, and jeopardizing the health of untold millions.

To preserve a livable planet, we need a strong global treaty to reduce climate-warming emissions. We also need an immense effort to transform our energy system, leave fossil fuels in the ground, and burn less of everything. Such changes will require the efforts of a broad cross-section of society. Investors and international institutions must provide the capital to power the transformation, and government and industry must provide the innovation, incentives, and investments to get the job done. We need bold public works programs that build on those President Obama has begun: the equivalent of a Manhattan Project to research new renewable-energy technologies, a Marshall Plan to finance them, an Apollo Plan to launch the transformation, and a Green New Deal to sustain it.

The scope of the undertaking is no doubt daunting, and the voices of resistance will again claim that we can't make the necessary changes. It threatens our rights, they'll say—threatens our jobs, threatens national sovereignty. They neglect to account for the mounting costs of inaction, however, and they neglect to remember that in tackling huge problems in the past, we've created huge opportunities. In tackling climate change, we'll create a clean and powerful engine for the twenty-first-century economy.

For hundreds of years, our global economy has been driven by endless growth and consumption. But in a world with finite resources and a limited capacity to recycle waste, endless consumption is not possible. Ultimately, we're going to have to consume less, generate much less waste, and create very different ways of doing business, in which companies are rewarded for conserving rather than consuming, for serving rather than growing. At long last we need a form of economic development that is truly sustainable.

Today it's fair to say that we have underestimated the rate at which climate and the Earth's ice cover would change, underestimated the breadth of biological responses to these changes, and underestimated how much these changes could harm our health and well-being. But perhaps, too, we have underestimated ourselves.

Speaking at the conclusion of the first Bretton Woods conference, American treasury secretary Henry Morgenthau, Jr., told the gathered delegates that "international cooperation" was the "wisest and most effective way to protect our national interests." He observed that "the great lesson of contemporary life is that the people of the earth are inseparably linked to one another by a deep underlying community of purpose."

Sixty-seven years have passed since that day of triumph and hope. Today, the greatest war we face is the one we have waged on our environment. The drivers of climate change—primarily fossil fuel combustion and deforestation—have caused a pervasive decline in the ecosystems upon which human life depends. We face an epidemic of epidemics. So far, impoverished nations have borne the brunt, but we are all seared by this fire. We have a chance to rewrite the rules. This time, we must write them together.

Epilogue

On July 15, 2010, eighty-six long days after the Deepwater Horizon oil rig exploded and crude oil began gushing from the Macondo well into mile-deep waters of the Gulf of Mexico, BP engineers placed a seventy-five-ton cap on the seafloor, at long last stanching the flow. By then eleven men had been killed, seventeen more had been injured, and oil sheens and floating tar balls had fouled the Gulf. BP had sprayed 800,000 gallons of potentially toxic oil-dispersing chemicals on the sea surface and spewed 360,000 gallons underwater, in effect turning the Gulf of Mexico into a massive science experiment, but salad-dressing-like suspensions of crude oil up to one kilometer (0.6 miles) deep and miles long nevertheless prowled the Gulf. Oil grounded family fishing businesses, diverted tourists to cleaner locales, and fouled salt marshes that nurture fish and crab larvae, tomorrow's seafood. Many called it the worst environmental disaster in the nation's history.

If there was ever an event that should have sparked an epiphany on the injurious effects of our fossil fuel addiction, the Gulf oil disaster was it. And Washington did respond, in a fashion. President Obama quickly placed a moratorium on deepwater drilling in the Gulf of Mexico until it was proven safe. An FBI-led team of investigators called "the BP squad" launched a criminal probe of BP and two other firms for violating environmental laws, falsifying safety reports, perjury, and obstruction of justice. A presidential oil spill commission began digging into the root causes of the Deepwater Horizon explosion and ways to modernize spill cleanup technology. And Congress passed so-called spill bills designed to tighten oversight of offshore oil and gas development, hold companies accountable for the spills they cause, and require drillers to pay royalties on the oil or gas they extract, closing a long-standing loophole.

But the larger lessons of the Gulf oil spill have so far gone unheeded. Before the ink was dry on Obama's moratorium, BP and Shell had announced plans for deepwater drilling in the Mediterranean Sea, placing more critical waters at risk. And despite three months of nonstop news of the Gulf oil disaster, in July 2010 the U.S. Senate, faced with determined opposition from conservatives and senators from coal and oil-producing states, many well funded by fossil fuel interests, gave up on their effort to pass a comprehensive climate and energy bill. Despite heart-tugging images of oil-soaked pelicans and out-of-work fishermen, we failed to grasp the larger lesson of the oil spill. The lesson was that we, as a society, had gone overboard.

We'd gone overboard with our fossil fuel addiction, which by 2010 had driven the industry to dangerous and marginal sources: deep-ocean drilling in the Gulf of Mexico, oil shale from Colorado, tar sands from Canada, coal to liquid fuel conversion. We'd gone overboard on our farms, where we drive petroleum-fueled machines over our once-rich land and dump millions of tons of fertilizer, created from fossil fuels, on our increasingly depleted soil. The fertilizer's breakdown products end up poisoning the Chesapeake Bay, the Gulf of Mexico, creating harmful algal blooms and dead zones, and other waste ends up in our atmosphere. We'd gone overboard with our forests, clear-cutting vast swaths, and polluting them with too much nitrogen and CO_2 from burning coal and oil. We'd gone overboard economically, drawn by a love of lucre and mesmerized by the false prophets of Wall Street.

Again and again, we want too much, waste too much, and fail to consider the consequences.

. . .

Why does this keep happening? We justify it with an outdated world-view, which holds that there are, or should be, boundless resources for the taking. U.S. politics promote it, as elected officials rely on wealthy contributors, often corporations, to pay for winning campaigns. Our mainstream economic theories promote it, too, via the dangerously misguided but mainstream idea that unfettered markets cure all ills, and the even more dangerous idea that endless growth and consumption are possible on a finite planet. We're dubbed "consumers," rather than human beings, who drive economic growth by buying ever more Gucci handbags, iPods, and McMansions. As the ecological economist Herman Daly has long argued, at a certain point a growing gross national product, our government-sanctioned measure of economic well-being, actually makes a society sicker. We moved past that point decades ago.

By the summer of 2010, a consensus had formed among health professionals that climate change endangered human health and well-being, and numerous professional groups had begun calling for action to ease the threat. The president of the American Medical Association, Dr. Cecil Wilson, stated at a congressional briefing that his group agrees with IPCC findings that the Earth is undergoing climate change, that human activity is accelerating it, and that "the potential exists for devastating events with serious health implications." The American Nurses Association, in a resolution, described the challenges of global climate change as "unprecedented in human history" and called for nurses to "speak out in a united voice and advocate for change on both individual and policy levels." The American Academy of Pediatrics stated in *Pediatrics,* its professional journal, that "children represent a particularly vulnerable group that is likely to suffer disproportionately from both direct and indirect adverse health effects of climate change." A joint commission by *The Lancet,* a top medical journal, and University College London's Institute for Global Health stated flatly that "climate change is the biggest global health threat of the 21st century."

Yet despite this knowledge, the world, and particularly the United States, continues on its dangerous course. To steer this ship away from the shoals, we must step up—as individuals, in choosing which house to live in, how far to commute, which technologies to buy, and what food to eat—and as members of workplaces, houses of worship, and communities. But the clean energy transformation won't happen unless our society elevates it to a top priority. We'll need to push our political leaders hard to implement the new rules we'll need to speed the shift. That means making much more happen than happened in 2009 and 2010, when developed nations, reeling from looming debt burdens, blinked when it came time to pony up the economic stimuli needed to boost—and green—the world economy.

. . .

In a perceptive 2009 essay, Dr. Peter C. Whybrow, a psychiatrist who directs the Semel Institute for Neuroscience and Human Behavior at the

University of California, Los Angeles, argued that there's a biological reason for "the debt fueled consumptive frenzy that has gripped the American psyche for the past two decades," which led to "epidemic rates of obesity, anxiety, depression and family dysfunction." (We can add polluted air, polluted water, ecosystem disruption, and climate change to Whybrow's list.) Our primitive, survival-oriented "lizard" brain houses the roots of the ancient dopamine reward pathways, which when over-stimulated by drugs, novel experience, or unlimited choice, he explains, "will trigger craving and insatiable desire." Normally these desires are kept in check by the prefrontal cortex, the executive, thinking part of the brain, and the limbic cortex, which fosters kinship behavior and nurturance.

In the eighteenth century, Adam Smith argued that harnessing the universal human traits of self-interest, novelty seeking, and social ambi-tion into markets creates an economic system that regulates itself. It does so, Whybrow relates, because "in a free society overweening self-interest is constrained by the wish to be loved by others (the limbic brain's drive for attachment) and by the 'social sentiment' (empathic and com-monsense behavior) that is learned by living in community." For a few brief decades, we forgot our civic traditions and deceived ourselves that unleashing our instinctual desires in an unfettered market would bring about a higher societal good. Instead we got SUVs, air pollution, an economic roller coaster, and worsened climate change.

To create a truly sustainable future, we must accept limits. These start with laws and regulations that change an economic system that thrives on overconsumption and that constrain damaging corporate behavior. That process is well underway. Besides tighter regulations on offshore oil drilling, we've seen a health care overhaul to rein in predatory health insurance practices; a tightened financial oversight system to rein in predatory lending; and stricter federal rules on food safety, working conditions, and pollution. But we must also accept that the pursuit of boundless economic growth no longer makes sense in today's world. The world's forests, oceans, and farms and freshwater supplies have their limits, and we've begun to surpass them.

To fight climate change as oil runs low, as a society we'll be forced to curb our own appetites. But in accepting limits, we'll reap tremendous health benefits. Living in smaller, more energy-efficient homes reduces electricity bills. Driving less and bicycling or walking more promotes physical fitness and prevents heart disease. Buying local, organically grown food cuts emissions from long-distance transport, while reducing pesticide use—plus, it tastes better. In fighting climate change, we'll save lives, and we'll restore much-needed balance to our society.

Looking back years from now, we may view April 20, 2010, the date the Deepwater Horizon rig exploded, as the 9/11 for the environment—a crossroads in the history of our species, when we collectively got a grip on our relationship with the environment, and just in the nick of time. It was a time, our grandchildren may say, when our society began acting on its long-thwarted instinct to care for the creatures with whom we share the planet, to care about future generations, to revere human life. If so, we'd veer from our tarnished history, creating a legacy of healing for both the Earth and ourselves.

Acknowledgments

Many people helped shape the ideas or took part in the experiences described in this book, and others made vital contributions by sharing expertise, time, and stories. We are indebted to all of them and apologize to anyone we have omitted.

We're grateful to all the scientists, doctors, policy makers, and businesspeople interviewed, whose knowledge and insights informed this book: Godofredo Andino, Kaid Benfield, Allan Carroll, Eric Chivian, Steve Dishart, James Elsner, Kerry Emanuel, Howard Frumkin, Chris Funk, Andy Goodman, Debarati Guha-Sapir, Ove Hoegh-Guldberg, Mark Jacobson, Dan Jaffe, Sadhu Johnston, Larry Kalkstein, Ronald Kessler, Patrick Kinney, Richard Levins, Ken Linthicum, Wally MacFarlane, Charles McNeil, Evan Mills, Stephen Morse, John Norquist, Federica Perera, Rob Pratt (of Cambridge Energy Alliance), Rob Pratt (of Pacific Northwest National Laboratory), Ken Raffa, Joan Rose, Steve Running, William Sprigg, Gary Szatkowski, Gary Tabor, Julie Trtanj, José Vasquez Bricaño, Mark Way, Tony Westerling, Louisa Willcox, Donald Wuebbles, X. B. Yang, Lewis Ziska.

We are indebted to everyone who shared their experiences, insights, and stories: Lucy Aguilera, Natividad Aguilera, Eva Ambrosia, Chris Ballas, Anna Barnes, Wafaa Bilal, Jerry Brous, Lettez Maribel Castro Basques, Ebony Clark, Rita Colwell, Latisha Doctor, Adelaida Flores Ihovares, Nelly Funes Calderon, Bernard Gikandi, Elena Maria Giron Lopez, Peter Mwangi Githeko, Robert Hoskins, Elizabeth James,

Margaret Kariuki, Ron Later, Laura McKeown, Bill Moomaw, Margot Moomaw, Catherine Muthoni Mukunya, Anne Mwangi, David Riecks, Chris Walker, Susan Wangiki, Susan Witheka, Bob Younger, Sandra Younger. Only some of their stories could be included, but all contributed.

Huge thanks to everyone who made Dan's on-the-scene reporting possible. In Kenya, Andrew Githeko was more than generous with his time and expertise, and he opened doors, shared wonderful stories, and patiently explained the connections between climate and malaria. In Honduras, three community leaders from Comayagüela—Doña Maritza Arévalo Amador, Doña Blanca Estela Serrano, and Doña Hilda Maradiaga Mejia—served as generous expert guides to their neighborhoods and to the challenges faced by residents. Three talented translators—Sophia Horwitz, Cheryl Ripley, and Laura Keresztesi—provided superb and spirited assistance. Luis Posadas of the Honduran Red Cross explained events and local lore in Choluteca, and members of the Valle Siria Environmental Committee provided expert guidance and hospitality in El Pedernal. Very special thanks go to Dr. Juan Almendares, whose generosity and introductions smoothed the way. Dr. Almendares provided an inspiring example through his wisdom, courage, and humanity.

In Harlem, Dr. Ben Ortiz and Dr. Vincent Hutchinson provided generous access to the pediatrics ward and outpatient clinic at Harlem Hospital. Dr. Ortiz went above and beyond in connecting Dan with patients to help him understand the links between air pollution and asthma. Dr. Hutchinson and the Harlem Children's Zone Asthma Initiative staff provided other key insights into asthma in Harlem. In Illinois, Evan DeLucia's good-natured hospitality enabled Dan's reporting at the SoyFACE plots run by the University of Illinois, Urbana-Champaign, and Clare Casteel put up with persistent questioning in the midst of a busy experiment. Also in Illinois, Bridget O'Neill offered insights into the lifestyles of Japanese beetles, and May Berenbaum provided an insect's eye view of her team's research on climate change and herbivory.

In Wyoming, Jesse Logan served up first-rate science on bark beetles and whitebark pines, as well as snowshoes, skis, and stories. We're grateful to him and everyone else at the June 2008 Whitebark Pine Citizen Science Workshop in Dubois, Wyoming, especially co-organizers Wally McFarlane and Louisa Willcox, for information and inspiration. And with Bruce Gordon of Ecoflight (www.ecoflight.info) at the helm, an overflight high in the Wind River Mountains provided a uniquely valuable view of a threatened alpine ecosystem.

Steve Dishart and Mark Way of Swiss Re patiently answered questions about the company's efforts to prevent and help adapt to climate change; Chris Walker provided an invaluable insider's view of the climate change–related business initiatives of this leading reinsurance company.

Many others contributed to this project in essential ways, and we are indebted to all of them. Philip Turner brought us together and got the project rolling. Ted Weinstein, our agent, offered enthusiasm and savvy advice that buoyed us on our journey. Mary Fergus and Sophia Horwitz provided key research assistance. Mason Inman, Kari Lydersen, and Erik Ness delivered research, editorial assistance, and other essential help in a pinch. Hillary Johnson provided crucial editorial assistance. Huge thanks go to Michael Denneny for his discerning reading of the manuscript, expert editorial advice, and support when the chips were down.

Jason Lindsey (perceptivevisions.com), photographer extraordinaire, accompanied Dan to Honduras and Harlem and generously donated eight powerful photographs that now grace these pages, and more on the book's Web site, www.changingplanetchanginghealth.com. Diana Asberry-Ferber provided expert photo-editing advice. Tom Dunne (tomdunneart.com) drew the delightful illustrations, and his life cycle diagrams were adapted from David Butler's.

Finally, we are grateful to the people at the University of California Press, especially our editor Hannah Love, whose verve, care, and wisdom helped elevate this book. We also thank Janet Reed Blake, Amy Cleary, Lou Doucette, Lynn Meinhardt, David Peattie, Sam Petersen, Leonard Rosenbaum, and Marilyn Schwartz for their professionalism and enthusiastic support.

FROM PAUL R. EPSTEIN, MD

My family's stay in Mozambique (1978–80) was a pivotal experience. Living outside the United States and working with those from many lands and cultures helped frame my views on how global forces shape national possibilities.

A dozen years later (June 1992), I sat in a sparse room inside a pastel-pink concrete building at the Federal University of Brazil, Rio de Janeiro. As the sun's rays glanced off the nearby bay, my colleagues and I were treated to a crash course in global climate change at a side event of the Rio Earth Summit. Tom Lovejoy, now president of the Heinz Center, and Bert Bolin, then head of the UN's Intergovernmental Panel on Climate Change, were our chief mentors.

I received graduate training in global change that fall in Woods Hole, Massachusetts, where George M. Woodwell, then director of the Woods Hole Research Center, conducted an electrifying weeklong workshop that looked comprehensively at how living organisms interact with the ocean, sky, ice, and soils to create Earth's life-supporting systems. My postgraduate training occurred in Woods Hole in November 1993, when our New Disease Group at the Harvard School of Public Health assembled researchers from multiple disciplines to tackle the wave of new and reemerging infectious diseases that had been building since the 1970s.

Several scientists and authors laid the foundation for my views on how things change. G.G. Simpson, with his 1950 book *The Meaning of Evolution,* introduced the term *quantum evolution,* precursor to evolutionary biologist Stephen Jay Gould's pivotal synthesis of "punctuated equilibrium," which pertains to sudden shifts in communities of species. (Climate, I would learn later, can also go through abrupt shifts after long periods of relative stability.) William Calvin argued simply, elegantly, and convincingly in *A Brain for All Seasons* that sudden climate shifts—rapid cooling events called "cold reversals" in the midst of warming trends—selected hominids who were best at communication and cooperation. And Rachel Carson wrote presciently in *The Sea Around Us* (1950) of ocean currents that drive Earth's climate and how marine species were undergoing redistribution due to global warming.

A grand cast of scientists have contributed key discoveries and ideas that have advanced the study of climate change and its health and ecological effects. In the climate and ecological sciences, they include James J. McCarthy, a biological oceanographer with whom I've had the great pleasure of teaching eager undergraduates for over a decade; Paul Mayewski and Ellen Mosley-Thompson, two intrepid ice-field explorers; JoAnn Burkholder, for her science and her courage in making it socially relevant; Cynthia Rosenzweig; and Ken Sherman. Others to thank include James Baker, the former head of NOAA, who understood health and climate connections early on, and Mike Hall, former chief of NOAA's Office of Global Programs, who strove to integrate analyses of natural climate variability with the current period of change.

Special acknowledgment is due Dr. Eric Chivian, who directs our Center for Health and the Global Environment at Harvard Medical

School (HMS), with whom I traveled to Rio and Kyoto and have the privilege of working each day. Another is due my dear friend, environmental epidemiologist Richard Clapp, who brings the highest quality science to communities most in need.

I'm indebted to my colleagues in the health and climate field: Anthony J. McMichael and Sir Andrew Haines, partners throughout; Jonathan Patz, John Balbus, Larry Kalkstein, and Howard Frumkin, who helped develop this discipline; Richard Levins, a key mentor to us all; Dr. Mary Wilson and Rita Colwell (a pioneering scientist who later headed up the NSF). Many thanks to our international colleagues, including Dr. Juan Almendares (Honduras), Drs. Ulisses Confalonieri and Manuel Cesario (Brazil), Andrew Githeko (Kenya), Pim Martens (Netherlands), Ricardo Thompson (Mozambique), and Bettina Menne, Roberto Bertollini, and Rudi Slooff (World Health Organization). Thanks, too, to our HMS teaching partners: Daniel Goodenough, Howard Hu, Tim Ford, Melissa Perry, and Aaron Bernstein.

In the corporate world, I appreciate the wise counsel of Franklin Nutter of the Reinsurance Association of America, Jacques DuBois (retired chair of the board and CEO, Swiss Re Holding Corp.), Stephen Dishart (for his insight and warmth), Mark Way, Peter Duerig, Adrienne Atwell, and Chris Walker (all with or formerly with Swiss Re), Amy Davidsen (formerly with JPMorgan Chase, now heading The Climate Group, U.S.), Jean Sweeny and Paul Narog (3M), Brian Boyd, Michael Bzdak, and Dennis Canavan (J&J), Mary Wenzel and Stephanie Rico (Wells Fargo Bank), Mindy Lubber (Ceres), Norman Myers, Frank Ackerman, Rob Pratt, and Jeffrey Sachs, director of Columbia's Earth Institute, author of this book's forward, and ever a passionate voice for righting the world's wrongs.

I have been fortunate to work with a wonderful, hardworking group: Eric Chivian, Kathleen Frith, Tracy Sachs, Margaret Katsumi, Emily Griffith, Natalie Wicklund, Susan Boa, Heather Foley, Kim Riek, Alyssa Manning, and Acacia Matheson, a dream team dedicated to bringing the best science, beautifully and professionally presented, to the general public and policy makers.

Special thanks to Ross Gelbspan, for our inspirational walks and talks, and Charles McNeill (United Nations Development Programme), a marvelous man who spreads warmth and encouragement across the globe.

Finally, I am deeply indebted to my wife, Andy, who's stood by me through thick and thicker, my staunchest critic and stalwart supporter.

And I am grateful to my two marvelous children—our daughter Jesse, a Brooklyn-based documentary filmmaker with an uncanny sense of the amazing, and Ben, a musician in New Orleans, dedicated father and soulful student of life. Finally, this book is dedicated to Isaiah Epstein Bagneris—Izzy—our dear, wondrous grandson.

FROM DAN FERBER

Dan is indebted to colleagues, friends, and family who offered logistical, practical, and moral support, each uniquely valuable in creating this book: Ron and Carol Asberry, Rich Asberry, Ben Ferber, Judy Ferber, Gloria and Peter Ferber, Scott Grubisich, Antonia Herbstreit, Anna Keck, David Kohn, Nancy Kriplen, Jason Lindsey (again), Cindy and Chris Marlow, Sandy Masur, Sue Russell, Vic Schuster, James Stewart, Jeff Tessler, Yang Yang, and Taiji friends far and wide. He offers thanks to his wise colleagues at the Writers Table of Indianapolis and the Writers Underground, and to the amazing Birders, whose savvy advice and wholehearted support helped him hang on through the bumps, twists, and turns of this long and sometimes arduous project. Finally, Dan is deeply grateful for the patience and loving support of his wife, Diana Asberry-Ferber, and daughters, Chelsea and Libby Asberry.

Notes

Page

1   *Larsen B ice shelf: Larsen Ice Shelf Has Progressively Thinned* and
photo captions from Scott Polar Research Institute, www.spri.cam.ac
.uk/research/projects/larseniceshelf, October 31, 2003; Paul R. Epstein
and James J. McCarthy, "Assessing Climate Stability," *Bulletin of the
American Meteorological Society,* December 2004, 1863–70; Eugene
Domack, et al., "Stability of the Larsen B Ice Shelf on the Antarctic Peninsula during the Holocene Epoch," *Nature* 436: 7051 (2005); Andrew
Shepherd, Duncan Wingham, Tony Payne, et al., "Larsen Ice Shelf Has
Progressively Thinned," *Science* 302: 5646 (2003); *Antarctic Ice Shelf
Breaks Apart,* BBC online, http://news.bbc.co.uk/1/hi/sci/tech/1880566
.stm, March 19, 2002.

2   *drowning polar bears:* Jim Carlton, "Is Global Warming Killing the
Polar Bears?" *Wall Street Journal,* December 14, 2005.

2   *melted 10 percent of the ice in the Alps:* Janet Larsen, "Setting the
Record Straight: More than 52,000 Europeans Died from Heat in Summer 2003," *Plan B Updates* (Washington, DC: Earth Policy Institute),
www.earth -policy.org/index.php?/plan_b_updates/2006/update56, July
28, 2006.

3   *150,000 additional deaths worldwide:* Jonathan A. Patz, Diarmid
Campbell-Lendrum, Tracey Holloway, et al., "Impact of Regional Climate Change on Human Health," *Nature* 438: 7066 (2005). Also, one
definition of *disability adjusted life years* can be found at World Health
Organization, Health Statistics and Health Information Systems, www
.who.int/healthinfo/boddaly/en, accessed October 27, 2010.

CHAPTER 1: MOZAMBIQUE

13   *absent from Africa for most of the twentieth century:* R.I. Glass, P.A. Blake, R.J. Waldman, M. Claeson, and N.F. Pierce, "Cholera in Africa: Lessons on Transmission and Control for Latin America," *Lancet* 338: 8770 (September 28, 1991): 791–95; C. Lam, S. Octavia, P. Reeves, L. Wang, and R. Lan, "Evolution of Seventh Cholera Pandemic and Origin of 1991 Epidemic, Latin America," *Emerging Infectious Disease* 16: 7 (July 2010): 130–32.

18   *United Nations report:* Gro Harlem Brundtland and the World Commission on Environment and Development, *Our Common Future* (New York: Oxford University Press, 1987).

19   *"Public health assumed that we'd licked it":* Interview with Dick Levins, April 2008.

20   *William Harvey:* R.E. Philips, Jr., *The Heart and the Circulatory System,* Access Excellence @ The National Health Museum, www.access-excellence.org/AE/AEC/CC/heart_background.php, accessed September 29, 2010.

21   *Ludwig von Bertalanffy:* For descriptions of systems theory and details of Bertalanffy's career, see F. Heylighen and C. Joslyn, "What Is Systems Theory?" *Principia Cybernetica Web* (Brussels: Principia Cybernetica), http://pespmc1.vub.ac.be/SYSTHEOR.html, November 1, 1992; T.E. Weckowicz, *Ludwig von Bertalanffy (1901–1972): A Pioneer of General Systems Theory,* Center for Systems Research Working Paper No. 89-2 (Edmonton: University of Alberta, February 1989). Bertalanffy spent the last years of his career at the University of Alberta; Weckowicz was his colleague. Also see *Ludwig von Bertalanffy (1901–1972),* International Society for the Systems Sciences, www.isss.org/lumLVB.htm, accessed September 29, 2010.

21   *intangible life force:* F. Capra, *The Web of Life* (New York: Anchor Books, 1997), 26.

22   *flame of a lit candle:* A. Scott, "Physicalism, Chaos, and Reductionism," in *The Emerging Physics of Consciousness,* ed. J.A. Tuszynski, 185 (New York: Springer, 2006).

23   *Feedback mechanisms:* Capra, *The Web of Life,* 57.

24   The Quark and the Jaguar: M. Gell-Mann, *The Quark and the Jaguar* (New York: Henry Holt and Company, 1994).

25   *"Epidemics . . . are like sign-posts":* R. Virchow, "Report on the typhus epidemic in Upper Silesia," in *Collected Essays on Public Health and Epidemiology,* ed. and trans. L.J. Rather (Canton, MA: Science History Publications, 1985).

25   *an epidemic of epidemics:* The recent flood of new and reemerging diseases, many moving from animals to humans, calls to mind the first centuries of the Holocene. As the continental ice sheets melted and climate warmed dramatically, many hunters and gatherers became farmers, and they domesticated animals, leading to modern civilization. At the time, a flood of new diseases leapt the species barrier, moving from animal to animal and on to humans.

26   *the groundbreaking research of Rita Colwell:* Accounts of Colwell's work are drawn from an interview and from the following papers: R.R. Colwell, "Global Climate and Infectious Disease: The Cholera Paradigm," *Science* 274: 5295 (1996); and R.R. Colwell, "An Unexpected Consequence: Re-emerging Human Diseases Driven by Climate Change," a presentation at the American Association for the Advancement of Science (AAAS) conference, February 2008.

27   *large cholera epidemics erupt as if on cue:* S.M. Faruque, et al., "Seasonal Epidemics of Cholera Inversely Correlate with the Prevalence of Environmental Cholera Phages," *Proceedings of the National Academy of Sciences* 102: 5 (2005).

28   V. cholerae *could hide out:* Interview with Rita Colwell, April 2008; P.R. Epstein, T.E. Ford, and R.R. Colwell, "Marine Ecosystems," *Lancet* 342 (1993).

28   *the cholera epidemic occurred during an El Niño:* Additional support for the climate–cholera link came later. During the massive 1997–98 El Niño event, Colwell and her colleagues showed that warmer sea surface temperatures in Peruvian coastal waters reliably predicted an increase in *V. cholerae* numbers in seawater. These results fortified the climate–cholera link and set the stage for a climate early warning system that could predict cholera outbreaks in the region during El Niño events. Normally, the seas along Peru's coastline are cold, with a strong upwelling current rich in nourishing plankton that feed fisheries. But the warm, stagnating surface waters that occur during El Niño events foster plankton proliferation.

CHAPTER 2: THE MOSQUITO'S BITE

31   *startling declines in frog populations:* Michael McCally, "Environment and Health: An Overview," *Canadian Medical Association Journal* 163: 5 (2000).

31   *James Hansen sounded the first major public alarm:* Quotes and details are from the hearing transcript and from Philip Shabecoff, "Global Warming Has Begun, Expert Tells Senate," *New York Times,* June 24, 1988.

33   *The IPCC authors concluded:* IPCC, *Climate Change: The IPCC Scientific Assessment* (Cambridge, U.K.: Cambridge University Press, 1990), also available at www.ipcc.ch/ipccreports/far/wg_I/ipcc_far_wg_I_full_report.pdf.

33   *nearly six billion tons of carbon:* Fiona Godlee and Alison Walker, "Importance of a Healthy Environment," *British Medical Journal* 303 (1991).

34   *the medical profession received its first warning:* Alexander Leaf, "Potential Health Effects of Global Climatic and Environmental Changes," *New England Journal of Medicine* 321: 23 (1989).

34   *the New Disease Group convened:* Mary E. Wilson, Richard Levins, and Andrew Spielman, "Disease in Evolution: Global Changes and

Emergence of Infectious Diseases," *Annals of the New York Academy of Sciences* 740 (1994): preface.

39   *Andrew Githeko's is the tale of a scientific underdog:* Material on Andrew Githeko's work and life are from coauthor Dan Ferber's interviews and observations during a trip to Kenya, March 19–25, 2008.

44   Anopheles gambiae: Material on malaria parasites, mosquito vectors, and their biological requirements are from interviews with Andrew Githeko, December 2007 and March 2008, and several journal articles and academic reports by Andrew Githeko, including "Malaria, Climate Change, and Possible Impacts on Populations in Africa," in *HIV, Resurgent Infections and Population Change in Africa,* ed. M. Carael and J. Glynn (New York: Springer, 2007); Andrew K. Githeko, et al., "Climate Change and Vector-Borne Diseases: A Regional Analysis," *Bulletin of the World Health Organization* 78: 9 (2000); and Andrew K. Githeko and William Ndegwa, "Predicting Malaria Epidemics in the Kenyan Highlands Using Climate Data: A Tool for Decision Makers," *Global Change and Human Health* 2: 1 (2001).

45   *tertiary prevention:* See any public health textbook, such as Mary-Jane Schneider, *Introduction to Public Health* (Sudbury, MA: Jones and Bartlett, 2006), 12.

46   *The model also made a startling new prediction:* Andrew Githeko's climate–malaria models are described in Githeko and Ndegwa, "Predicting Malaria Epidemics in the Kenyan Highlands Using Climate Data."

49   *IPCC's third assessment report:* IPCC, *Climate Change 2007 Synthesis Report* (Geneva: Intergovernmental Panel on Climate Change, 2008), www.ipcc.ch/publications_and_data/publications_ipcc_fourth_assessment_report_synthesis_report.htm.

50   *turned into a slugfest:* The studies by climate–health naysayers that informed this section include David J. Rogers and Sarah E. Randolph, "The Global Spread of Malaria in a Future, Warmer World," *Science* 289: 5485 (2000); Chris Dye and Paul Reiter, "Temperatures without Fevers?" *Science* 289: 5485 (2000); Simon I. Hay, et al., "Climate Change and the Resurgence of Malaria in the East African Highlands," *Nature* 415: 6874 (2002); and S.I. Hay, D.J. Rogers, S.E. Randolph, et al., "Hot Topic or Hot Air? Climate Change and Malaria Resurgence in East African Highlands," *Trends in Parasitology* 18: 12 (2002).

Articles by Andrew Githeko and like-minded colleagues that informed this section include Githeko, Lindsay, Confanlonieri, et al., "Climate Change and Vector-Borne Diseases"; Jonathan Patz, Mike Hulme, Cynthia Rosenzweig, et al., "Regional Warming and Malaria Resurgence," *Nature* 420: 6916 (2002); Guofa Zhou, Noboru Minakawa, Andrew K. Githeko, et al., "Association between Climate Variability and Malaria Epidemics in the East African Highlands," *Proceedings of the National Academy of Sciences* 101: 8 (2004); Mercedes Pascual, J.A. Ahumada, L.F. Chaves, et al., "Malaria Resurgence in the East African Highlands: Temperature Trends Revisited," *Proceedings of the National Academy of Sciences* 103: 15 (2006); Hong Chen, Andrew K. Githeko, Guofa

Zhou, et al., "New Records of *Anopheles arabiensis* Breeding on the Mount Kenya Highlands Indicate Indigenous Malaria Transmission," *Malaria Journal* 5: 17 (2006).

53 *"It really changed opinions"*: More recently, a researcher named Kevin Lafferty presented a more sophisticated challenge to climate's role in the spread of infectious diseases, writing in an ecology journal that warming might shift rather than expand the range of such ills and that the whole problem is overblown. But the Earth is warming more at higher latitudes than in the tropics, and winter temperatures are rising in higher latitudes as well. These effects would combine to expand the range of temperature-dependent infectious diseases like malaria. See Kevin D. Lafferty, "The Ecology of Climate Change and Infectious Diseases," *Ecology* 90 (2009).

56 *For people here, Mount Kenya is the only mountain:* Information on the geography and biodiversity of the Mount Kenya region is from *Mt. Kenya Environment,* Mountain Club of Kenya, http://boxproductions.com/cms/mck/Environment.19.0.html, accessed October 3, 2010; *EEIU Mt. Kenya—Eastern and Central Kenya Integrated Conservation and Development Plan—Operation Plan 2004–2007,* Eco-Ethics International Union, www.eeiu.org/chapters/mtkenya/index.html, accessed October 3, 2010.

CHAPTER 3: SOBERING PREDICTIONS

62 *the United Nations Conference on Environment and Development:* For a capsule history of the event, see United Nations Conference on Environment and Development (1992), www.un.org/geninfo/bp/enviro.html, accessed October 4, 2010.

63 *My colleague and I were stunned:* Paul Epstein's recollections, and an interview with Eric Chivian, December 2, 2008.

63 *slammed vice presidential candidate Gore:* David Remnick, "Ozone Man," *New Yorker,* April 24, 2006.

64 *"on a collision course": 1992 World Scientists' Warning to Humanity,* Union of Concerned Scientists, www.ucsusa.org/about/1992-world-scientists.html, accessed October 4, 2010.

64 The Lancet *published the eight-part series:* The eight-part series on the health effects of global environmental change ran in *The Lancet* in successive issues from October 23 to November 27, 1993. Reports referred to include P.R. Epstein and D. Sharp, "Medicine in a Warmer World"; K. Maskell, I.M. Mintzer, and B.A. Callander, "Basic Science of Climate Change"; Andrew Dobson and Robin Carper, "Biodiversity"; Steven A. Lloyd, "Stratospheric Ozone Depletion"; P.R. Epstein, et al., "Marine Ecosystems"; Jerome E. Freier, "Eastern Equine Encephalitis"; David J. Rogers and Michael J. Packer, "Vector-Borne Diseases, Models and Global Change"; Neville Nicholls, "El Niño-Southern Oscillation and Vector-Borne Disease"; and Martin L. Parry and Cynthia Rosenzweig, "Food Supply and Risk of Hunger."

65 *the view from twenty-five thousand feet:* Epstein and Sharp, "Medicine in a Warmer World."

66 *"Climate change is likely to have wide-ranging":* IPCC, *Climate Change 1995: Impacts, Adaptations and Mitigation of Climate Change: Scientific-Technical Analyses* (Cambridge, U.K.: Cambridge University Press, 1996), also available at www.ipcc.ch/ipccreports/sar/wg_II/ipcc_sar_wg_II_full_report.pdf.

66 *observations and mathematical models:* Sophisticated climate naysayers often dismiss scientific forecasts of continuing climate change by attacking climate models. Such models take data from observations of the environment (e.g., carbon dioxide levels, air temperature, ocean temperature, or rainfall) and feed them into sophisticated computer programs (models) that simulate the known behavior of the Earth's climate system. Because the Earth's climate system is complex, different climate models sometimes give slightly different forecasts of what the future will look like—a fact that naysayers use to harp on uncertainties and question the reality of climate change. The naysayers neglect to mention, however, that climate predictions are also based on bedrock scientific principles of physics, chemistry, and biology. The truth is that there is much we are certain of, and that body of knowledge should inform what we do.

67 *a Swedish physicist and chemist named Svante Arrhenius: Svante Arrhenius, 1859–1927,* Earth Observatory, http://earthobservatory.nasa.gov/Features/Arrhenius, accessed October 4, 2010.

67 *thirty-six billion metric tons:* IPCC, *Climate Change 2007: The Physical Science Basis* (Cambridge, U.K.: Cambridge University Press, 2008), 238, also available at www.ipcc.ch/publications_and_data/publications_ipcc_fourth_assessment_report_wg1_report_the_physical_science_basis.htm.

68 *black soot:* V. Ramanathan and G. Carmichael, "Global and Regional Climate Changes Due to Black Carbon," *Nature Geoscience* 1 (2008).

68 *This heat has warmed the atmosphere:* Atmospheric warming of 0.7°C from IPCC, *Climate Change 2007 Synthesis Report* (Geneva: Intergovernmental Panel on Climate Change, 2008), www.ipcc.ch/publications_and_data/publications_ipcc_fourth_assessment_report_synthesis_report.htm. For more on warming of the oceans, see Sydney Levitus, John Antonov, Timothy Boyer, et al., "Warming of the World Ocean, 1955–2003," *Geophysical Research Letters* 32 (2005); IPCC, *Climate Change 2007: The Physical Science Basis,* 243.

68 *These effects combine to rev up the global water cycle:* P. Y. Groisman, et al., "Contemporary Changes of the Hydrological Cycle over the Contiguous United States," *Journal of Hydrometeorology* 5: 1 (2004); IPCC, *Climate Change 2007: The Physical Science Basis,* 238.

68 *more intense downpours:* Rain increased 7 percent overall in the continental United States between 1970 and 2009. Rains over two inches a day are up 14 percent. Rains over four inches are up 20 percent, and rains over six inches a day are up 27 percent. The increase in heavy rains has significant public health implications because rains over two

inches a day tend to trigger flooding that sets the stage for outbreaks of waterborne disease like those caused by *E. coli* and *Cryptosporidium,* while lighter rains don't.

70   *dengue fever infected fifty million people: Dengue and Dengue Haemorrhagic Fever,* Fact Sheet 117, World Health Organization, www.who .int/mediacentre/factsheets/fs117/en, March 2009.

71   *the disease once appeared:* Information on dengue fever and dengue hemorrhagic fever is from the following sources: Mark Stevenson, "Hemorrhagic Dengue Fever Surges in Mexico with Climate Change, Migration, Urbanization," Associated Press Worldstream, March 31, 2007; Tony Pugh, "Tropical Disease Headed toward U.S., Health Officials Warn," McClatchy Washington Bureau, January 10, 2008; "Is Climate Change Affecting Dengue in the Americas?" *Lancet* 371: (2008): 9617; *2007: Number of Reported Cases of Dengue and Dengue Hemorrhagic Fever (DHF), Region of the Americas,* data table, Pan American Health Organization, www.paho.org/common/Display.asp?Lang=E&RecID=10389, accessed October 4, 2010.

71   *Asian tiger mosquito: U.S. Distribution, Asian Tiger Mosquito,* a map from the University of Florida Institute of Food and Agricultural Sciences, http://entnemdept.ufl.edu/creatures/aquatic/Asian-tiger2000.htm, accessed October 4, 2010.

72   *Rift Valley fever:* Denise Grady, "An Outbreak of Rift Valley Fever Kills Dozens in Kenya," *New York Times,* January 8, 2007; information on Ken Linthicum's studies is from the Global Warming: More Than Just Hot Air! conference held April 17, 2008, at the National Institutes of Health, Bethesda, Maryland, and from these journal articles: Assaf Anyamba, Jean-Paul Chretien, Jennifer Small, et al., "Developing Global Climate Anomalies Suggest Potential Disease Risks for 2006–2007," *International Journal of Health Geographics* 5 (2006); Paul R. Epstein, "Climate and Health," *Science* (July 16, 1999); Kenneth J. Linthicum, Assaf Anyamba, Compton J. Tucker, et al., "Climate and Satellite Indicators to Forecast Rift Valley Fever Epidemics in Kenya," *Science* 285: 5426 (1999).

77   *complications make Lyme disease very expensive to treat:* John R. Brownstein, "Lyme Disease: Implications of Climate Change," in *Climate Change Futures,* ed. Paul R. Epstein and Evan Mills (Boston: Center for Health and the Global Environment, Harvard Medical School, 2005).

77   *Lyme cases like Ballas's have become more common: Reported Cases of Lyme Disease by Year, United States, 1994–2008,* Centers for Disease Control and Prevention, www.cdc.gov/ncidod/dvbid/lyme/ld_Up ClimbLymeDis.htm, accessed October 4, 2010.

78   *the black-legged tick feeds just three times:* Life cycle of black-legged tick, *Ixodes scapularis,* including feeding times, is from Michael R. Patnaude and Thomas N. Mather, Featured Creatures Web site, http://entnemdept.ufl.edu/creatures/urban/medical/deer_tick.htm, accessed October 4, 2010.

78   *John S. Brownstein, an epidemiologist:* Information on John S. Brown-
      stein and Durland Fish's research is from these sources: Durland Fish,
      "Environmental Determinants of Lyme Disease Risk," lecture at the
      annual conference of the American Institute of Biological Sciences,
      May 2008, available online at https://live.blueskybroadcast.com/bsb/
      client/CL_DEFAULT.asp?Client=26&PCAT=813&CAT=813, accessed
      October 4, 2010; John S. Brownstein, Theodore R. Holford, and Dur-
      land Fish, "Effect of Climate Change on Lyme Disease Risk in North
      America," *Ecohealth* 2: 1 (March 2005); John S. Brownstein, Theodore
      R. Holford, and Durland Fish, "A Climate-Based Model Predicts the
      Spatial Distribution of the Lyme Disease Vector *Ixodes scapularis* in the
      United States," *Environmental Health Perspectives* 111: 9 (2003); Wil-
      liam Speed Weed, "Tick Dragger," *Popular Science,* www.popsci.com/
      scitech/article/2004-11/tick-dragger, accessed October 4, 2010.

CHAPTER 4: EVERY BREATH YOU TAKE

80   *industries had begun to spend millions of dollars:* Material on spending
      by industrial companies and their PR firms is from Ross Gelbspan, *The
      Heat Is On* (New York: Perseus Books, 1998), 33–57; Sheldon Rampton
      and John Stauber, *Trust Us We're Experts: How Industry Manipulates
      Science and Gambles with Your Future* (New York: Tarcher/Putnam,
      2001), 267–88.

83   *what we health and environmental advocates were up against:* For a
      contemporaneous description of the *Times* ad insert, see www.well
      .com/user/sfflier/Savio-warming.html, December 4, 1997. Details of
      the energy industry's misinformation campaign are from Rampton and
      Stauber, *Trust Us We're Experts: How Industry Manipulates Science
      and Gambles with Your Future,* 276.

85   *Foster planted ragweed seeds:* Peter Wayne, Susannah Foster, John Con-
      nolly, Fakhri Bazzaz, and Paul Epstein, "Production of Allergenic Pollen
      by Ragweed (*Ambrosia artemisiifolia* L.) Is Increased in CO_2-Enriched
      Atmospheres," *Annals of Allergy, Asthma and Immunology* 8 (2002).

86   *Ziska's team planted ragweed:* Descriptions of Lew Ziska's work are
      from an interview in April 2008 and other communications, and these
      journal articles: Ben D. Singer, Lewis H. Ziska, David A. Frenz, Dennis
      E. Gebhard, and James G. Straka, "Increasing Amb a 1 Content in
      Common Ragweed *(Ambrosia artemisiifolia)* Pollen as a Function of
      Rising Atmospheric CO_2 Concentration," *Functional Plant Biology* 32:
      7 (2005); Lewis Ziska, K. George, and David A. Frenz, "Establishment
      and Persistence of Common Ragweed (*Ambrosia artemisiifolia* L.) in
      Disturbed Soil as a Function of an Urban–Rural Macro-environment,"
      Global Change Biology 13 (2007).

86   *Although many factors affect allergies, several of them come from
      burning fossil fuels:* Christine A. Rogers, "Carbon Dioxide and Aero-
      allergens," in *Climate Change Futures,* ed. Paul R. Epstein and Evan

Mills (Boston: Center for Health and the Global Environment, Harvard Medical School, 2005).

87 *urban heat island effect:* For information on heat islands, including increased vulnerability to heat exhaustion and heat stroke, see the links listed at the U.S. Environmental Protection Agency Web site, www.epa .gov/hiri.

87 *Jaquan Doctor:* Material on Jaquan Doctor and Latisha Doctor is from reporting by Dan Ferber based on time spent with Jaquan Doctor and interviews with Latisha Doctor, Dr. Vincent Hutchinson, and Dr. Ben Ortiz, the attending doctor in charge of the inpatient unit, during July 88 at Harlem Hospital, New York City.

88 *asthma is the top cause of serious chronic illness:* Asthma facts and statistics are from these sources: Lara J. Akinbami, "The State of Childhood Asthma, United States, 1980–2005," *Advance Data* (a publication of the Centers for Disease Control and Prevention) 381 (2006); *Asthma,* Fact Sheet 307, World Health Organization, www.who.int/mediacentre/ factsheets/fs307/en/index.html, May 2008; Jeanne E. Moorman, et al., "National Surveillance for Asthma—United States, 1980–2004," *Morbidity and Mortality Weekly Reports* 56: 8 (2007); *About Childhood Asthma,* American Lung Association, www.lungusa.org/lung-disease/ asthma/about-asthma; Epstein and Mills, *Climate Change Futures.*

88 *Asthma has also been on the rise:* Asthma facts and statistics are from Paul R. Epstein, "Fossil Fuels, Allergies, and a Host of Other Ills," *Journal of Allergy and Clinical Immunology* 122: 3 (2008); Akinbami, "The State of Childhood Asthma"; Epstein and Mills, *Climate Change Futures.*

89 *a remarkable one in five people admitted to Harlem Hospital:* Personal communications with Dr. Ben Ortiz, July 2008.

90 *When allergies kick off an asthma attack:* "Asthma In-Depth Report," *New York Times Health Guide,* http://health.nytimes.com/health/ guides/disease/asthma/causes.html, accessed October 7, 2010.

92 *horrible pollution events like the Great Smog of London:* Information on the Great Smog of London is from Bert Brunekreef and Stephen T. Holgate, "Air Pollution and Health," *Lancet* 360: 9341 (2002). Details on air pollutants and their health effects are from Brunekreef and Holgate, "Air Pollution and Health," and from the U.S. Environmental Protection Agency, www.epa.gov/air/urbanair.

92 *heart attacks, stroke, irregular heartbeats:* Robert D. Brook, "Cardiovascular Effects of Air Pollution," *Clinical Science* 115 (2008); Robert D. Brook, et al., "Air Pollution and Cardiovascular Disease: A Statement for Healthcare Professionals . . . ," *Circulation* 109: 21 (2004); Robert D. Brook, "Air Pollution: What Is Bad for the Arteries Might Be Bad for the Veins," *Archives of Internal Medicine* 168: 9 (2008).

92 *One of the most dangerous types of particle pollution:* On cancer and black soot, see Erica Weir, "Diesel Exhaust, School Buses and Children's Health," *Canadian Medical Association Journal* 167: 5 (September 3, 2002): 505. Information on asthma studies in the South Bronx

and Harlem is from personal communications with Dr. Ben Ortiz, July 2008; "Asthma Symptoms Linked to Soot Particles from Diesel Trucks in South Bronx," press release, New York University, October 16, 2006, www.nyu.edu/about/news-publications/news/2006/10/16/asthma_symptoms_linked_to_soot.html, accessed October 7, 2010; and "Black Soot and Asthma," *New York Times,* November 19, 2006.

95 *a dust cloud the size of the continental United States:* Information on mobile pollution is from S.T. Ott, et al., "Analysis of a Transatlantic Sub-Saharan Dust Outbreak Based on Satellite and GATE Data," *Monthly Weather Review* 119 (1991); Epstein and Mills, *Climate Change Futures.*

96 *Saudi refugee camps plagued by clouds of powdery dirt:* Interview with former Saudi refugee camp resident Wafaa Bilal, September 2007.

96 *southern United States has gotten drier in recent years:* William Sprigg, Institute of Atmospheric Physics, University of Arizona, presentation at the American Association for the Advancement of Science conference in February 2008.

96 *pathogen-laden dust from new deserts:* Epstein and Mills, *Climate Change Futures.*

96 *William Sprigg, an atmospheric physicist . . . who studies dust storms:* Sprigg interview and presentation at the annual conference of the American Association for the Advancement of Science, February 2008; Heidi Rowley, "UA's Dust Research Has Public Health, Safety Applications," *Tucson Citizen,* November 3, 2008.

97 *a heat wave of unprecedented intensity:* Janet Larsen, "Setting the Record Straight: More than 52,000 Europeans Died from Heat in Summer 2003," *Plan B Updates* (Washington, DC: Earth Policy Institute), www.earth-policy.org/index.php?/plan_b_updates/2006/update56, July 28, 2006; Bruce Crumley, "Elder Careless: With Perhaps 10,000 Dead from the Heat, France Confronts Its Inaction and Neglect of the Aged," *Time International,* September 1, 2003.

97 *a trend toward increased heat waves worldwide:* Kevin Trenberth, National Center for Atmospheric Research, *Climate Change and Extreme Weather Events,* PowerPoint presentation available at www.cgd.ucar.edu/cas/Trenberth/trenberth_presentations.html; Peter A. Stott, D.A. Stone, and M.R. Allen, "Human Contribution to the European Heatwave of 2003," *Nature* 432: 7017 (2004); Epstein and Mills, *Climate Change Futures.*

98 *stifling heat wave that slammed Chicago in 1995:* Information on heat wave dangers is from *Extreme Heat: A Prevention Guide to Promote Your Personal Health and Safety,* U.S. Centers for Disease Control and Prevention, www.bt.cdc.gov/disasters/extremeheat/heat_guide.asp, accessed October 7, 2010. Details of the Chicago heat wave are from interviews with Larry Kalkstein and Eric Klinenberg, and from Joel Kaplan and Sharman Stein, "City Deaths in Heat Wave Triple Normal: Toll Now at 436 and Still Rising," *Chicago Tribune,* July 20, 1995;

Eric Klinenberg, *Heat Wave: A Social Autopsy of Disaster in Chicago* (Chicago: University of Chicago Press, 2002), 8.

99   *In India, a 2003 heat wave:* "Encephalitis Kills 110 Children," *Chicago Tribune,* July 16, 2003.

99   *poster child for Kalkstein's warning system is Philadelphia:* Material on the 1993 heat wave is from Judy Pasternak and Stephen Braun, "Heat Wave: Why Did So Many Die?" *Los Angeles Times,* July 21, 1995. Material on the heat wave early warning system is from interviews with Larry Kalkstein and National Weather Service meteorologist Gary Szatkowski, November 2008; Kristie L. Ebi, et al., "Heat Watch/Warning Systems Save Lives: Estimated Costs and Benefits for Philadelphia 1995–98," *Bulletin of the American Meteorological Society* (August 2004): 1067–72.

100   *unacceptable someone should die from excessive heat:* Interview with Gary Szatkowski, November 2008.

100   *Kalkstein's warning system was used in forty U.S. municipalities:* Kalkstein personal communications and PowerPoint presentations, November and December 2008.

CHAPTER 5: HARVEST OF TROUBLE

101   *Not far offshore, the warm Gulf Stream:* Information on the Marine Biological Laboratory is from Lewis Thomas, *Lives of a Cell: Notes of a Biology Watcher* (New York: Penguin Group, 1978), 69 and 72, and from the Marine Biological Laboratory, www.mbl.edu/about/discovery/breakthrough.html, accessed October 7, 2010.

103   *Strains of bean plants that resist the virus at ordinary temperatures become vulnerable:* P.K. Anderson and F.J. Morales, "The Emergence of New Plant Diseases: The Case of Insect-Transmitted Plant Viruses," *Annals of the New York Academy of Sciences* 740 (1994).

103   *on a hot midsummer morning in an experimental Illinois farm field:* Material on soybean experiments is from reporting by coauthor Dan Ferber based on a visit to SoyFACE facilities and Evan DeLucia's laboratory, University of Illinois, Urbana-Champaign, July 2008 and interviews and e-mails with Clare Casteel, Bridget O'Neill, Evan DeLucia, and May Berenbaum in April, May, and July of 2008. Information on Evan DeLucia's and May Berenbaum's research is available at their respective home pages: www.life.uiuc.edu/delucia/delucia.htm and www.life.uiuc.edu/entomology/faculty/berenbaum.html.

105   *What plants, algae, and blue-green algae (cyanobacteria) gain:* Peter H. Raven and George B. Johnson, *Biology,* 4th ed. (New York: McGraw-Hill, 1996), 216–26.

106   *led early agronomists to predict:* Evan DeLucia interview, Urbana, Illinois, July 2008.

107   *Global warming naysayers . . . did not miss the significance of this news:* "Greening Earth Society," *SourceWatch,* an online encyclopedia of the Center for Media and Democracy, www.sourcewatch.org/index

.php?title=Greening_Earth_Society, accessed October 7, 2010; *Fact sheet: Greening Earth Society,* an ExxonSecrets.org documentation of industry-funded climate change skeptics compiled by Greenpeace, www.exxonsecrets.org/html/orgfactsheet.php?id=88, accessed October 7, 2010; Ross Gelbspan, *U.S. Coal Industry: Global Warming Is Good for Us,* The Heat Is Online, www.heatisonline.org/contentserver/object handlers/index.cfm?id=3046&method=full, accessed October 7, 2010.

109 *real-world conditions cut the CO_2 fertilization boon in half:* Evan DeLucia interviews; Stephen P. Long, et al., "Food for Thought: Lower-Than-Expected Crop Yield Stimulation with Rising CO_2 Concentrations," *Science* 312: 5782 (2006); David Schimel, "Climate Change and Crop Yields: Beyond Cassandra," *Science* 312 (2006).

109 *soybean aphids:* The soybean aphid description is from CropWatch, University of Nebraska-Lincoln, http://cropwatch.unl.edu/archives/2008/crop15/soybean_aphids.htm, accessed October 7, 2010.

112 *some of these pheromones draw parasitic wasps:* An example of a parasitic insect that lays eggs that develop inside plant-eating caterpillars is the citrus leafminer parasitoid. See Featured Creatures Web site, http://entomology.ifas.ufl.edu/creatures/beneficial/a_citricola.htm, accessed October 7, 2010.

112 *Plant defenses such as these:* Gregory J. Retallack, et al., "Early Forest Soils and Their Role in Devonian Global Change," *Science* 276 (1997).

114 *Geological evidence indicates:* James C. Zachos, et al., "An Early Cenozoic Perspective on Greenhouse Warming and Carbon-Cycle Dynamics," *Nature* 451: 7176 (2008).

114 *Ellen Currano of the Smithsonian Institution:* Information on events at the Paleocene–Eocene Thermal Maximum are from Ellen D. Currano, et al., "Sharply Increased Insect Herbivory during the Paleocene–Eocene Thermal Maximum," *Proceedings of the National Academy of Sciences* 105: 6 (2008); Evan H. DeLucia, et al., "Insects Take a Bigger Bite Out of Plants in a Warmer, Higher Carbon Dioxide World," *Proceedings of the National Academy of Sciences* 105: 6 (2008).

115 *strikes dread in the hearts of farmers:* Information on the threat of soybean rust and its movement into the United States is from Erik Stokstad, "Plant Pathologists Gear Up for Battle with Dread Fungus," *Science* 306: 5702 (2004) and the Web site www.stopsoybeanrust.com.

115 *"charcoal rot":* Paul R. Epstein and Evan Mills, ed., *Climate Change Futures* (Boston: Center for Health and the Global Environment, Harvard Medical School, 2005).

117 *the heat itself may kill them:* Cynthia Rosenzweig, *Climate Change and U.S. Agriculture: The Impacts of Warming and Extreme Weather Events on Productivity, Plant Diseases and Pests* (Boston: Center for Health and the Global Environment, Harvard Medical School, 2000).

118 *It is estimated that 35–50 percent of the world's wheat crops:* Chee Chee Leung, "GM Wheat Yields Hope for Big Dry; Bid for 50 Drought Tolerant Crops," *The Age* (Australia), June 18, 2008.

118 *farmers in the bush had committed suicide:* Sarah Wotherspoon, "Bush

Suicide Shock; Drought Exacts Deadly Toll on Victorian Farmers," *Herald Sun* (Australia), October 8, 2007.

118 *In the developing world, the toll of drought can be even harsher:* Information on the Sahel drought in the 1970s that killed an estimated 250,000 people is from *Millennium Ecosystem Assessment, Ecosystems and Human Well-Being: Synthesis* (Washington, DC: Island Press, 2005) and from *Poverty Eradication, MDGs and Climate Change* (New York: United Nations Development Programme [UNDP], Energy and Environment), www.undp.org/climatechange/adap01.htm, accessed December 18, 2008.

118 *Ethiopia's ability to grow food for its citizens:* Interview with Chris Funk, February 2008; Chris Funk's presentation at the annual meeting of the American Association for the Advancement of Science, Boston, Massachusetts, February 2008; James Verdin, et al., "Climate Science and Famine Early Warning," *Philosophical Transactions of the Royal Society* 360: 1463 (2005).

119 *these dire projections do not even take into account a changing climate:* Epstein and Mills, *Climate Change Futures.*

120 *In the United States, water conflicts have already begun:* Jay Reeves, "Wet December Eases Drought in the South," Associated Press, December 19, 2008; U.S. Drought Monitor, www.drought.unl.edu/dm/monitor .html, accessed December 19, 2008; Abraham Lustgarten and David Hasemyer, "Colorado River May Face Fight of Its Life," U-T Special Report, *San Diego Union Tribune,* December 21, 2008.

120 *civil society institutions can play crucial roles:* Molly E. Brown and Christopher C. Funk, "Food Security under Climate Change," *Science* 319: 5863 (2008).

CHAPTER 6: SEA CHANGE

122 *Sherman is an innovative ecologist:* Kenneth Sherman, "Coastal Ecosystem Health: A Global Perspective," *Annals of the New York Academy of Sciences* 740 (1994); *Large Marine Ecosystems: A Breakthrough Concept for Ecosystem Management,* National Oceanic and Atmospheric Administration, http://celebrating200years.noaa.gov/breakthroughs/ ecosystems, accessed October 8, 2010.

125 *The three broad questions we asked broke down into dozens of smaller ones:* Marine scientists blame the bulk of coastal ecosystem decline on three main factors: loss of coastal wetlands ("nature's kidneys"); pollution from sewage, fertilizer, and nitrates deposited from burning coal; and diseases that decimate species such as oysters and sea urchins that play essential ecological roles in marine ecosystems (filtering water and scavenging waste, respectively). In our study, we asked if these factors contributed to disease in humans and marine organisms. We also asked whether extreme weather, a known consequence of climate change, triggers disease outbreaks via coastal flooding.

126 *We had no shortage of material:* Information on die-offs of mammals,

plants, invertebrates, and vertebrates is from C.D. Harvell, et al., "Emerging Marine Diseases—Climate Links and Anthropogenic Factors," *Science* 285: 5433 (1999) and from our full Health, Ecological and Economic Dimensions of Global Change Program report: P.R. Epstein, B.H. Sherman, E.S. Siegfried, et al., ed., *Marine Ecosystems: Emerging Diseases as Indicators of Change: Health of the Oceans from Labrador to Venezuela* (Boston: Center for Conservation Medicine and Center for Health and the Global Environment, Harvard Medical School, 1998), http://heedmd.org/report.html.

128   *The Prince Edward Island cases: Diatoms Ecology,* Monterey Bay Aquarium Research Institute, www.mbari.org/staff/conn/botany/diatoms/john/basics/eco.htm, accessed October 8, 2010. Information on the effects on the brain is from IPCC, *Climate Change 2001: Impacts, Adaptation and Vulnerability* (Cambridge, U.K.: Cambridge University Press, 2001), also available at http://www.grida.no/publications/other/ipcc _tar; IPCC, "North America," chapter 8 in *Regional Impacts of Climate Change: An Assessment of Vulnerability* (Cambridge, U.K.: Cambridge University Press, 1997), also available at www.ipcc.ch/ipcc reports/sres/regional/index.php?idp=0; S.S. Bates, "Amnesic Shellfish Poisoning: Domoic Acid Production by *Pseudonitzschia* Diatoms," *Aqua Info Aquaculture Notes* 16: 4 (2004).

128   *domoic acid poisoning next appeared in the Pacific off the West Coast of the United States: Crisis Off Our Coast,* International Bird Rescue Research Center, www.ibrrc.org/pelican_domoic.html, accessed October 8, 2010.

129   *dinoflagellates:* Information on dinoflagellates and red tides is from Andrew MacRae, *Dinoflagellates,* www.geo.ucalgary.ca/~macrae/paly nology/dinoflagellates/dinoflagellates.html, accessed October 8, 2010; Centers for Disease Control and Prevention (CDC), Marine Toxins Web page, www.cdc.gov/ncidod/dbmd/diseaseinfo/marinetoxins_g .htm, accessed October 8, 2010; Interagency Working Group on Harmful Algal Blooms, Hypoxia, and Human Health, *Interagency Oceans Human Heath Annual Report 2004–2006* (Washington, DC: Council on Environmental Quality and Office of Science and Technology Policy, Executive Office of the President, September 2008), www.eol.ucar.edu/projects/ohhi/documents/jsost_iohh0908.pdf, accessed October 8, 2010; Epstein, et al., *Marine Ecosystems: Emerging Diseases as Indicators of Change;* CDC, Ciguatera Fish Poisoning Web page, www.cdc.gov/nceh/ciguatera/default.htm, accessed October 8, 2010.

129   *harmful algal blooms:* Like other real-world phenomena, harmful algal blooms have multiple causes. These include fallout from coal-fired power plants. (More than half the nitrogen coming into U.S. East Coast waters arrives as fallout from coal-fired power plants in the Midwest and the Northeast.) Loss of coastal wetlands to shrimp farming and development removes the filters that would otherwise take up waste. Rains flush lots of nutrients to trigger blooms, and warm stagnant waters allow the nasty algae to dominate.

131   *Dermo and MSX:* "Oysters shrivel and become unpalatable" is from Paul R. Epstein and Evan Mills, ed., *Climate Change Futures* (Boston: Center for Health and the Global Environment, Harvard Medical School, 2005); "oysters were unable to shake the invaders" is from Harvell, et al., "Emerging Marine Diseases."

132   *Marine life . . . could provide us with pharmaceuticals:* Committee on the Ocean's Role in Human Health, National Research Council, *From Monsoons to Microbes: Understanding the Ocean's Role in Human Health* (Washington, DC: National Academy Press, 1999); "Painkiller from Cone Snail to Go Commercial," press release, University of Melbourne, November 17, 2003, http://uninews.unimelb.edu.au/news/1038, accessed October 8, 2010; Marc Kaufman, "Snail Sting 1000 Times Stronger than Morphine," *Washington Post,* December 30, 2004; Interagency Working Group on Harmful Algal Blooms, Hypoxia, and Human Health, *Interagency Oceans Human Heath Annual Report 2004–2006.*

132   *Marine organisms also are used as biomedical models:* Information on zebra fish treated with polycyclic aromatic hydrocarbons is from Interagency Working Group on Harmful Algal Blooms, Hypoxia, and Human Health, *Interagency Oceans Human Heath Annual Report 2004–2006,* 49.

133   *Reefs provide habitat:* Joan Kleypas, et al., *Impacts of Ocean 133 on Coral Reefs and Other Marine Calcifiers,* report of a workshop sponsored by NSF, NOAA, and USGS, www.ucar.edu/communications/Final_acidification.pdf, June 2006.

133   *reefs' ability to protect shorelines from storm damage:* Epstein and Mills, *Climate Change Futures.*

135   *As fisheries decline:* Information on fisheries, and how climate change could harm them, is from *Report of the FAO Expert Workshop on Climate Change Implications for Fisheries and Aquaculture,* FAO Fisheries Report 870 (Rome: Food and Agriculture Organization of the United Nations, 2008), ftp://ftp.fao.org/docrep/fao/010/i0203e/i0203e00.pdf; Edward Allison, et al., *Effects of Climate Change on the Sustainability of Capture and Enhancement Fisheries Important to the Poor* (London: U.K. Department for International Development, 2005), www.fmsp.org.uk/Documents/r4778j/r4778j_1.pdf, accessed October 8, 2010.

136   *Approximately 20 percent has gone into the oceans and 30 percent into trees and soils:* The ocean and terrestrial sinks have forestalled far greater atmospheric warming on Earth, perhaps lulling us into complacency. But as vast as the oceans are, they are becoming saturated with the CO_2 they already hold. And, just as warming a soda makes it hold less carbonation and go flat, warming an ocean also reduces the amount of CO_2 the seas can hold, leaving more in the atmosphere to warm the planet.

136   *pH declines of at least 0.3 pH units by midcentury:* Peter Brewer, et al., "Designing Experiments to Predict the Impact of Rapidly Changing

Ocean Chemistry," a presentation at the American Association for the Advancement of Science (AAAS) conference, February 2008.

137 *slight increases in acidity slow shell building:* Elizabeth Kolbert, "The Darkening Sea," *New Yorker,* November 20, 2006.

137 *"the rise of slime":* Kolbert, "The Darkening Sea."

CHAPTER 7: FORESTS IN TROUBLE

140 *Persistent warming can kill off vegetation, turning grasslands into deserts:* Rudoph Kuper and Stefan Kröpelin, "Climate-Controlled Holocene Occupation in the Sahara: Motor of Africa's Evolution," *Science* 313 (August 11, 2006).

141 *For thirty years, Logan had investigated:* Material on Jesse Logan, whitebark pine, and mountain pine beetles is from interviews with Logan and others before and during Dan Ferber's reporting trip to Wyoming and Montana, June 2008.

141 *Whitebark pines . . . anchor the high-altitude forest:* Jesse Logan interviews, May and June 2008; Logan presentation at the Whitebark Pine Citizen Scientists Project Training Workshop held by the Natural Resources Defense Council, June 16–17, 2008; Charles Petit, "In the Rockies, Pines Die and Bears Feel It," *New York Times,* January 30, 2007; Jim Robbins, "Bark Beetles Kill Millions of Acres of Trees in West," *New York Times,* November 18, 2008.

141 *On a cool, windswept mountainside:* Description of the flora and fauna in the whitebark pine ecosystem, and their workings, is from interviews with Logan, Louisa Willcox of the Natural Resources Defense Council, and others, and these: *Whitebark Pine Communities,* U.S. Geological Survey, www.nrmsc.usgs.gov/research/whitebar.htm, accessed October 9, 2010; *Grizzly Bear and Black Bear Ecology,* U.S. Geological Survey, www.nrmsc.usgs.gov/research/grizzlyb.htm, accessed October 9, 2010; Dana L. Perkins, *Ecology,* Whitebark Pine Ecosystem Foundation, www.whitebarkfound.org/ecology.html, accessed October 9, 2010; *Clark's Nutcracker,* Field Guide to Birds of North America, online data base of Wild Birds Unlimited, http://whatbird.wbu.com/obj/83/_/Clarks _Nutcracker.aspx, accessed October 9, 2010.

144 *the beetles have ways of fighting back:* Bark beetle life cycle and defense strategies are from *Bark Beetle Management Guidebook,* Province of British Columbia, www.for.gov.bc.ca/tasb/legsregs/fpc/fpcguide/beetle/ chap1.htm, October 1995; Kenneth Raffa interview, November 2008.

147 *entomologists saw the potential of Logan's modeling method:* Logan's backstory and science are from interviews in May and June 2008 and from Jesse Logan and James Powell, "Ghost Forests, Global Warming and the Mountain Pine Beetle," *American Entomologist* 47: 3 (2001).

150 *For example, when chestnut blight was introduced to North America:* Scott Brennan and Jay Withgott, *Environment: The Science behind the Stories* (New York: Pearson/Benjamin Cummings, 2005), 140, 152.

150 *But ecosystems, like all systems, have their limits:* The comma but-

terfly story is summarized from Elizabeth Kolbert, *Field Notes from a Catastrophe* (New York: Bloomsbury, 2006), 67–71. The golden toad extinction story is summarized from Brennan and Withgott, *Environment: The Science behind the Stories*, 146.

151 *"ecosystem services are indispensable":* Millennium Ecosystem Assessment, *Ecosystems and Human Well-Being: Synthesis* (Washington, DC: Island Press, 2005), 2.

152 *this uneasy equilibrium was broken:* Allan Carroll interview, December 2008; Wendy Stueck, "The Mighty Are Falling," *Globe and Mail,* September 1, 2007; Logan and Powell, "Ghost Forests, Global Warming and the Mountain Pine Beetle."

153 *The foresters tried:* The story of the bark beetle battle in British Columbia is from Mark Hume, "Poisoned Bark Worse Than Beetles' Bite?" *Globe and Mail,* December 21, 2004; "Beetles Invade BC," *Edmonton Sun,* May 11, 2004; Larry Pynn, "BC Parks, Forests Ministry Will Combat Bark Beetles," *Vancouver Sun,* September 30, 2003.

153 *scientists and foresters realized the battle was lost:* Information on the bark beetle battle is from Timothy Egan, "On Hot Trail of Tiny Killer in Alaska," *New York Times,* June 25, 2002; Robbins, "Bark Beetles Kill Millions of Acres of Trees in West"; Margaret Munro, "Insects Wage 'Mass Attack': Climate Changes Have Made Valuable BC Forests 'Just What the Mountain Pine Beetle Wants,' Biologist Says," *National Post* (Canada), July 11, 2003; Allan Carroll interview, December 2008. Information on the aftermath is from Rob Granatstein, "Bark Beetle Won't Bug Off," *Toronto Sun,* August 31, 2008; Nathan Vanderklippe, "Winter Cold May End Pine Beetle's Push Eastward: 'More Optimistic,'" *Financial Post,* June 10, 2008; Stueck, "The Mighty Are Falling."

153 *Such failure has ramifications for the planet:* Werner A. Kurz, et al., "Mountain Pine Beetle and Forest Carbon Feedback to Climate Change," *Nature* 452: 7190 (2008); Werner Kurz, "Making the Paper," *Nature* 452 (2008).

154 *Today, the beetles have infested:* The story of the bark beetles' attack on Alberta, and their threat to the boreal forest, is from the following: Munro, "Insects Wage 'Mass Attack,'"; Allan Carroll interview, December 2008; Vanderklippe, "Winter Cold May End Pine Beetle's Push Eastward."

154 *Global warming has hit the mountains of the U.S. West:* Stephen Saunders, Charles Montgomery, and Tom Easley, *Hotter and Drier: The West's Changed Climate,* Rocky Mountain Climate Organization and the Natural Resources Defense Council, www.rockymountainclimate .org/website%20pictures/Hotter%20and%20Drier.pdf, March 2008.

155 *Trees that have perished in the previous year or two are still full of highly combustible resin:* Logan and Powell, "Ghost Forests, Global Warming and the Mountain Pine Beetle."

155 *beetle infestations may actually reduce fire risk:* A new study in Yellowstone by Monica Turner of the University of Wisconsin, Madison, and colleagues was in press at *Ecological Monographs* when this book went

to press; it was cited in Bettina Boxall, "Bark Beetles Kill Trees, but May Not Raise Fire Risk," *Los Angeles Times,* September 26, 2010.

155 *a beetle-ravaged forest is like a field of roman candles:* Bark beetle 155 Kenneth Raffa describes this whole situation as a vicious cycle, a positive feedback loop wherein a warming climate means bark beetle infestations, which mean fierce fires that release tons of carbon dioxide, which in turn mean more warming. "I know some critics of climate change scenarios talk about tipping points as if they are some hypothetical construct," Raffa said. "Here's a classic example."

155 *Fires have grown larger and more frequent:* Information on global warming and the U.S. West is from Saunders, Montgomery, and Easley, *Hotter and Drier: The West's Changed Climate.* Information on the increase in fires is from Tim Molloy, "Fire Season Rages Early in California," *The Gazette* (Montreal, Quebec), May 5, 2004; Egan, "On Hot Trail of Tiny Killer in Alaska"; Dan Glaister, "Plague of Beetles Raises Climate Change Fears for American Beauty," *Guardian,* March 19, 2007.

155 *the infamous Biscuit Fire:* Dominique Bachelet, James M. Lenihan, and Ronald P. Neilson, *Wildfires and Global Climate Change: The Importance of Climate Change for Future Wildfire Scenarios in the Western United States,* Pew Center on Global Climate Change, www.pew climate.org/docUploads/Regional-Impacts-West.pdf, December 2007; Stephen Pyne, "Passing the Torch: Why the Eons-Old Truce between Humans and Fire Has Burst into an Age of Megafires, and What Can Be Done about It," *American Scholar* 77: 2 (2008): 22–23.

155 *California . . . has been particularly hard hit:* Media reports, including Chris Ayres, "Celebrities among 250,000 Forced to Flee California Wild Fires," *Ottawa Citizen,* October 23, 2007; "Calif. Fires Rage Despite Calmer Winds," CBS/AP, November 17, 2008; Andrew Bridges, "Firefighter Dies Trying to Save Historic Town from Wildfires," Associated Press, October 30, 2003.

156 *Tony Westerling, a scientist with the Sierra Nevada Research Institute:* The story of Westerling's research on climate change–induced western fires comes from interviews with Westerling and Stephen Running, November and December 2008; Stephen Running, "Is Global Warming Causing More, Larger Wildfires?" *Science* 313 (2006): 927; Anthony L. Westerling, et al., "Warming and Earlier Spring Increase Western U.S. Forest Wildfire Activity," *Science* 313 (2006).

156 *they just went home and had a beer:* Interview with Steven Running, November 2008.

157 *None of this bodes well for human health:* Interview with Dan Jaffe, December 5, 2008. Information on smoke-induced heart or respiratory problems is from *Wildfires,* fact sheet, U.S. Centers for Disease Control and Prevention, www.bt.cdc.gov/disasters/wildfires, accessed October 9, 2010. Information on air quality impacts is from Vishal Verma, et al., "Physicochemical and Toxicological Profiles of Particulate Matter in Los Angeles during the October 2007 Southern California Wildfires,"

Environmental Science and Technology 43: 3 (2009); Harish C. Phuleria, et al., "Air Quality Impacts of the October 2003 Southern California Wildfires," *Journal of Geophysical Research* 110 (2005). Information on toxic smoke is from Paul Epstein and Gary Tabor, "Climate Change Is Really Bugging Our Forests," *Washington Post,* September 7, 2003.

158 *woke up to the acrid smell of smoke:* The story of Bob and Sandra Younger's ordeal with the fire is from an interview with Sandra Younger, November 2008.

159 *The victims of the Cedar Fire included:* Ryan Kim, "Flames Ravage Tiny Community," *San Francisco Chronicle,* October 29, 2003; Kevin Caruso, "Tribute to Steven Rucker," www.firefighterheroes.com/tribute -to-steven-rucker.html, accessed November 30, 2008; "James, Solange and Randy Shohara, 3 Victims of Cedar fire," *San Diego Union Tribune,* November 12, 2003.

CHAPTER 8: STORMS AND SICKNESS

161 *in the spring of 1997, conditions in the Pacific Ocean portended trouble:* Details on the 1997–98 El Niño and El Niño prediction are from Curt Suplee, "El Niño/La Niña: Nature's Vicious Cycle," *National Geographic* online, www.nationalgeographic.com/elnino/mainpage.html, accessed October 10, 2010; Richard M. Todaro and David M. Legler, *Predicting El Niño and Other Climate Variations: Successes and Remaining Challenges,* U.S. Global Change Research Program, www.usgcrp.gov/usgcrp/ Library/highlights/highlight-june2002.htm, June 20, 2002.

163 *My presentation described:* Paul Epstein's presentation and those of Godfrey Chikwenhere and Julie Cliff covered results reported in these studies: A. Barreto, M. Aragon, and Paul R. Epstein, "Bubonic Plague in Mozambique, 1994," *Lancet* 345 (1995); Paul R. Epstein and Godfrey P. Chikwenhere, "Environmental Factors in Disease Surveillance," *Lancet* 343 (1994); Paul R. Epstein and Godfrey P. Chikwenhere, "Biodiversity Questions," letter, *Science* 265 (1994); Julie Cliff and J. Coutinho, "Acute Intoxication from Newly Introduced Cassava during Drought in Mozambique," *Tropical Doctor* 25 (1995).

163 *Our failure to predict rains in Mozambique:* Mozambican officials were skeptical of regional climate predictions—that is, until they were hit in 2000 with six weeks straight of incessant downpours, including three tropical cyclones, which caused widespread floods that took the lives of many, followed by outbreaks of cholera and a fivefold spike in malaria. Officials in Mozambique and elsewhere now accept the reality of climate change and recognize it as a public health threat.

167 *it was bedlam all over Tegucigalpa:* Godofredo Andino, chief of the Program of Emergencies and Disasters, Secretary of Health, Republic of Honduras, personal communication, January 2008; Juan Almendares interviews, January 2008.

168 *Hurricane Mitch generated many short-term horror stories:* Andino,

personal communication, January 2008; José Angel Vasquez Briceño, director general of Epidemiology, Republic of Honduras, personal communication, January 2008. Cholera and leptospirosis facts are from *Extreme Weather Events: The Health and Economic Consequences of the 1997/98 El Niño and La Niña* (Boston: Center for Health and the Global Environment, Harvard Medical School, 1999).

168   *Such outbreaks typically follow major hurricanes and flooding:* John T. Watson, Michelle Gayer, and Maire A. Connolly, "Epidemics after Natural Disasters," *Emerging Infectious Diseases* 13: 1 (2007).

168   *images of Hurricane Katrina:* The fatality count of more than 1,450 is from Dan Barry, "Tracing the Path of a Corpse, from the Street to Dignity," *New York Times*, August 27, 2006; information on the 2,000 left missing is from Shaila Dewan, "Storm's Missing: Lives Not Lost but Disconnected," *New York Times*, March 1, 2006.

169   *Other health problems emerged later:* Material on outbreaks of *Vibrio* bacteria is from the U.S. Centers for Disease Control and Prevention, "*Vibrio* Illnesses after Hurricane Katrina: Multiple States," *Morbidity and Mortality Weekly Reports* 54: 37 (September 23, 2005). Material on "Katrina cough" is from an interview with Elizabeth James, May 2008, and from "Researcher Seeks Truth about Katrina Cough," press release, Tulane School of Medicine, June 2, 2008. Information on the contaminated trailers is from Sheila Kaplan, "FEMA Covered Up Cancer Risks to Katrina Victims," *Salon* online journal, www.salon.com, January 29, 2008. Information on mental health issues is from Howard Osofsky, "Katrina's Children Revisited: A Longitudinal Study," presentation at the Association for the Advancement of Science conference, February 2008; Hurricane Katrina Community Advisory Group, "Mental Illness and Suicidality after Hurricane Katrina," *Bulletin of the World Health Organization* 84: 12 (2006); Sandro Galea, et al., "Exposure to Hurricane-Related Stressors and Mental Illness after Hurricane Katrina," *Archives of General Psychiatry* 64: 12 (2007).

170   *Such storms begin:* Robert Henson, *The Rough Guide to Weather* (London: Rough Guides, 2007), 86; IPCC, *Climate Change 2007 Synthesis Report* (Geneva: Intergovernmental Panel on Climate Change, 2008), www.ipcc.ch/publications_and_data/publications_ipcc_fourth_assessment_report_synthesis_report.htm, 24.

170   *As the globe warms, warmer oceans store more energy:* Greenhouse gas emissions are responsible for most of the ocean heat buildup in the past five decades, according to studies by Tim Barnett of the Scripps Institution of Oceanography and his colleagues. Indeed, the pattern of warming is precisely what one would predict from the buildup of greenhouse gases. There is a tight link as well between ocean surface temperatures and the formation and power of hurricanes.

170   *These effects may have contributed to the devastating power of hurricanes Mitch and Katrina: Hurricanes and Climate Change,* Union of Concerned Scientists, www.ucsusa.org/global_warming/science_and_impacts/science/hurricanes-and-climate-change.html, accessed October 10, 2010.

170 *Studies over the last five years:* Kerry Emanuel, Judith Curry, Kevin Trenberth, Tom Knutson, personal communications, September 2008. Information on early studies on hurricanes from the past thirty years is from Kerry Emanuel, "Increasing Destructiveness of Tropical Cyclones over the Past 30 Years," *Nature* 436 (2005); Peter J. Webster, et al., "Changes in Tropical Cyclone Number, Duration, and Intensity in a Warming Environment," *Science* 309 (2005). For more recent studies on the projections for future hurricanes, see Morris A. Bender, Thomas R. Knutson, Robert E. Tuleya, Joseph J. Sirutis, Gabriel A. Vecchi, Stephen T. Garner, and Isaac M. Held, "Modeled Impact of Anthropogenic Warming on the Frequency of Intense Atlantic Hurricanes," *Science* 327 (January 22, 2010) and Thomas Knutson, *Global Warming and Hurricanes: An Overview of Current Research Results,* National Oceanic and Atmospheric Administration, www.gfdl.noaa.gov/~tk/glob_warm_hurr_webpage.html, May 5, 2010.

172 *When the waters subsided in Cedar Rapids:* Information on postflood Cedar Rapids, Iowa, is from reporting by Kari Lydersen, June 2008; *Hurricane and Flood Recovery,* U.S. Centers for Disease Control and Prevention, Emergency Preparedness and Response Web page, www.bt.cdc.gov/disasters/hurricanes/recovery.asp, accessed October 10, 2010; *Flooding,* Environmental Protection Agency, Natural Disasters and Weather Emergencies Web page, www.epa.gov/naturalevents/flooding.html, accessed October 10, 2010.

173 *epidemic of cryptosporidiosis:* William R. Mac Kenzie, Neil J. Hoxie, Mary E. Proctor, et al., "A Massive Outbreak in Milwaukee of Cryptosporidium Infection Transmitted through the Public Water Supply," *New England Journal of Medicine* 331: 3 (1994).

173 *To prevent such outbreaks in the stormier, flood-prone years:* Interview with Joan Rose, a waterborne disease and climate expert at Michigan State University, October 2008.

174 *Of the hundreds of neighborhoods in Tegucigalpa that had lost access to public drinking water:* Godofredo Andino, personal communication, January 2008; Vasquez Briceño, personal communication, January 2008.

177 *"Of course we're worried":* Interview with Teddy Thomas, July 2008.

178 *"I try not to think about it":* Interview with Marguerite Burke, July 2008.

CHAPTER 9: THE AILING EARTH

179 *The 1997–98 El Niño:* For an overview see Curt Suplee, "El Niño/La Niña: Nature's Vicious Cycle," *National Geographic* online, www.national geographic.com/elnino/mainpage.html, accessed October 12, 2010. Statistics on cholera in East Africa are from World Health Organization, "El Niño and Its Health Impacts," *Weekly Epidemiological Record* 73: 20 (1998); Paul R. Epstein, "Climate and Health," *Science* 285: 5 (1999).

180 *I began to urge my colleagues to test early health warning systems:*

For example, see some of the early health warning system recommendations from the Center for Health and the Global Environment, *Extreme Weather Events: The Health and Economic Consequences of the 1997/98 El Niño and La Niña* (Boston: Center for Health and the Global Environment, Harvard Medical School, 1999), http://chge.med .harvard.edu/publications/documents/enso.pdf.

182 *We'd also need to monitor the health of animals:* The need to monitor the status of animal reservoirs, and not just vector organisms like mosquitoes and rodents, was underscored by the explosive 1999 U.S. debut of West Nile virus, which is a disease of birds that spilled over to humans. The 2009 outbreak of pandemic swine flu (type A H1N1) is another case in point. The virus first emerged in pigs and contained genetic material from four types of organisms: Eurasian and American pigs, birds, and humans. Besides preventing human epidemics, monitoring animal populations for such developments can be helpful in preventing animal epidemics.

183 *Just a few days before I was scheduled to speak:* My article, "Climate and Health," appeared in *Science* on July 16, 1999. In that same issue was Kenneth J. Linthicum, Assaf Anyamba, Compton J. Tucker, et al., "Climate and Satellite Indicators to Forecast Rift Valley Fever Epidemics in Kenya," *Science* 285: 5426 (1999). The second meeting of the panel, when I gave my talk, occurred July 20–21, 1999.

184 *Cheney was an old Washington hand:* Newspaper coverage includes "Bush Chooses Cheney as Running Mate," *USA Today,* July 25, 2000. Cheney biographical details are from *Angler: The Cheney Vice Presidency,* a multimedia piece on the *Washington Post* Web site, http:// blog.washingtonpost.com/cheney, accessed October 12, 2010; *The Life and Career of Dick Cheney,* a 2008 slide show on the *Washington Post* Web site, www.washingtonpost.com/wp-srv/photo/gallery/070622/ GAL-07Jun22-78888, accessed October 12, 2010.

184 *Bush reversed his campaign pledge:* Andrew C. Revkin, "Despite Opposition in Party, Bush to Seek Emissions Cuts," *New York Times,* March 10, 2001; Douglas Jehl with Andrew C. Revkin, "Bush, in Reversal, Won't Seek Cut in Emissions of Carbon Dioxide," *New York Times,* March 14, 2001.

184 *Bush charged Cheney with developing a new energy policy:* Cheney's roles are from the *Washington Post* multimedia piece *Angler: The Cheney Vice Presidency.*

184 *remarkable series of lawsuits and investigations:* Details of who the task force met with come from these articles: Dana Milbank and Justin Blum, "Document Says Oil Chiefs Met with Cheney Task Force," *Washington Post,* November 16, 2005; Michael Abramowitz and Steven Mufson, "Papers Detail Industry's Role in Cheney's Energy Report," *Washington Post,* July 18, 2007; Jeff Gerth, "Enron and Cheney Aides Met 4 Times," *New York Times,* January 9, 2002. "Conservation . . . a personal virtue" was quoted in Richard Benedetto, "Cheney's Energy Plan Focuses on Production," *USA Today,* May 1, 2001.

185 *The next year, 2003, our center launched a large two-and-a-half-year study:* UNDP, Swiss Re and the Harvard Center for Health and the Global Environment Embark on Groundbreaking Study, Swiss Re, www.swissre.com/media/media_information/undp_swiss_re_and_the _harvard_center_for_health_and_the_global_environment_embark_on _groundbreaking_study.html, September 22, 2003.

187 *an eye-opening report:* "Office of Net Assessment," *SourceWatch,* an online encyclopedia of the Center for Media and Democracy, www .sourcewatch.org/index.php?title=Office_of_Net_Assessment, accessed October 12, 2010; the statement of Stephen P. Rosen was cited by Gary J. Schmitt of the Project for a New American Century, a neoconservative think tank, www.newamericancentury.org/defnov1097.htm, November 10, 1997.

187 *Just as President George W. Bush was launching his campaign:* The report was dated October 2003, the *Fortune* magazine article was January 2004, and Bush's campaign ran through 2004; *About GBN,* Global Business Network, www.gbn.com/about/index.php, accessed October 12, 2010.

187 *evidence for steady change: Keeling Curve Lessons,* Scripps Institution of Oceanography, http://scrippsco2.ucsd.edu/program_history/keeling _curve_lessons_4.html, accessed October 12, 2010.

188 *an extremely sobering report: Abrupt Climate Change: Inevitable Surprises (Report in Brief)* (Washington, DC: National Academy of Sciences, 2002), 1, www.nap.edu/html/climatechange-brief/abruptclimate change-brief.pdf.

188 *when the world's warming climate hurtled over a precipice:* Peter Schwartz and Doug Randall, *An Abrupt Climate Change Scenario and Its Implications for United States National Security* (San Francisco: Global Business Network, October 2003).

188 *same latitude as Anchorage:* William H. Calvin, "The Great Climate Flip-Flop," *Atlantic Monthly,* January 1998.

188 *This actually happened about 12,700 years ago:* The description of abrupt conveyor change is from Harry L. Bryden, Hannah R. Longworth, and Stuart A. Cunningham, "Slowing of the Atlantic Meridional Overturning Circulation at 25°N," *Nature* 438: 7068 (2005); Wallace Broecker, "Was the Younger Dryas Triggered by a Flood?" *Science* 312: 5777 (2006).

190 *Researchers determine a temperature for a particular year:* W. Dansgaard, S. J. Johnsen, H. B. Claussen, et al., "Evidence for General Instability of Past Climate from a 250-kyr Ice-Core Record," *Nature* 364: 6434 (1993).

190 *The temperature was amazingly constant for the past 10,000 years:* Holocene climate has been mostly stable, but there was an abrupt cooling 8,200 years ago that lasted 150 years, an abrupt shift in precipitation that turned the Sahara from grassland to desert, and other shifts in temperatures and rainfall.

191 *a series of sharp dips and rises:* Dansgaard, et al., "Evidence for Gen-

eral Instability of Past Climate from a 250-kyr Ice-Core Record"; NAS, *Abrupt Climate Change*, 1.

191 *climate can change surprisingly fast:* Strictly speaking, we are still in an ice age and have been for two million years. What we call "the last ice age" (known by scientists as the last glacial maximum), refers to a period about eighteen thousand years ago, when continental ice sheets extended their farthest away from the poles. Viewed another way, it was a period of large polar ice caps. Today we have medium-sized ice caps and may be heading toward a small (or absent) north polar cap. We might also be heading toward a cold reversal like that depicted in the Pentagon scenario, in which gradual warming is punctuated by a sudden shift back to a colder climate. Cold reversals have been much more common than scientists previously appreciated, according to studies of ancient temperature records from ice cores in Greenland and Antarctica.

192 *the idea of a living Earth was more than a metaphor:* James Lovelock, *Homage to Gaia* (Oxford: Oxford University Press, 2001); Oliver Morton's book on photosynthesis, *Eating the Sun* (New York: HarperCollins, 2008), is an excellent source on Lovelock's personal history and that of the Gaia hypothesis.

193 *When they examined Mars:* Dian R. Hitchcock and James E. Lovelock, "Life Detection by Atmospheric Analysis," *Icarus* 7: 149 (1967).

195 *Lovelock's hypothesis was ridiculed and attacked:* The long years of ridicule and ultimate vindication of the Gaia hypothesis are described in detail here: Jim Gillon, "Feedback on Gaia," *Nature* 406 (2000); Drake Bennett, "Dark Green," *Boston Globe,* January 11, 2009; Stephen H. Schneider and Penelope J. Boston, *Scientists on Gaia* (Cambridge, MA: MIT Press, 1993).

196 *A third positive feedback involves water vapor:* Gavin Schmidt, "Water Vapour: Feedback or Forcing?" *RealClimate,* www.realclimate.org/index.php?p=142, April 6, 2005.

197 *We may be headed for a third stable state:* William James Burroughs, *Climate Change* (Cambridge, U.K.: Cambridge University Press, 2001), 225; Jonathan Cowie, *Climate Change: Biological and Human Aspects* (New York: Cambridge University Press, 2007). Also, in the warm period that began just 130,000 years ago, an eyeblink in geological time, the temperature was just 1°C warmer, but sea levels were at least three meters (almost ten feet) and possibly up to eight meters (twenty-six feet) higher.

198 *changing more, and more quickly:* Richard P. Allan and Brian J. Soden, "Atmospheric Warming and the Amplification of Precipitation Extremes," *Science* 321: 5 (2008); Quirin Schiermeier, "A Rising Tide," *Nature* 428: 6979 (2004). Information on the cryosphere is from the National Snow and Ice Data Center, www.nsidc.org/sotc/iceshelves.html, accessed October 12, 2010; Krishna Ramanujan, *Fastest Glacier in Greenland Doubles Speed,* NASA news, www.nasa.gov/vision/earth/lookingatearth/jakobshavn.html, December 1, 2004.

198   *how likely it is to change abruptly:* Paul R. Epstein and James J. McCarthy, "Assessing Climate Stability," *Bulletin of the American Meteorological Society* 85 (2004).

198   *Also on the rise are novel events and outliers:* The hottest European summer since at least 1500 is described in P. A. Stott, et al., "Human Contribution to the European Heatwave of 2003," *Nature* 432: 7017 (2004), and the South Atlantic hurricane is described in Chris Mooney, *Storm World* (Orlando: Harcourt Books, 2007), 111.

CHAPTER 10: GAINING GREEN BY GOING GREEN

200   *For years afterward, Swiss Re employees at the conference that day:* Paul Epstein's recollections and an interview with Stephen Dishart, who was present at the event, May 2008.

201   *Coomber was a trained actuary:* Coomber's career trajectory is from a biographical sketch he provided for the program of a different event associated with the Climate Change Futures project, our September 2005 presentation of the results at the American Museum of Natural History; Stephen Dishart, personal communication, May 1, 2008; the Coomber biographical sketch from the Swiss Re Web site, www.swissre.com/pws/about%20us/corporate%20governance/governing%20bodies/board%20of%20directors/john%20r.%20coomber.html, accessed December 10, 2008.

203   *Hurricane Andrew:* Accounts of the storm and short-term damage are from Craig Pittman, "Storm's Howl Fills the Ears of Survivors," *St. Petersburg Times,* August 18, 2002; "Hurricane Andrew, After the Storm: Ten Years Later," *St. Petersburg Times,* www.sptimes.com/2002/webspecials02/andrew, accessed October 13, 2010; *History of the Homestead Air Reserve Base,* Homestead Air Reserve Base, www.homestead.afrc.af.mil/library/factsheets/factsheet.asp?id=3401, accessed October 13, 2010.

204   *Eleven insurers became insolvent, abandoning thousands of homeowners:* Jeff Harrington, "Insurance Customers Still Paying the Price," *St. Petersburg Times,* August 18, 2002.

204   *When the restrictions expired, companies reacted:* Insurance company actions and Florida's response are from Harrington, "Insurance Customers Still Paying the Price," and from Evan Mills, *Responding to Climate Change—The Insurance Industry Perspective,* Climate Action Programme, http://evanmills.lbl.gov/pubs/pdf/climate-action-insurance.pdf, November 26, 2007.

204   *insured property losses topped $61 billion:* Robert Hartwig, *Catastrophic Loss in Mississippi: The Aftermath of Katrina,* Insurance Information Institute, www.iii.org/Catastrophic-Loss-in-Mississippi-The-Aftermath-of-Katrina, March 2008.

205   *"caught unprepared for ostensibly 'inconceivable' disasters":* Evan Mills, "Insurance in a Climate of Change," *Science* 309: 5737 (August 12, 2005): 1040–44.

206   *"climate change will drive up insured losses 37 percent per year"*: Mills, *Responding to Climate Change.*

206   *How much economic pain will climate change deliver*: Swiss Re's thinking is from Stephen Dishart, personal communication, May 1, 2008. Jacques Dubois's remarks are from "Harvard Study Shows Escalating Climate Change Impacts on Human Health, the Environment, and the Economy," press release from Harvard Medical School, Swiss Re, and the UN Development Programme, November 1, 2005.

207   *what happened around Kuala Lumpur, Malaysia, in 2005*: "Smog Blankets Kuala Lumpur," *Guardian* (U.K.), August 11, 2005.

208   The Stern Review *warned*: Nicholas Stern, *The Economics of Climate Change: The Stern Review* (Cambridge, U.K.: Cambridge University Press, 2007). We should also acknowledge that *The Stern Review* was criticized by U.S. economists for using too small a discount rate—a measure used to account for the loss of value of commodities over time in comparison with their present value. Stern used a discount rate of 1.6 percent, while the critical U.S. economists wanted to use 6 percent. Either rate assumes the value of the thing in question goes to zero over 100 years, which holds true for commodities like cars and refrigerators. But ecological systems, whose value is measured by the goods and services they provide and the sustainability they offer, gain value in a crowded world. Since it's nonsensical to say that a forest will be worthless after 101 years, a discount rate does not apply to living systems. Rather than using a discount rate, one might use a negative discount rate—or better yet, an appreciation rate—to help determine the value of an ecological system.

209   *The mission of public health*: Institute of Medicine, *The Future of Public Health* (Washington, DC: National Academy Press, 1988), 40.

210   *Swiss Re gave Chris Walker a groundbreaking assignment*: Accounts of Chris Walker's work at Swiss Re, and examples used, are from interviews with Chris Walker, March 2009, and an interview with Stephen Dishart, formerly of Swiss Re, May 2008. Background on insurers' roles in fighting climate change is from Evan Mills, *From Risk to Opportunity: Insurer Responses to Climate Change* (Boston: CERES, 2009).

214   *increase investment in flood defenses*: Mills, *From Risk to Opportunity.*

214   *42 percent less severe*: The Benefits of Modern Wind Resistant Building Codes on Hurricane Claim Frequency and Severity, Institute for Business and Home Safety, www.ibhs.org/newsroom/down loads/20070810_102941_10167.pdf, August 2007.

214   *new pay-as-you-drive auto policies*: Jason E. Bordoff, Brookings Institution, "Pay-As-You-Drive Car Insurance," *Democracy Journal* 8 (spring 2008); Mills, *From Risk to Opportunity.*

215   *I was introduced to Charles O. "Chuck" Prince*: Paul Epstein's recollections of Citi visit; Dominic Rushe, "Banking on Going Green," *Times* (London), May 13, 2007.

215 *the company had pledged to invest $50 billion:* Dominic Rushe, "Banking on Going Green."

216 *In March 2007, Bank of America committed $20 billion:* Rushe, "Banking on Going Green"; "Bank of America Announces $20 Billion Environmental Initiative," Bank of America press release, www.sustainable-business.com/index.cfm/go/news.display/id/12475, March 6, 2007.

216 *Equator Principles:* See the Equator Principles Web site, www.equator principles.com.

217 *the Bush administration kept up its role:* The account of the Bush administration actions draws from Andrew C. Revkin, "Bush Aide Softened Greenhouse Gas Links to Global Warming," *New York Times,* June 8, 2005; Andrew C. Revkin, "Climate Expert Says NASA Tried to Silence Him," *New York Times,* January 29, 2006; Andrew C. Revkin, "NASA's Goals Delete Mention of Home Planet," *New York Times,* July 22, 2006.

217 *ten U.S. states, frustrated by the Bush administration's intransigence:* This account of the states' lawsuit, the response of the Environmental Protection Agency, and the Supreme Court's decision are from *U.S. Supreme Court to Consider Greenhouse Gas Rules (Update 1),* Bloomberg, www.bloomberg.com, June 26, 2006. The oral argument transcript in the case is available from the Supreme Court of the United States at www.supremecourtus.gov/oral_arguments/argument_tran scripts/05-1120.pdf. Additional material is from *Massachusetts et al. vs. EPA et al.,* Pew Center on Global Climate Change, www.pewclimate .org/epavsma.cfm, accessed October 13, 2010; *Letter from Jason Kestrel Burnett to Senator Barbara Boxer,* Center for American Progress Action Fund, Think Progress Web site, http://thinkprogress.org/wonkroom/ wp-content/uploads/2008/07/burnett_epw.pdf, July 6, 2008; Danny Hakim, "10 States Sue E.P.A. on Emissions," *New York Times,* April 28, 2006; Linda Greenhouse, "Justices Say EPA Has Power to Act on Harmful Gases," *New York Times,* April 3, 2007.

218 *the White House had censored it:* On the censoring of Gerberding's testimony, see H. Josef Hebert, "Heavy Editing Is Alleged in Climate Testimony," Associated Press, October 24, 2007; Julie L. Gerberding, "Climate Change and Public Health," Testimony before the Committee on Environment and Public Works, U.S. Senate, testimony obtained by the authors, now available online at www.desmogblog.com/full-version -of-white-house-edited-cdc-climate-report-with-highlights, accessed October 13, 2010.

219 *the science in the deleted testimony:* Alison Young, "Deleted CDC Material Similar to Panel Data," *Atlanta Journal-Constitution,* October 26, 2007.

219 *Gerberding's full testimony was finally presented to Congress:* For the account of Gerberding's original testimony and the Congressional hearing, see *Select Committee Hearing Takes Pulse of Health and Climate Issues* on the House Select Committee on Energy Independence and Global Warming Web site, http://globalwarming.house

.gov/pubs/archives_110?id=0036#main_content, accessed October 13, 2010. Also see Howard Frumkin's testimony at this hearing, *Climate Change and Public Health*, http://globalwarming.house.gov/tools/assets/files/0486.pdf, April 9, 2008. Also see *CDC House Hearing Witness Frumkin Submits CDC Director Gerberding's Previously Censored Testimony*, Climate Science Watch, www.climatesciencewatch.org/2008/04/12/cdc-house-hearing-witness-frumkin-submits-cdc-director-gerberding%E2%80%99s-previously-censored-testimony, accessed October 13, 2010; Richard Simon, "Cheney's Office Tried to Alter Greenhouse Gas Testimony, Former Official Says," *Los Angeles Times*, July 9, 2008. The letter from Jason Kestrel Burnett to U.S. Senator Barbara Boxer, July 6, 2008, was discussed at a hearing called An Update on the Science of Global Warming and Its Implications, at which Burnett testified, on July 22, 2008; it is available from the Senate Environment and Public Works Committee Web site, www.epw.senate.gov/public/index.cfm?FuseAction=Files.View&FileStore_id=7d0fb527-d157-4353-b7ce-b72daee6aeed, accessed October 13, 2010.

219 *The 2008 Ceres Conference promised far more:* The Ceres Conference description is from coauthor Dan Ferber's firsthand observation and interviews with participants. Background information on Ceres and conference participants is from biographical sketches in the Ceres Conference 2008 brochure and from the Ceres Web site, www.ceres.org, and related sources.

220 *Global Reporting Initiative:* Global Reporting Initiative, www.globalreporting.org/Home, and links within; Global Reporting Initiative's *G3 Guidelines* on sustainability reporting, www.globalreporting.org/ReportingFramework/ReportingFrameworkDownloads, accessed October 13, 2010. For background on socially responsible investing, see the Social Investment Forum, www.socialinvest.org.

222 *"Stonyfield is my definition of a sustainable business":* The Stonyfield Farm and Hirshberg history is from Andy Savitz's and Gary Hirshberg's talks at the April 2008 Ceres meeting; from Stonyfield's Web sites, www.stonyfield.com/about_us/making_a_difference_since_1983/our_story/index.jsp, accessed October 13, 2010, and www.stonyfield.com/about_us/meet_our_ceyo_and_his_team/meet_gary_our_ceyo/index.jsp, accessed October 13, 2010; and from Neil Shister, "Manufacturer of the Year of Supply Chain Excellence: Stonyfield Farm," *World Trade*, May 1, 2008.

CHAPTER 11: HEALTHY SOLUTIONS

223 *David Riecks and Anna Barnes faced a dilemma:* The story of their decision to install a heat pump is based on interviews conducted in September 2008.

223 *Just a few feet below the ground:* "What Is a Ground Source Heat Pump?" in *What Is Geothermal?* at International Ground Source Heat

Pump Association, www.igshpa.okstate.edu/geothermal/geothermal.htm, accessed October 22, 2010.

224 *Ron Later:* The story of his windmill is based on a phone interview with Later in October 2008.

225 *Jerry Brous and his wife, Pat:* The story of their experience with Grid-Wise is based on phone interviews in September 2008.

226 *"the most expensive power you can buy":* Our explanation of how Grid-Wise works is based on a phone interview with Rob Pratt in September 2008.

227 *The ground source heat pump industry:* Liz Galst, "With Energy in Focus, Heat Pumps Win Fans," *New York Times,* August 14, 2008.

227 *An ambitious deployment of . . . wind turbines:* Paul R. Epstein, William Moomaw, and Christopher Walker, *Healthy Solutions for the Low Carbon Economy* (Boston: Center for Health and the Global Environment, Harvard Medical School, 2008); *GridWise Demonstration Project Fast Facts,* Pacific Northwest National Lab, http://gridwise.pnl.gov/docs/pnnl_gridwiseoverview.pdf, December 2007.

227 *the stabilization wedge:* Stephen Pacala and Robert Socolow, "Stabilization Wedges: Solving the Climate Problem for the Next 50 Years with Current Technologies," *Science* 305: 5 (2004); Fred Wellington, et al., *Scaling Up: Global Technology Deployment to Stabilize Emissions* (Washington, DC: World Resources Institute, April 2007), 3.

229 *Asbestos provides a good example:* The history of asbestos mining, asbestosis, lung cancer, mesothelioma, and five hundred thousand deaths is from *Richard Ennals, Asbestos and Public Health: The International Dimension,* proceedings of a meeting of the International Commission on Occupational Health, June 5, 2001, www.icohweb.org/news/asbestos.asp. Asbestos uses are from *Asbestos Exposure and Cancer Risk,* National Cancer Institute fact sheet, www.cancer.gov/cancertopics/factsheet/Risk/asbestos, accessed October 20, 2010. Information on how industry fought asbestos bans is from Rick Barrett, "Asbestos Suits Costs U.S. Businesses up to $275 Billion, Report Says," *Milwaukee Journal Sentinel,* December 5, 2002.

229 *Consider, too, compact fluorescent lightbulbs:* CFL background is from *Light Bulbs (CFLs),* Energy Star, www.energystar.gov/index.cfm?c=cfls.pr_cfls, accessed October 20, 2010; information on mercury toxicity is from *ToxFAQs for Mercury,* Agency for Toxic Substances and Disease Registry, www.atsdr.cdc.gov/toxfaqs/tf.asp?id=113&tid=24, April 1999.

230 *nanoparticles made from petroleum:* Epstein, et al., *Healthy Solutions for the Low Carbon Economy.*

231 *The precautionary principle:* Epstein, et al., *Healthy Solutions for the Low Carbon Economy.*

231 *an earthen retaining wall collapsed in Harriman, Tennessee:* Shaila Dewan, "Tennessee Ash Flood Larger Than Initial Estimate," *New York Times,* December 26, 2008. Sampling information is from *Preliminary Summary Report from Water, Sediment and Fish samples collected*

at the TVA Ash Spill by Appalachian State University, Appalachian Voices, Tennessee Aquarium and Wake Forest University, a study conducted by scientists at Appalachian State University and Wake Forest University, from the Appalachian Voices Web site, www.appvoices.org/resources/AppVoices_TVA_Ash_Spill_Report_May15.pdf, accessed October 20, 2010. Additional details were compiled from press reports and Tennessee Valley Authority press releases. Except where noted otherwise, further details in this section are from Epstein, et al., *Healthy Solutions for the Low Carbon Economy.*

231 *At current burn rates our stores could last:* The amount of usable coal left in the ground has recently become contentious. The U.S. Energy Information Administration says 138 years of worldwide reserves: *Coal Reserves,* www.eia.doe.gov/neic/infosheets/coalreserves.html, accessed October 20, 2010. The coal industry projects that by 2050 U.S. coal use will double. But Richard Heinberg argues in his book *Blackout* (Gabriola Island, BC, Canada: New Society Publishers, 2009) that much of that coal is uneconomical to mine and burn, and therefore the peak production of coal is just two decades away.

232 *more than 1,300 similar dumps:* Shaila Dewan, "Hundreds of Coal Ash Dumps Lack Regulation," *New York Times,* January 6, 2009.

232 *coal extracts a blood price:* R.G. Dunlop and Laura Ungar, "Miners Keep Dying Despite Federal Laws," *The Courier-Journal* (Louisville, KT), www.courier-journal.com/cjextra/blacklung/index.html, accessed October 20, 2010. Information on China's coal miner deaths is from Jim Yardley, "As Most of China Celebrates New Year, a Scramble Continues in Coal Country," *New York Times,* February 9, 2008.

233 *First, forests and soils are bulldozed:* Information on the miles of streams buried is from "Environmentalists Fight E.P.A. Rule," *New York Times,* December 3, 2008. The damage from mountaintop removal mining is even more extensive than indicated here. The shattered mountain is vulnerable to landslides. Coal companies claim that they conduct "restoration," but in truth mountaintop removal turns forested mountain ranges into disaster zones: piles of rocks and cleared trees, with just a thin stubble of grasses.

233 *coal transport trains make up as much as 70 percent:* Epstein, et al., *Healthy Solutions for the Low Carbon Economy.*

233 *Finally, burning coal releases:* Effects of pollutants released by coal are from *National Lake Fish Tissue Study: Basic Information,* Environmental Protection Agency, www.epa.gov/waterscience/fish/study/overview .htm, accessed October 20, 2010; effects of particle pollution, including premature deaths, are from Conrad G. Schneider, *Death, Disease, and Dirty Power. Mortality and Health Damage Due to Air Pollution from Power Plants* (Bethesda, MD: Abt Associates, October 2000).

234 *how clean "clean coal" really is:* Bush's promise of clean coal is from *National Energy Policy,* National Energy Policy Development Group, www.ne.doe.gov/pdfFiles/nationalEnergyPolicy.pdf, May 2001, 5–15;

see also Andrew Revkin, "A 'Bold' Step to Capture an Elusive Gas Falters," *New York Times,* February 3, 2008.

234 *coal gasification: How Coal Gasification Power Plants Work,* U.S. Department of Energy, www.fossil.energy.gov/programs/powersystems/gasification/howgasificationworks.html, accessed October 20, 2010.

235 *The infrastructure needed would be enormous:* Material on carbon capture and sequestration is from the author's work and from Revkin, "A 'Bold' Step to Capture an Elusive Gas Falters." Risk of acidifying aquifers is from IPCC, *Carbon Dioxide Capture and Storage* (Cambridge, U.K.: Cambridge University Press, 2005), also available at www.ipcc.ch/publications_and_data/publications_and_data_reports_carbon_dioxide.htm.

236 *And oil is no better:* Paul Purpura, "Supply Vessel Watched Rig Burst into Fireball: Ship Scrambles to Pull Survivors out of Gulf," *New Orleans Times-Picayune,* May 8, 2010; Peter Grier, "BP Oil Spill: Harrowing Escapes of Deepwater Horizon Survivors," *Christian Science Monitor,* May 27, 2010; Pam Radtke Russell, "Regulations for Offshore Drilling Rigs Are Inadequate: Coast Guard Official," *Inside Energy/with Federal Lands,* May 17, 2010; Noel L. Grise, "Resolving BP Spill Will Take Years," *Atlanta Journal-Constitution,* June 17, 2010.

236 *peak oil:* Information on oil supplies and possible manifestations of peak oil is from Mark Hertsgaard, "Running on Empty," *Nation,* April 24, 2008; Elizabeth Kolbert, "Unconventional Crude," *New Yorker,* November 12, 2007; Lester R. Brown, "Is World Oil Production Peaking?" *Plan B Updates* (Washington, DC: Earth Policy Institute), www.earth-policy.org/index.php?/plan_b_updates, November 15, 2007; and the documentary *The End of Suburbia: Oil Depletion and the Collapse of the American Dream,* available at www.endofsuburbia.com. This documentary features authors like James Howard Kunstler *(The Geography of Nowhere)* and energy experts such as Matthew Simmons, member of Dick Cheney's 2001 energy task force and chair of Simmons and Company, an investment banking firm serving the oil industry. (See www.simmonsco-intl.com/research.aspx?Type=msspeeches for Simmons's speeches on energy and gas supplies.)

238 *life cycle analyses:* Our oil life cycle analysis is published as Paul R. Epstein and Jesse Selber, ed., *Oil: A Life Cycle Analysis of Its Health and Environmental Impacts* (Boston: Center for Health and the Global Environment, Harvard Medical School, 2002), http://chge.med.harvard.edu/publications/documents/oilfullreport.pdf. Material on Mark Jacobson's alternative energy life cycle analysis is from an interview with Mark Jacobson, November 2008, and from Mark Z. Jacobson, "Review of Solutions to Global Warming, Air Pollution, and Energy Security," *Energy and Environmental Science* 2 (2009).

239 *The nation's designated permanent nuclear waste repository:* Yucca Mountain background is from media reports, including "Milestones on the Road to Yucca Mountain," *Las Vegas Review-Journal,* www

.reviewjournal.com/news/yuccamtn/milestones.jpg, accessed October 20, 2010.

240 *Bill Moomaw:* The story of the Moomaws' Berkshires home is based on interviews conducted in October and November 2008.

242 *When Rob Pratt . . . talks to audiences:* Rob Pratt's story is based on personal communications in September 2008.

243 *Chicago, like any large city: Regional Snapshot,* Chicago Metropolitan Agency for Planning, www.chicagoareaplanning.org/snapshot, accessed October 20, 2010; Katharine Hayhoe and Donald Wuebbles, *Climate Change and Chicago: Projections and Potential Impacts* (Chicago: Department of Environment, September 2008), www.chicago climateaction.org/filebin/pdf/report/Chicago_climate_impacts_report _Executive_Summary.pdf, 5.

244 *evolving green agenda:* Personal communication with Sadhu Johnston, July 2007, about Chicago's green efforts; *Bike 2015 Plan, City of Chicago,* Mayor's Bicycle Advisory Council, www.bike2015plan.org, January 2006.

244 *the green roof:* For benefits of green roofs, see Karen Liu, *Energy Efficiency and Environmental Benefits of Rooftop Gardens* (Ottawa: National Research Council of Canada, March 2002), www.nrc-cnrc.gc .ca/obj/irc/doc/pubs/nrcc45345/nrcc45345.pdf.

245 *In a Minnesota study:* Weidt Group, *Top 6 Benefits of High-Performance Buildings,* brochure summarizing study, Minnesota Pollution Control Agency, www.pca.state.mn.us/oea/publications/highperformance-bro chure.pdf, June 2005. The study *High Performance Building Design in Minnesota* is available at the Minnesota Pollution Control Agency's Web site, www.pca.state.mn.us/oea/publications/highperformance-weidt.pdf, June 30, 2005.

245 *better land use planning:* John Norquist, personal communication, April 2007.

246 *turning itself into a green city:* Our center offers a clear path toward greener, healthier cities, as we outline here for Chicago. Programs we call "healthy cities initiatives" use a coordinated planning effort to achieve some or all of these elements: green buildings, rooftop gardens, walking paths, biking lanes, tree-lined streets, open space, congestion control, smart growth, and enhanced public transport. The payoff can be immense. By making cities walkable, with spokes to an outer ring of walkable communities and open land between the spokes, city planners can preserve wilderness, promote exercise, and increase social cohesion, thereby protecting the environment and promoting good health. Green cities, particularly when connected by electric light-rail, reduce airway traffic and miles driven, both in town and on highways, and this conserves energy and fights climate change. Green cities also save money, create jobs, and help launch climate-stabilizing technologies into the global marketplace.

CHAPTER 12: OF RICE AND TRACTORS

251 *On a hot, sun-baked afternoon in Choluteca, Honduras:* The material in this section is from coauthor Dan Ferber's trip to Choluteca with Juan Almendares, photographer Jason Lindsey, and two translators in January 2008.

253 *In 1972, Choluteca's climate:* Juan Almendares, et al., "Critical Regions, a Profile of Honduras," *Lancet* 342 (1993); "Gulf of Fonseca Mangroves," *National Geographic* online, www.nationalgeographic.com/wildworld/profiles/terrestrial/nt/nt1412.html, accessed October 20, 2010.

253 *General Mining Law:* Interviews with Juan Almendares in January 2008; Juan Almendares, Letter to the [Honduran] National Congress, May 15, 2006; Juan Almendares, "Impact of Climatic Changes: Natural Disasters in Honduras," in *Precaution, Environmental Science, and Preventive Public Policy,* ed. Joel Tickner (Washington, DC: Island Press, 2003).

253 *to encourage the companies to begin their open-pit mining practices:* Today's mines are part of a long history of devastation, as depicted in the classic book by Eduardo Galeano, *Open Veins of Latin America,* which depicts how colonial powers extracted riches from the region and left behind a scarred landscape. Today, the scarring continues.

253 *Although Goldcorp denies that the water is contaminated: Honduras Imposes Fine on GOLDCORP Subsidiary for Irresponsible Environmental Pollution,* Development and Peace, www.devp.org/devpme/eng/pressroom/2007/pf-comm2007-07-26-eng.html, July 26, 2007; Dina Meza, *Lead, Mercury, Arsenic Found in Siria Valley Population* (in Spanish), Revistazo.com, www.revistazo.biz/cms, January 22, 2008; Juan Almendares, personal communication.

258 *In their three weeks at Bretton Woods, Keynes and White:* John Cassidy, "New World Disorder," *New Yorker,* October 26, 1998.

259 *the three decades following World War II:* Cassidy, "New World Disorder."

261 *In his eye-opening book:* John Perkins, *Confessions of an Economic Hit Man* (San Francisco: Berrett-Koehler, 2004), 15 and xvii.

264 *the Washington Consensus:* Paul Epstein and Greg Guest, "International Architecture for Sustainable Development and Global Health," in *Globalization, Health and the Environment: An Integrated Perspective,* ed. Greg Guest (Lanham, MD: Altamira Press, 2005), 239–58.

264 *money, much of it interest, collectively being paid by developing nations:* Epstein and Guest, "International Architecture for Sustainable Development and Global Health."

266 *a rapid rise in currency speculation:* Bernard Lietaer, "From the Real Economy to the Speculative," *IFG News* (International Forum on Globalization) 2 (summer 1997).

266 *As long as different national currencies exist:* Currency speculation background and examples are adapted from Christian Weller, "Currency Speculation: How Great a Danger?" *Dollars and Sense,* May–June 1998, www.dollarsandsense.org/archives/1998/0598weller.html.

266   *$3 trillion a day:* Niall Ferguson, television series *The Ascent of Money,*
       Chimerica Media Limited and WNET.org, first aired January 13, 2009;
       Weller, "Currency Speculation: How Great a Danger?"

267   *Asian financial crisis of the late 1990s:* "The Crash: Timeline of the
       Panic," *Frontline,* WGBH (Boston), www.pbs.org/wgbh/pages/frontline/
       shows/crash/etc/cron.html, accessed October 21, 2010; Amiya Kumar
       Bagchi, "A Turnaround in South Korea" *Frontline* (India's national
       magazine), July 17–30, 1999, www.hindu.com/fline/fl1615/16150620
       .htm; Uday Mohan, *The East Asia Crisis and the Poor: What Role for
       Agriculture?* International Food Policy Research Institute, www.ifpri
       .org/2020/NEWSLET/Nv_0698/NV0698A.HTM, June 1998.

267   *that's what happened in Thailand:* "The Crash: Timeline of the Panic";
       George Soros, *The Crisis of Global Capitalism* (New York: Public
       Affairs, 1998), 245.

267   *The value of South Korea's currency, the won, plunged:* The South
       Korean financial crisis was clearly also caused by Western pressure to
       liberalize capital flow. In the 1980s, the rapidly growing nation taxed
       withdrawals from abroad of recently deposited money. But under pres-
       sure, it dismantled controls on capital movements in the early 1990s,
       which made its economy far more susceptible to currency speculation.

269   *Democrats and Republicans alike had dismantled regulations:* Robert
       Kuttner, "The Great Crash," *American Prospect,* June 2003.

269   *Glass-Steagall Act:* Reem Heakal, "What Was the Glass-Steagall Act?"
       Investopedia, www.investopedia.com/articles/03/071603.asp, accessed
       October 21, 2010.

270   *Then the house of cards collapsed:* Sources for this account of the finan-
       cial crisis include Jonathan Rauch, "Capitalism's Fault Lines," *New
       York Times,* May 17, 2009, which is a review of a book by Richard A.
       Posner, *A Failure of Capitalism: The Crisis of '08 and the Descent into
       Depression.* Information on Bear Stearns is from Gary Shorter, *Bear
       Stearns: Crisis and "Rescue" for a Major Provider of Mortgage-Related
       Products,* Congressional Research Service Report for Congress, http://
       assets.opencrs.com/rpts/RL34420_20080326.pdf, March 26, 2008.
       Information on failed financial firms, including IndyMac Bancorp, is
       from Louise Story, "Regulators Seize Mortgage Lender," *New York
       Times,* July 12, 2008. Information on insurance company failures is
       from Eric Dash and Diana B. Henriques, "Six Insurers Named to Get
       U.S. Taxpayer Aid," *New York Times,* May 14, 2009.

271   *By early 2009, the economic contagion had spread:* World Bank's
       background paper prepared for the G-20 finance ministers meet-
       ing in São Paulo, Brazil, November 8, 2008, *The Financial Crisis:
       Implications for Developing Countries,* http://web.worldbank.org/
       WBSITE/EXTERNAL/COUNTRIES/SOUTHASIAEXT/0,,content
       MDK:21974412~menuPK:158937~pagePK:2865106~piPK:2865128~
       theSitePK:223547,00.html, November 11, 2008. Information on Ice-
       land's collapse is from Andrew Pierce, "Financial Crisis: Iceland's
       Dreams Go Up in Smoke," *Telegraph* (U.K.), October 7, 2008, www

.telegraph.co.uk/finance/financetopics/financialcrisis/3147866/Financial -crisis-Icelands-dreams-go-up-in-smoke.html; *Open Letter from the Prime Minister of Iceland,* Government of Iceland, www.iceland.org/ info/iceland-crisis/timeline/nr/6986, December 1, 2008. Information on Latvia's woes is from *Latvian Economy in Rapid Decline,* BBC online, http://news.bbc.co.uk/2/hi/business/8043972.stm, May 11, 2009.

271 *By unhinging exchange rates and constraints on capital flow:* Paul Epstein, "Yes, We May All Be Keynesians—But He Has More to Teach Us," *Financial Times,* January 5, 2009.

CHAPTER 13: REWRITING THE RULES

272 *the Earth's climate system was also in critical condition:* The state of the climate system is from Paul R. Epstein, William Moomaw, and Christopher Walker, *Healthy Solutions for the Low Carbon Economy* (Boston: Center for Health and the Global Environment, Harvard Medical School, 2008). The time since CO_2 has been so high is from Dieter Lüthi, Martine Le Floch, Bernhard Bereiter, et al., "High-Resolution Carbon Dioxide Concentration Record 650,000–800,000 Years before Present," *Nature* 453: 7193 (2008).

272 *In February of that year:* A report on the speech by preeminent climate scientist John Holdren at United Nations Headquarters, New York, February 14, 2008, is in *Investor Summit on Climate Risk Final Report* (Boston: Ceres, 2008).

273 *which would swamp coastal regions worldwide:* Low-lying land that is home to 10 percent of the world's population would be inundated by a one-meter (three-foot) rise, according to an analysis in the *Financial Times;* scientists are now projecting up to two meters (over six feet) by century's end (S. Knight, "The Human Tsunami," *Financial Times,* June 20–21, 2009).

274 *"Here in the early years of the twenty-first century":* Bill Moomaw was quoted in Andrew C. Revkin, "Climate Panel Reaches Consensus on the Need to Reduce Harmful Emissions," *New York Times,* May 4, 2007.

274 *But Darbee had plenty of company:* For example, see the description of the 2008 CERES conference in chapter 10. USCAP is another example: Timothy Gardner, "U.S. Industry Plan to Cut Greenhouse Gases Criticized," *New York Times,* May 2, 2007.

275 *States, too, had demonstrated their impatience:* Information on the battle between California and the feds to curb CO_2 emissions and impose auto efficiency standards is from Felicity Barringer, "California, Taking Big Gamble, Tries to Curb Greenhouse Gases," *New York Times,* September 15, 2006; John M. Broder and Felicity Barringer, "E.P.A. Says 17 States Can't Set Emission Rules," *New York Times,* December 20, 2007; "Arrogance and Warming," editorial, *New York Times,* December 21, 2007.

276 *President George W. Bush planted a Shumard oak tree:* "Global Warming: Do 'Green' Initiatives Matter?" *The Week,* May 2, 2008; *Fact Sheet:*

Earth Day 2008, Protecting Our Environment, Achieving Results, a White House backgrounder, April 22, 2008, http://georgewbush-white house.archives.gov/ceq/factsheet_earthday_2008.pdf, accessed June 2, 2009.

276   *In a cap-and-trade system:* Alan Durning, et al., *Cap and Trade 101: A Climate Policy Primer,* September 2008, and *Climate Pricing 101 fact sheet,* both available from Sightline Institute, http://sightline.org/research/energy/ res_pubs/cap-and-trade-101?gclid=CNSJ17bj45cCFQrFGgodShAGCw, accessed October 29, 2010; John M. Broder, "From a Theory to a Consensus on Emissions," *New York Times,* May 17, 2009.

276   *The cap-and-trade system was pioneered:* James Gustave Speth, *The Bridge at the Edge of the World* (New Haven, CT: Yale University Press, 2008), 94.

278   *"creation care":* Richard Cizik is quoted from *Evangelical, Scientific Leaders Launch Effort to Protect Creation,* National Association of Evangelicals, www.nae.net/index.cfm?FUSEACTION=editor.page&page ID=413&IDcategory=1, January 17, 2007. Background is from Calvin DeWitt, "Climate Ethics: Winds of Change," from the PlanetU Conference, University of Illinois, Urbana-Champaign, April 9, 2009; *Calvin DeWitt on Science, Evangelicals, and Climate Change,* Step It Up 2007 Web site, http://april.stepitup2007.org/article.php?id=57, January 19, 2007; Evangelical Climate Initiative, *Climate Change: An Evangelical Call to Action,* NPR online, www.npr.org/documents/2006/feb/evangel ical/calltoaction.pdf, accessed October 22, 2010; and media accounts, including Laurie Goodstein, "Evangelical Leaders Join Global Warming Initiative," *New York Times,* February 8, 2006.

278   *Al Gore . . . gave a major speech:* David Stout, "Gore Calls for Carbon-Free Electric Power," *New York Times,* July 18, 2008.

279   *Senator John McCain:* McCain's and Obama's views on climate change are from media accounts, especially David Kestenbaum, "Candidates Call Climate Change an 'Urgent' Priority," NPR, www.npr.org/tem plates/story/story.php?storyId=93562705, August 13, 2008; *The Candidates on Climate Change,* Council on Foreign Relations, www.cfr.org/ publication/14765, September 11, 2008; Edmund L. Andrews, "Candidates Offer Different Views on Energy Policy," *New York Times,* November 28, 2007.

279   *Sarah Palin: Alaska Gov. Palin Appears to Deny Global Warming Is Due to Human Activity,* Climate Science Watch, www.climatescience watch.org/2008/08/30/alaska-gov-palin-appears-to-deny-global-warm ing-is-due-to-human-activity, August 30, 2008.

279   *after hurricanes Ivan, Katrina, Rita, and Wilma disrupted seafloor pipelines:* Ray Tyson, "Gulf Suffers Ivan Hangover," *Petroleum News* 9: 42 (October 17, 2004); *Hurricane Impacts on the U.S. Oil and Natural Gas Markets,* Energy Information Administration, http://tonto.eia.doe .gov/oog/special/eia1_katrina.html, accessed October 22, 2010; Steve Quinn, "Rita Socked Oil Rigs Hard," Associated Press via *San Jose Mercury News,* September 29, 2005.

279   *A global food crisis also worsened in 2008:* Information on the global food crisis is from UNEP, *The Environmental Food Crisis* (UN Environment Programme, 2009), 13, www.unep.org/pdf/FoodCrisis_lores .pdf.

279   *a mulityear drought in this major wheat exporter:* Australia's drought information is from *Australian Agriculture, Fisheries and Forestry,* Australian Government Department of Foreign Affairs and Trade, www .dfat.gov.au/facts/affaoverview.html, accessed October 22, 2010; *Climate Change "Will Prolong" Drought Conditions,* ABC News online, www.abc.net.au/news/newsitems/200506/s1389858.htm, June 11, 2005; Kathy Marks, "Australia's Epic Drought: The Situation Is Grim," *Independent* (U.K.), April 20, 2007, www.independent.co.uk/news/ world/australasia/australias-epic-drought-the-situation-is-grim-445450 .html.

280   *As prices of staple foods soared, riots broke out:* Information on global food riots is from Vivienne Walt, "The World's Growing Food-Price Crisis," *Time,* February 27, 2008; Joseph Guyler Delva and Jim Loney, "Haiti's Government Falls after Food Riots," Reuters, April 12, 2008; "Food Riots in Egypt," Al Jazeera, March 13, 2008.

280   *the price of corn tortillas:* The account of the tortilla protests of 2007 is from Elisabeth Malkin, "Thousands in Mexico City Protest Rising Food Prices," *New York Times,* February 1, 2007; "Mexicans Stage Tortilla Protest," BBC News, February 1, 2007; Jerome Taylor, "How the Rising Price of Corn Made Mexicans Take to the Streets," *Independent* (U.K.), June 23, 2007. Information on $5 billion a year in subsidies is from *Biofuels: At What Cost?—Government Support for Ethanol and Biodiesel in the United States,* Global Subsidies Initiative, www.global subsidies.org/en/research/biofuel-subsidies-united-states, October 2006, 38, 56–61.

281   *U.S. energy and climate policy had turned a corner:* Accounts of Obama's speech are from media reports, including Margot Roosevelt, "Obama's Video Message Energizes Climate Conference," *Los Angeles Times,* November 19, 2008; Samantha Young, "Governors Pledge to Fight Global Warming Together," SFGate (Web site of the *San Francisco Chronicle*), www.sfgate.com/cgi-bin/article.cgi?f=/n/a/2008/11/17/state/ n130925S65.DTL, November 19, 2008; Samantha Young, "Schwarzenegger Opens Climate Summit with Obama," *USA Today,* November 18, 2009; John M. Broder, "Obama Affirms Climate Change Goals," *New York Times,* November 19, 2008.

282   *Then he followed through:* Information is from media accounts of Obama's early climate-preserving initiatives, including Kent Garber, "The Green Energy Economy: What It Will Take to Get There," *U.S. News and World Report,* March 20, 2009; John M. Broder, "Obama to Toughen Rules on Emissions and Mileage," *New York Times,* May 19, 2009; *Van Jones: The Face of Green Jobs,* The Big Money, a Slate Group Web site, www.thebigmoney.com/articles/mothers-milk/2009/04/19/ van-jones-face-green-jobs, April 19, 2009; *Stimulus Package En Route*

to Obama's Desk, CNN online, www.cnn.com/2009/POLITICS/02/13/stimulus/index.html, February 13, 2009.

282   *His $787 billion fiscal stimulus bill:* Michael A. Fletcher, "Obama Leaves D.C. to Sign Stimulus Bill," *Washington Post,* February 18, 2009; Kate Galbraith, "Obama Signs Stimulus Packed with Clean Energy Provisions," *New York Times,* February 17, 2009.

282   *Obama's stagecraft communicated his priorities:* Kate Galbraith, "Obama Touts Clean Energy Achievements," *New York Times,* May 27, 2009.

282   *Global Green New Deal:* "Realizing a 'Green New Deal,'" press release from the UN Environment Programme, February 16, 2009; Edward B. Barbier, *Rethinking the Economic Recovery: A Global Green New Deal* (UN Environment Programme, April 2009), www.unep.org/greeneconomy/portals/30/docs/GGND-Report-April2009.pdf.

283   *climate-warming emissions were rising faster than anyone had anticipated:* Dan Ferber, "Climate Change Worst-Case Scenarios: Not Worst Enough," Findings, the *Science* magazine news blog, http://blogs.sciencemag.org/newsblog/2009/02/climate-change-worst-case-scen.html, February 14, 2009.

283   *Obama sought to influence the world's nations on climate policy:* Kent Garber, "Browner: Climate Change Law Would Bolster U.S. Role at Global Warming Talks," *U.S. News and World Report,* April 13, 2009.

283   *Then the horse-trading started:* Darren Samuelsohn, "Power Companies Bring Fragile Coalition to Climate Debate," *New York Times,* March 19, 2009; "Business and Environmental Leaders Release Landmark Blueprint for Climate Protection Legislation," press release from the U.S. Climate Action Partnership, January 15, 2009; *Summary Overview: USCAP Blueprint for Legislative Action,* backgrounder from U.S. Climate Action Partnership, www.us-cap.org/blueprint, accessed October 22, 2010; Garber, "The Green Energy Economy: What It Will Take to Get There."

284   *Theatrics followed anyway:* Mitch Daniels, "Indiana Says 'No Thanks' to Cap and Trade," *Wall Street Journal,* May 15, 2009; Mike Pence, "Bill Is an Economic Declaration of War," *Indianapolis Star,* May 31, 2009; "GOP Called Democratic Climate Proposal 'Unwise,'" *Wall Street Journal,* May 30, 2009.

284   *Climategate!:* Bryan Walsh, "Has 'Climategate' Been Overblown?" *Time,* December 7, 2009; "'ClimateGate' Doesn't Show Global Warming Was Faked, AP Reports," *Huffington Post,* www.huffingtonpost.com, December 12, 2009; Raphael G. Satter, "'Climategate' Inquiry Largely Clears Scientists," Associated Press via *Seattle Times,* March 30, 2010; *The Disclosure of Climate Data from the Climatic Research Unit at the University of East Anglia* (London: House of Commons Science and Technology Committee, March 31, 2010); *Report of the International Panel Set Up by the University of East Anglia to Examine the Research of the Climatic Research Unit,* University of East Anglia, www.uea.ac.uk/mac/comm/media/press/CRUstatements/SAP, April 12, 2010.

290   *Tobin tax: Tobin Taxes: How to Tame Hot Money and Fund Urgent Global Priorities,* fact sheet, Center for Environmental Economic Development, www.ceedweb.org/iirp/factsheet.htm, accessed October 22, 2010.

291   *one of Keynes's ideas that never came to pass:* George Monbiot, "Keynes Is Innocent: The Toxic Spawn of Bretton Woods Was No Plan of His," *Guardian,* November 18, 2008.

293   *when London treated the River Thames like an open sewer:* Michelle Allen, *Good Intentions, Unexpected Consequences: Thames Pollution of and the Great Stink of 1858* (an excerpt from Michelle Allen, *Cleansing the City: Sanitary Geographers in Victorian London* [Athens, OH: Ohio University Press, 2008]), Victorian Web, www.victorianweb.org/science/health/thames1.html, accessed October 22, 2010; George P. Landow, "'A Fine Corrective': A Review of Michelle Allen's *Cleansing the City: Sanitary Geographers in Victorian London,*" Victorian Web, www.victorianweb.org/science/health/allen1.html, accessed October 22, 2010; Laurelyn Douglas, *Health and Hygiene in the Nineteenth Century,* Victorian Web, www.victorianweb.org/science/health/health10.html, accessed October 22, 2010; Steven J. Burian, Stephan J. Nix, Robert E. Pitt, et al., "Urban Wastewater Management in the United States: Past, Present, and Future," *Journal of Urban Technology* 7: 3 (December 2000).

EPILOGUE

295   *eighty-six long days after the Deepwater Horizon oil rig exploded:* Details of the Gulf oil spill are from media reports: Robert Lee Hotz, "Scientists Try to Gauge Potential Long-Term Environmental Impact," *Wall Street Journal,* June 12, 2010; Noel L. Griese, "Resolving BP Spill Will Take Years," *Atlanta Journal-Constitution,* June 17, 2010; Joel Achenbach, "Cap May Erase Sense of Hopelessness in Oil Spill," *Washington Post,* July 18, 2010; Justin Gillis and Campbell Robertson, "On the Surface, Gulf Oil Spill Is Vanishing Fast; Concerns Stay," *New York Times,* July 27, 2010; "Barab Testifies before Senate on Protection of Oil Refinery Workers," *OSHA Quick Takes* 9: 12 (June 15, 2010), www.osha.gov/as/opa/quicktakes/qto6152010.html.

295   *An FBI-led . . . probe:* Jerry Markon, "Criminal Probe of Oil Spill to Focus on 3 Firms and Their Ties to Regulators," *Washington Post,* July 28, 2010; Seth Borenstein, "Gulf Oil Spill Panel to Look at Root Causes," Associated Press via the *Christian Science Monitor,* July 9, 2010.

296   *spill bills:* Kate Sheppard, "It's Spill Bill Time," *MotherJones* online, http://MotherJones.com, July 27, 2010; John M. Broder, "House and Senate Roll Out Spill Bills," *New York Times,* July 27, 2010.

297   *numerous professional groups had begun calling for action:* Cecil Wilson, MD, address to the November 20, 2009, Senate hearing sponsored by Senator Tom Udall, quoted in "The Health Community Speaks

Out about Climate Change: Implications for the U.S. States and for the Nation," press release, Center for Health and the Global Environment, November 20, 2009; American Nurses Association, "Resolution: Global Climate Change," to the 2008 House of Delegates, in *Resolutions: Global Climate Change: Summary of Proceedings,* www.nursing world.org/MainMenuCategories/OccupationalandEnvironmental/ environmentalhealth/PolicyIssues/GlobalClimateChangeandHuman Health.aspx, accessed October 22, 2010; Katherine M. Shea and the Committee on Environmental Health (of the American Academy of Pediatrics), "Global Climate Change and Children's Health," *Pediatrics* 120: 5 (November 2007); *The Lancet* and University College London Institute for Global Health Commission, "Managing the Health Effects of Climate Change," *Lancet* 373 (May 16, 2009).

297 *In a perceptive 2009 essay:* Peter C. Whybrow, "Dangerously Addictive: Why We Are Biologically Ill-Suited to the Riches of Modern America," *Chronicle of Higher Education,* March 13, 2009.

298 *That process is well under way:* Eric Lipton, "With Obama, Regulations Are Back in Fashion," *New York Times,* May 12, 2010.

Index

TEXT
10/13 Sabon

DISPLAY
DIN

COMPOSITOR
BookMatters, Berkeley

PRINTER AND BINDER
Sheridan Books, Inc.